Springer Proceedings in Complexity

Springer Complexity

Springer Complexity is an interdisciplinary program publishing the best research and academic-level teaching on both fundamental and applied aspects of complex systems—cutting across all traditional disciplines of the natural and life sciences, engineering, economics, medicine, neuroscience, social, and computer science.

Complex Systems are systems that comprise many interacting parts with the ability to generate a new quality of macroscopic collective behavior the manifestations of which are the spontaneous formation of distinctive temporal, spatial, or functional structures. Models of such systems can be successfully mapped onto quite diverse "real-life" situations like the climate, the coherent emission of light from lasers, chemical reaction–diffusion systems, biological cellular networks, the dynamics of stock markets and of the Internet, earthquake statistics and prediction, freeway traffic, the human brain, or the formation of opinions in social systems, to name just some of the popular applications.

Although their scope and methodologies overlap somewhat, one can distinguish the following main concepts and tools: self-organization, nonlinear dynamics, synergetics, turbulence, dynamical systems, catastrophes, instabilities, stochastic processes, chaos, graphs and networks, cellular automata, adaptive systems, genetic algorithms, and computational intelligence.

The three major book publication platforms of the Springer Complexity program are the monograph series "Understanding Complex Systems" focusing on the various applications of complexity, the "Springer Series in Synergetics", which is devoted to the quantitative theoretical and methodological foundations, and the "SpringerBriefs in Complexity" which are concise and topical working reports, case-studies, surveys, essays, and lecture notes of relevance to the field. In addition to the books in these two core series, the program also incorporates individual titles ranging from textbooks to major reference works.

More information about this series at http://www.springer.com/series/11637

Stefano Battiston · Francesco De Pellegrini
Guido Caldarelli · Emanuela Merelli
Editors

Proceedings of ECCS 2014

European Conference on Complex Systems

Springer

Editors
Stefano Battiston
RAF
University of Zürich
Zürich
Switzerland

Francesco De Pellegrini
CREATE-NET
Trento
Italy

Guido Caldarelli
IMT Lucca
Lucca
Italy

Emanuela Merelli
School of Science and Technology
University of Camerino
Camerino
Italy

ISSN 2213-8684 ISSN 2213-8692 (electronic)
Springer Proceedings in Complexity
ISBN 978-3-319-29226-7 ISBN 978-3-319-29228-1 (eBook)
DOI 10.1007/978-3-319-29228-1

Library of Congress Control Number: 2016934958

Printed on acid-free paper

This Springer imprint is published by Springer Nature
The registered company is Springer International Publishing AG Switzerland

Foreword

In the past decade the field of Complexity Science has moved into a new stage of its life. The big data and information technology revolutions are finally providing the necessary data, numerical experiments and validation tests to the many conceptual and theoretical advances that complex systems science has already provided to a large number of scientific disciplines. These fast paced developments are augmenting complex systems science with an "applied" dimension. Our increasing capability to solve many open problems, in a large diversity of scientific fields, has made it possible that Complex Systems Science becomes one of the conceptual and methodological keys to understand and deal with important real-world challenges that range from epidemics and traffic congestions, to systemic risks and cultural evolution, to cite a few.

In this framework, it is no wonder that the Complex Systems Society, gathering all researchers engaged in complex systems research has grown and developed along the same lines. The general Society conference is annually gathering about 1,000 scientists from all disciplines and it is a meeting point where every scientist interested in complex systems research can network the collective with a vibrant research community.

The annual conference on Complex Systems of 2014, organized at the IMT School for advanced studies in Lucca, was a smashing success, breaking many records for attendance, number of presentations—more than 200—and parallel workshops. The Lucca conference is certainly a milestone in the life of the field and the Complex Systems Society. We are extremely glad to see that the chairmen of the conference Guido Caldarelli and Stefano Battiston—Chairmen of the Lucca's conference—have teamed up with Francesco De Pellegrini and Emanuela Merelli to edit a book that collects a selection of 27 papers presented at the conference. The final result is a proceedings volume that is truly representative of the wide range of problems addressed by the community and the depth of the technical approaches used to tackle them. It speaks loudly for itself and we are sure that it will also become a reference for those that want to grasp what the community is doing nowadays.

On behalf of the Complex Systems Society and its members we thank the organizers of the conference and the editors of this Proceedings of ECCS 2014 for all their work, the exemplary engagement and their service to Complex Systems Science.

Alessandro Vespignani
President of the Complex Systems Society 2012–2015
Boston, MA, USA

Yamir Moreno
President of the Complex Systems Society 2016–2019
Zaragoza, Spain

Preface

This volume collects a series of multidisciplinary contributions in the field of complex systems science. Several works presented in this collection pivot on the theory and applications of formal and computational approaches. These methods are suitable to construct and simulate models of complex systems so as to analyse their properties. This is indeed an emerging research area encompassing a broad range of fields including—but not limited to—physics, computer science and mathematics, economics, business, political science, biology, sociology, neuroscience and medicine. The collection is thus addressed to the new generation of transdisciplinary researchers.

The work contains contributions which have been initially discussed in the *European Conference on Complex Systems* (ECCS'14) held at IMT, Lucca from 22 to 26 September 2014, under the sponsorship of the Complex Systems Society. ECCS'14 is a major international conference in the area of Complex Systems and interdisciplinary science in general. The main aim is to offer unique opportunities to study novel foundational approaches in a multitude of application areas. Thus, it spans from Complexity in ICT and Social Systems, to Complexity in Infrastructures, Complexity in Environment and Cities, Complexity in Natural Sciences, Complexity in Humanities, Linguistics and Society Complexity in Economics and Finance.

The project had an internal call for papers presented at the ECCS14 Conference. It contains a selection of 27 papers which originated from the conference oral presentations and poster sessions. All the manuscripts are extended versions of the contributions presented there and went through an independent review process.

The editors express their thanks to all authors of the articles submitted to this special issue. They also acknowledge the efforts of our many reviewers for their help in selecting the papers published in this special issue.

Zürich, Switzerland — Stefano Battiston
Trento, Italy — Francesco De Pellegrini
Lucca, Italy — Guido Caldarelli
Camerino, Italy — Emanuela Merelli

Contents

Contributors

Madalena Alarcão Faculty of Psychology and Education Sciences, University of Coimbra, Coimbra, Portugal

Talayeh Aledavood Aalto University School of Science, Espoo, Finland

Hideaki Aoyama Graduate School of Sciences, Kyoto University, Kyoto, Japan

Edmund Barter Department of Engineering Mathematics, University of Bristol, University Walk, Bristol, UK

Timothy Bates Physics Department, Assumption College, Worcester, MA, USA

Stefano Battiston Department of Banking and Finance, University of Zurich, Zürich, Switzerland

Igor Borovikov Nekkar.net. International Labs, Foster City, CA, USA

Dmitry Bratsun Theoretical Physics Department, Perm State Pedagogical University, Perm, Russia

Markus Brede Institute for Complex Systems Simulation, University of Southampton, Southampton, UK

F. Caravelli Invenia Technical Computing, Winnipeg, Canada; Department of Computer Science, UCL, London, UK; London Institute of Mathematical Sciences, London, UK

T. Carletti Department of Mathematics and naXys, University of Namur, Namur, Belgium

Alexander Casey Physics Department, Assumption College, Worcester, MA, USA; University of Notre Dame, Notre Dame, IN, USA

Filippo Castiglione Institute for Applied Mathematics (IAC) CNR, Rome, Italy

A. Chakraborti School of Computational and Integrative Sciences (SCIS), Jawaharlal Nehru University, New Delhi, India

Anna Chernyshova Siberian Federal University, Krasnoyarsk, Russia

Michael Daly Physics Department, Assumption College, Worcester, MA, USA; Meditech, Framingham, MA, USA

Ana Teixeira de Melo Faculty of Psychology and Education Sciences of the University of Coimbra, Coimbra, Portugal

Antonio De Nicola ENEA-CR Casaccia, Rome, Italy; University of Rome Tor Vergata, Rome, Italy

Francesco De Pellegrini CREATE-NET, Trento, Italy

Alonso Espinosa Mireles de Villafranca Department of Engineering Mathematics, University of Bristol, University Walk, Bristol, UK

Lei Dong School of Architecture, Tsinghua University, Beijing, China

Robin I.M. Dunbar Department of Experimental Psychology, University of Oxford, Oxford, UK

Gregorio D'Agostino ENEA-CR Casaccia, Rome, Italy; Center for Polymer Studies, Boston University, Boston, MA, USA

Yoshi Fujiwara Graduate School of Simulation Studies, University of Hyogo, Kobe, Japan

Georgi Yordanov Georgiev Physics Department, Assumption College, Worcester, MA, USA; Physics Department, Tufts University, Medford, MA, USA; Department of Physics, Worcester Polytechnic Institute, Worcester, MA, USA

Erin Gombos Physics Department, Assumption College, Worcester, MA, USA; National Cancer Institute, Bethesda, MD, USA

E. Heinsalu National Institute of Chemical Physics and Biophysics (NICPB), Tallinn, Estonia

Kaitlin Henry Physics Department, Assumption College, Worcester, MA, USA

Jorge Hidalgo Departamento de Electromagnetismo y Física de la Materia e Instituto Carlos I de Física Teórica y Computacional, Universidad de Granada, Granada, Spain

Masafumi Hino NEC Corporation, Minato-ku, Tokyo, Japan

Masato Hisakado Financial Services Agency, Chiyoda-ku, Tokyo, Japan

Johanne Hizanidis National Center for Scientific Research "Demokritos", Athens, Greece; Crete Center for Quantum Complexity and Nanotechnology, Department of Physics, University of Crete, Heraklion, Greece

Philipp Hövel Institut für Theoretische Physik, Technische Universität Berlin, Berlin, Germany; Bernstein Center for Computational Neuroscience Berlin, Humboldt-Universität zu Berlin, Berlin, Germany

Hiroshi Iyetomi Department of Mathematics, Niigata University, Niigata, Japan

Mark P. Johnson Ryan Institute, National University of Ireland Galway, Galway, Ireland

Andreas Karpf Université Paris 1 Panthéon-Sorbonne, Centre d'Économie de la Sorbonne/Paris School of Economics, Paris, France

K. Kaski Department of Computer Science, Aalto University School of Science, Aalto, Finland

Espen Knoop Department of Engineering Mathematics, University of Bristol, University Walk, Bristol, UK; Bristol Robotics Laboratory, Bristol, UK

Ruiqi Li School of Systems Science, Beijing Normal University, Beijing, China

Eduardo López CABDyN Complexity Center, Saïd Business School, University of Oxford, Oxford, UK

Antoine Mandel Université Paris 1 Panthéon-Sorbonne, Centre d'Économie de la Sorbonne, Paris, France

Dimitris Maniadakis Department of Informatics and Telecommunications, University of Athens, Athens, Greece

Ramona Marguta Departamento de Física, Biosystems Integrative Sciences Institute (BioISI), Faculdade de Ciências da Universidade de Lisboa, Campo Grande, Lisbon, Portugal

L. Marzola National Institute of Chemical Physics and Biophysics (NICPB), Tallinn, Estonia; Laboratory of Theoretical Physics, Institute of Physics, University of Tartu, Tartu, Estonia

Antoni Matyjaszkiewicz Department of Engineering Mathematics, University of Bristol, University Walk, Bristol, UK

Christopher McWilliams Department of Engineering Mathematics, University of Bristol, University Walk, Bristol, UK

F. Medda QASER Lab, UCL, London, UK

Emanuela Merelli School of Science and Technology, University of Camerino, Camerino, Italy

Daniele Miorandi CREATE-NET, Trento, Italy

Paolo Moretti Institute of Materials Simulation (WW8), Friedrich-Alexander-University, Erlangen-Nünberg, Fürth, Germany

Shintaro Mori Department of Physics, Kitasato University, Sagamihara, Kanagawa, Japan

Esteban Moro Departamento de Matemáticas & GISC, Universidad Carlos III de Madrid, Leganés, Spain

Igor Mozetič Jožef Stefan Institute, Ljubljana, Slovenia

Miguel A. Muñoz Departamento de Electromagnetismo y Física de la Materia e Instituto Carlos I de Física Teórica y Computacional, Universidad de Granada, Granada, Spain

James T. Murphy Sorbonne Universités, UPMC Univ Paris 6, UMR 7144, Station Biologique de Roscoff, Roscoff, France; CNRS, UMR 7144, Equipe Div&Co, Station Biologique de Roscoff, Roscoff, France; Ryan Institute, National University of Ireland Galway, Galway, Ireland

D. Nicolay Department of Mathematics and naXys, University of Namur, Namur, Belgium

Iryna Omelchenko Institut für Theoretische Physik, Technische Universität Berlin, Berlin, Germany

Andrea Parisi Departamento de Física, Biosystems Integrative Sciences Institute (BioISI), Faculdade de Ciências da Universidade de Lisboa, Campo Grande, Lisbon, Portugal

M. Patriarca National Institute of Chemical Physics and Biophysics (NICPB), Tallinn, Estonia

Marco Pettini Centre de Physique Théorique, Aix-Marseille University, Marseille, France

Thong Pham Osaka University, Osaka, Japan

Felix Reed-Tsochas CABDyN Complexity Center, Saïd Business School, University of Oxford, Oxford, UK; Department of Sociology, University of Oxford, Oxford, UK

R. Reinanda Faculty of Science, University of Amsterdam, Amsterdam, The Netherlands

Sam G.B. Roberts Department of Psychology, University of Chester, Chester, UK

Lewis Roberts Department of Engineering Mathematics, University of Bristol, University Walk, Bristol, UK

A. Roli Department of Computer Science and Engineering (DISI), University of Bologna, Campus of Cesena, Bologna, Italy

Matteo Rucco School of Science and Technology, University of Camerino, Camerino, Italy

Michael Sadovsky Institute of Computational Modelling SB RAS, Krasnoyarsk, Russia

Jari Saramäki Aalto University School of Science, Espoo, Finland

Luís F. Seoane ICREA-Complex Systems Lab, Universitat Pompeu Fabra-PRBB, Barcelona, Spain; Institut de Biologia Evolutiva, CSIC-UPF, Barcelona, Spain

Paul Sheridan The University of Tokyo, Tokyo, Japan

Hidetoshi Shimodaira Osaka University, Osaka, Japan

Borut Sluban Jožef Stefan Institute, Ljubljana, Slovenia

Jasmina Smailović Jožef Stefan Institute, Ljubljana, Slovenia

Ricard Solé ICREA-Complex Systems Lab, Universitat Pompeu Fabra-PRBB, Barcelona, Spain; Institut de Biologia Evolutiva, CSIC-UPF, Barcelona, Spain; Santa Fe Institute, Santa Fe, NM, USA

Massimo Stella Institute for Complex Systems Simulation, University of Southampton, Southampton, UK

Taiki Takahashi Department of Behavioral Science, Faculty of Letters, Hokkaido University, Sapporo, Japan; Center for Experimental Research in Social Sciences, Hokkaido University, Sapporo, Hokkaido, Japan

V.A. Traag CWTS, Leiden University, Leiden, The Netherlands

G. van Klinken KITLV, Leiden, The Netherlands

Dimitris Varoutas Department of Informatics and Telecommunications, University of Athens, Athens, Greece

Frédérique Viard Sorbonne Universités, UPMC Univ Paris 6, UMR 7144, Station Biologique de Roscoff, Roscoff, France; CNRS, UMR 7144, Equipe Div&Co, Station Biologique de Roscoff, Roscoff, France

Pablo Villegas Departamento de Electromagnetismo y Física de la Materia e Instituto Carlos I de Física Teórica y Computacional, Universidad de Granada, Granada, Spain

Andrea Vüllings Institut für Theoretische Physik, Technische Universität Berlin, Berlin, Germany

Xinran Wang College of Resources Science and Technology, Beijing Normal University, Beijing, China

Hiroshi Yoshikawa Graduate School of Economics, The University of Tokyo, Tokyo, Japan

Andrey Zakharov Theoretical Physics Department, Perm State Pedagogical University, Perm, Russia

Mattia Zeni Department of Information Engineering and Computer Science, University of Trento, Trento, Italy

Jiang Zhang School of Systems Science, Beijing Normal University, Beijing, China

Chapter 1
Detection of Non-self-correcting Nature of Information Cascade

Shintaro Mori, Masafumi Hino, Masato Hisakado and Taiki Takahashi

Abstract We propose a method of detecting non-self-correcting information cascades in experiments in which subjects choose an option sequentially by observing the choices of previous subjects. The method uses the correlation function $C(t)$ between the first and the $t+1$th subject's choices. $C(t)$ measures the strength of the domino effect, and the limit value $c \equiv \lim_{t\to\infty} C(t)$ determines whether the domino effect lasts forever ($c > 0$) or not ($c = 0$). The condition $c > 0$ is an adequate condition for a non-self-correcting system, and the probability that the majority's choice remains wrong in the limit $t \to \infty$ is positive. We apply the method to data from two experiments in which T subjects answered two-choice questions: (i) general knowledge questions ($T_{avg} = 60$) and (ii) urn-choice questions ($T = 63$). We find $c > 0$ for difficult questions in (i) and all cases in (ii), and the systems are not self-correcting.

1.1 Introduction

Herding phenomena are ubiquitous in human and animal behavior [1, 2]. An example is an information cascade, in which a person observes others' choices and chooses the majority's choice even though the person's private signal contradicts it [3, 4]. It is a rational behavior for people who are uncertain about choosing. If an information

S. Mori (✉)
Department of Physics, Kitasato University, 1-15-1 Kitasato, Sagamihara, Kanagawa 252-0373, Japan
e-mail: mori@sci.kitasato-u.ac.jp

M. Hino
NEC Corporation, Siba 5-7-1, Minato-ku, Tokyo 108-8001, Japan

M. Hisakado
Financial Services Agency, Kasumigaseki 3-2-1, Chiyoda-ku, Tokyo 100-8967, Japan

T. Takahashi
Department of Behavioral Science, Faculty of Letters, Hokkaido University, Sapporo, Japan

T. Takahashi
Center for Experimental Research in Social Sciences, Hokkaido University, Kita 10, Nishi 7, Kita-ku, Sapporo, Hokkaido 060-0810, Japan

S. Battiston et al. (eds.), *Proceedings of ECCS 2014*, Springer Proceedings in Complexity, DOI 10.1007/978-3-319-29228-1_1

cascade occurs, the same mechanism applies to later decision-makers, and the majority's choice tends to prevail. In some cases, the successive choices are wrong, and the cascade leads to irrational herding behavior [5].

An experimental setup demonstrates a situation in which an information cascade occurs [6]. There are two urns, A and B, and urn A (B) contains two a (b) balls and one b (a) ball. In each run of the experiment, an urn is randomly chosen initially and called X. Then, the subjects guess whether urn X is A or B and choose sequentially. They get a reward for the correct choice. In the course of the experiment, each subject draws a ball from X, which is his private signal. If the ball is a (b), urn X is more likely to be A (B). He also observes the choices of the previous subjects. If the difference between the numbers of subjects who choose each urn exceeds two, the private signal cannot overcome the majority's choice. An information cascade starts if someone chooses the majority's choice although his private signal suggests the minority's one. As the probability that the first two persons both choose the wrong option is non-zero, the probability for the onset of a cascade where the majority's choice is wrong is positive.

We now consider whether the wrong cascade continues [5]. If it continues forever, the majority's choice converges to the wrong option. Information cascades were initially considered to be fragile phenomena. As the trigger of the cascade is a small imbalance, people can be dissuaded from following the majority's choice [3]. In addition, an agent model with a Bayesian update of the private belief showed that the information cascade is self-correcting [8]. As the number of agents tends toward infinity, the wrong cascade disappears, and the majority's choice converges to the optimal option.

Using an information cascade experiment with a general knowledge two-choice quiz, we have shown that a phase transition occurs between a one-peak phase and a two-peak phase [9]. If the questions are easy, the ratio $z(t)$ of the correct choices of t subjects converges to a value $z_+ > 1/2$ in the limit $t \to \infty$. As there is only one peak in the probability distribution function of $z(t)$, we call the corresponding phase the one-peak phase [10, 11]. If the questions are difficult and most people do not know the answers, $z(t)$ converges to $z_+ > 1/2$ or $z_- < 1/2$. One cannot predict the value in $\{z_+, z_-\}$ to which $z(t)$ converges. We call the corresponding phase the two-peak phase. In the two-peak phase, the wrong cascade does not necessarily disappear, and the system is not self-correcting.

It was recently shown that the limit value of the normalized correlation function is the order parameter of the phase transition [14]. The normalized correlation function shows how the first subject's choice propagates to later subjects. It provides a measure of the domino effect. In addition, the positiveness of the limit value is a sufficient condition for a non-self-correcting system. By extrapolating the results for a finite system to infinity, we can determine whether the system is self-correcting. We report on the application of the method to data from two types of information cascade experiments. In Sect. 1.2, we define the normalized correlation function. We also explain the behavior of the function in each phase and the extrapolation method used to estimate its limit. We present the results of the data analysis in Sect. 1.3. Section 1.4 summarizes the results.

1.2 Correlation Function and Asymptotic Behaviors

We consider a typical information cascade experiment. T subjects answer a two-choice question sequentially in each run. We denote the order of the subjects as t, where $t = 1, 2, \ldots, T$. We denote the choice of subject t by $X(t) \in \{0, 1\}, t = 1, 2, \ldots, T$. If the choice is true (false), $X(t)$ takes 1 (0).

The correlation function $C(t)$ is defined as the covariance between $X(1)$ and $X(t+1)$ divided by the variance of $X(1)$:

$$C(t) \equiv \mathrm{Cov}(X(1), X(t+1))/\mathrm{Var}(X(1)).$$

$C(t)$ can be expressed as the difference of two conditional probabilities.

$$C(t) = \Pr(X(t+1) = 1|X(1) = 1) - \Pr(X(t+1) = 1|X(1) = 0). \quad (1.1)$$

$C(t)$ shows the degree to which the first subject's choice is transmitted to later subjects. It is a measure of the domino effect in an information cascade.

$C(t)$ is generally positive, and its asymptotic behavior depends on the phase of the system and the shape of the response function $q(z)$. Here $q(z)$ represents the dependence of the probability of the correct choice by subject $t+1$ on the ratio $z(t)$ of the correct choices of the previous t subjects.

$$q(z) \equiv \Pr(X(t+1) = 1|z(t) = z), \quad z(t) = \frac{1}{t}\sum_{s=1}^{t} X(s).$$

With the definition of $q(z)$, the stochastic process $\{X(t)\}, t = 1, 2 \ldots$ becomes a generalized Pólya urn process [12]. If there is one solution for $z = q(z)$ at z_+ (left panel in Fig. 1.1), $z(t)$ converges to z_+. $C(t)$ shows power-law decay for large t with two constants, c' and l, as

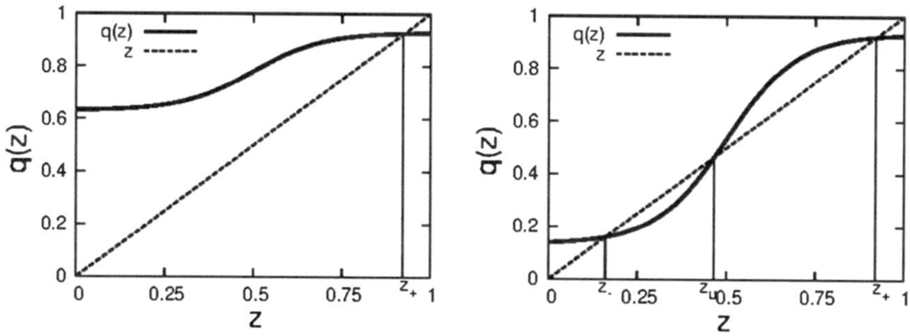

Fig. 1.1 Response function $q(z)$ versus z. *Left* panel shows the one-peak phase, in which there is one solution, z_+, for $z = q(z)$. *Right* panel shows the two-peak phase, in which there are three solutions, $z_- < z_u < z_+$, for $z = q(z)$

$$C(t) \simeq c' \cdot t^{l-1} \quad l < 1.$$

Here, l is the exponent for the power-law decay and is less than 1. The value of l is given by $g'(z_+)$ [11, 13]. If there are three solutions for $z = q(z)$ at $z_- < z_u < z_+$ (right panel in Fig. 1.1), the system is in the two-peak phase; $\lim_{t\to\infty} z(t) = z_+$ or z_- [12]. The limit value $c \equiv \lim_{t\to\infty} C(t)$ is positive, and the first subject's choice propagates to an infinite number of later subjects [14]. $C(t)$ behaves asymptotically as

$$C(t) \sim c + c' \cdot t^{l-1}. \tag{1.2}$$

Here $c' \cdot t^{l-1}$ is the subleading term of $C(t)$, and l is given by the larger value among $\{g'(z_+), g'(z-)\}$. Further, c acts as an order parameter of the phase transition, and (1.2) is the general asymptotic behavior of $C(t)$ [15].

As it is difficult to estimate c using $c \equiv \lim_{t\to\infty} C(t)$ with empirical data, where the system size and number of samples are strictly limited, we introduce two quantities for the estimation. First, we define the nth moment $m_n(t)$ for $C(t)$ as $m_n(t) \equiv \sum_{s=0}^{t-1} C(s)(s/t)^n$. We define the integrated correlation time $\tau(t)$ as $\tau(t) = m_0(t)$. We also define the second moment correlation time $\xi(t)$ as $\xi(t) \equiv t \cdot \sqrt{m_2(t)/m_0(t)}$. Using the asymptotic behavior of $C(t)$, we estimate the subsequent asymptotic behavior of $\tau(t)/t$ and $\xi(t)/t$.

$$\tau(t)/t \simeq c + \frac{c'}{l} \cdot t^{l-1} \tag{1.3}$$

$$\xi(t)/t \to \begin{cases} \sqrt{l/l+2} & c = 0 \\ \sqrt{1/3} & c > 0 \end{cases} \tag{1.4}$$

As $\tau(t)/t$ is defined as the summation of $C(s)$ over $0 \le s < t$ divided by t, the standard error becomes smaller than that of $C(t)$. The asymptotic behavior of $\tau(t)/t$ in (1.3) provides a more reliable estimate of c and l than the fitting of $C(t)$ to (1.2). $\xi(t)/t$ also provides a reliable estimate for l [15]. If $c > 0$, the leading term of $C(t)$ is the constant c, and l should be interpreted as $l = 1$.

We define whether the system is self-correcting according to whether $z(t)$ always converges to z_+. In the one-peak (two-peak) phase, the system is (non-)self-correcting. If $c > 0$, the system is in the two-peak phase and is non-self-correcting. However, $c = 0$ does not necessarily mean that the system is self-correcting. For the system to be self-correcting, $q(z) = z$ has to have only one solution, z_+.

1.3 Domino Effect and Detection of Non-self-correcting Nature

We study the domino effect and non-self-correction in information cascades. We discuss two types of information cascade experiments.

In experiment 1 (EXP-I), subjects answered a general knowledge two-choice quiz. First, the subjects answered using only their own knowledge. Then, they observed the choices of previous subjects and answered the question again. The average length of the sequence of subjects is $T = 60$, and the number of choice sequences is 240. The choice sequences are classified into four bins according to the ratio of correct choices $z_0(T)$ of the first answers without observation as $z_0(T) = 50 \pm 5, 60 \pm 5, 70 \pm 5$, and $80 \pm 5\,\%$, and the number of samples in each bin is $38(50 \pm 5\,\%), 52(60 \pm 5\,\%), 38(70 \pm 5\,\%)$, and $38(80 \pm 5\,\%)$, respectively [16].

Experiment 2 (EXP-II) is similar to the situation explained in the Introduction. There are two urns, A and B, which contain a and b balls in different configurations. We use two configuration patterns: (i) two a balls and one b ball in urn A versus one a ball and two b balls in urn B and (ii) five a balls and four b balls in urn A versus four a balls and five b balls in urn B. Urn $X \in \{$A,B$\}$ is chosen at random at the beginning of each run, and subjects are asked to choose between A or B. Each subject draws one ball from X and checks whether it is a or b. The ball corresponds to the type of urn X with probability $q = 2/3(5/9)$ for (i) [(ii)]. In addition, the subject also observes the choices of previous subjects. Our results, unlike those of previous experiments [6–8], show the summary statistics of the number of subjects who have chosen each urn. The length T and number of questions I are 63 and 200, respectively, for $q \in \{2/3, 5/9\}$ [17].

We denote the choice sequences in each bin as $\{X(i,t)\}, i = 1, \ldots, I, t = 1, \ldots, T(i)$. Here, the length of the sequence depends on question i in EXP-I; we denote it as $T(i)$. The number of samples I also depends on the bins. In EXP-II, $T(i) = 63$, and $I = 200$. First, we estimate $C(t)$ and its standard error $\Delta C(t)$ using (1.1). We denote the estimate and standard error of the probabilities as $q_x(t+1) = \Pr(X(t+1) = 1|X(1) = x)$ and $\Delta q_x(t+1)$, respectively. They are estimated from experimental data $\{X(i,t)\}$ as

$$q_x(t+1) = \frac{1 + \sum_{i=1}^{I} X(i,t+1)\delta_{X(i,1),x}}{N_x + 2},$$

$$N_x = \sum_{i=1}^{I} \delta_{X(i,1),x},$$

$$\Delta q_x(t+1) = \sqrt{\frac{q(x,t+1)(1 - q_x(t+1))}{N_x + 3}}.$$

Here, we use the expectation value and standard deviation obtained from the posterior probability distribution for the probabilities. $C(t)$ is then estimated as

$$C(t) = q_1(t+1) - q_0(t+1).$$

The error bars of $C(t)$ are given as

$$\Delta C(t) = \sqrt{\Delta q_1(t+1)^2 + \Delta q_0(t+1)^2}. \qquad (1.5)$$

Using $C(t)$ and $\Delta C(t)$, we estimate the error bars of $m_n(t)$ as

$$\Delta m_n(t) = \sqrt{\sum_{s=1}^{t-1} \Delta C(s)^2 (s/t)^{2n}}.$$

Here we assume that $\Delta C(s)$ and $\Delta C(s')$ are independent of each other if $s \neq s'$. We estimate the error bars of $\tau_t(t)$ and $\xi_t(t)$ as

$$\begin{aligned} \Delta \tau_t &= \frac{1}{t} \Delta m_0(t), \\ \Delta \xi_t &= \sqrt{\xi_t} (\Delta m_2(t)/2m_2(t) + \Delta m_0(t)/2m_0(t)). \end{aligned} \tag{1.6}$$

In the estimation of $\Delta \xi_t$, we assume that $\Delta m_2(t)$ and Δm_0 are completely correlated.

1.3.1 *EXP-I: General Knowledge Quiz Case*

Figure 1.2 plots $C(t)$ versus t. The value of $C(t)$ generally decreases from its initial value of 1 with increasing t. Because the sample number is restricted, $\Delta C(t)$ is large. We see that for difficult questions with $z_0(T) = 50 \pm 5$ and 60 ± 5 %, $C(t)$ is positive for large values of t. On the other hand, for easy questions with $z_0(T) = 70 \pm 5$ and 80 ± 5 %, $C(t)$ decreases to zero with increasing t. These results suggest that the system is in the two-peak phase for difficult questions. For $z_0(T) = 70 \pm 5$ and 80 ± 5 %, an analysis of $q(z)$ showed that the system was in the one-peak phase [16].

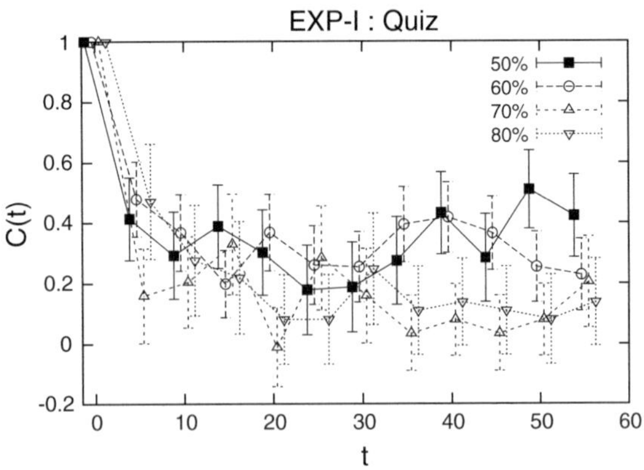

Fig. 1.2 $C(t)$ versus t for EXP-I. The sample choice sequences are classified according to the value of $z_0(T)$ as $z_0(T) = 50 \pm 5$ % (*filled quare*), 60 ± 5 % (*opened circle*), 70 ± 5 % (*opened triangle*), and 80 ± 5 %(*opened down triangle*). We plot only data with the interval $\Delta t = 5$. To see the behavior clearly, we slightly shift the data horizontally

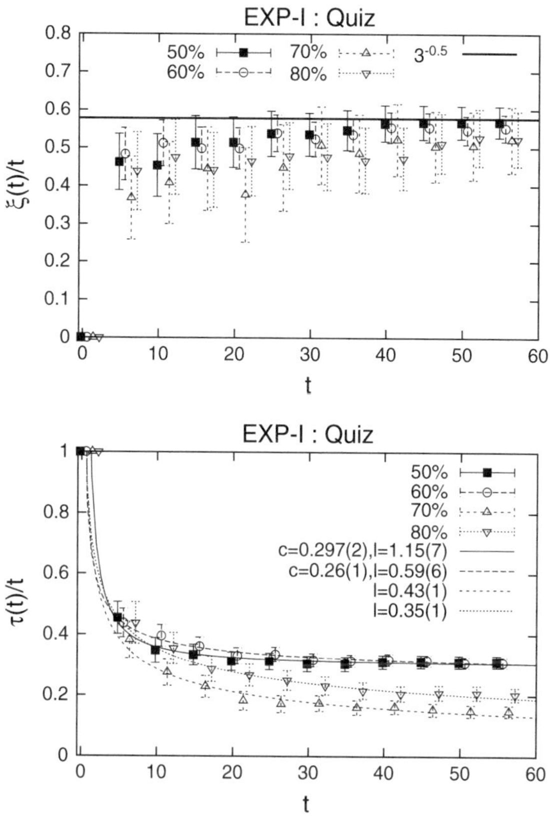

Fig. 1.3 $\xi(t)/t$ and $\tau(t)/t$ versus t for EXP-I with the interval $\Delta t = 5$. We also plot the fitted results for $\tau(t)/t$

Figure 1.3 shows plots of $\xi(t)/t$ and $\tau(t)/t$ versus t. The standard errors for $\xi(t)/t$ are larger than those for $\tau(t)/t$ because $\xi(t)$ is calculated with the second moment $m_2(t)$. For large values of t, $\xi(t)/t$ takes $\sqrt{1/3}$ for difficult questions with $z_0(T) = 50 \pm 5$ and $60 \pm 5\,\%$. The results suggest that the system is in the two-peak phase. For easy questions with $z_0(T) = 70 \pm 5$ and $80 \pm 5\,\%$, $\xi(t)/t \simeq 0.5$ for large values of t. As $\xi(t)/t \simeq \sqrt{l/l+2}$, $l \simeq 0.7$ for easy questions. As l is smaller than 1, the system is in the one-peak phase.

As the system is considered to be in the two-peak phase for $z_0(T) = 50 \pm 5$ and $60 \pm 5\,\%$, we assume $\tau(t)/t = c + d \cdot t^{l-1}$ and estimate c, l, d using the least square fit. We find that $c = 0.297(2)$ for $z_0(T) = 50 \pm 5\,\%$ and $c = 0.26(1)$ for $z_0(T) = 60 \pm 5\,\%$. For $z_0(T) = 70 \pm 5$ and $80 \pm 5\,\%$, we assume $\tau(t)/t = d \cdot t^{l-1}$ and estimate l and d. We find that $l = 0.43(1)$ for $z_0(T) = 70 \pm 5\,\%$ and $l = 0.35(1)$ for $z_0(T) = 80 \pm 5\,\%$, which differ slightly from the value of $l \simeq 0.7$ estimated from $\xi(t)/t$.

1.3.2 EXP-II: Urn Choice Case

Figure 1.4 shows plots of $C(t)$, $\xi(t)/t$, and $\tau(t)/t$ versus t for $q \in \{2/3, 5/9\}$. As the number of samples is larger than that in EXP-I, the standard errors are smaller than the symbols' size for $\tau(t)/t$ and large t. We see that $C(t)$ is positive for large values of t for both cases of q, where $q \in \{2/3, 5/9\}$. In addition, $\xi(t)/t$ for large values of t converges to $\sqrt{1/3}$, and the exponent l for $C(t) \sim t^{l-1}$ is almost one. These results suggest that the system is in the two-peak phase for both values of q. We assume $\tau(t)/t = c + d \cdot t^{l-1}$ and estimate c, l, d using the least square fit. We find that $c = 0.261(1)$ for $q = 2/3$ and $c = 0.207(1)$ for $q = 5/9$.

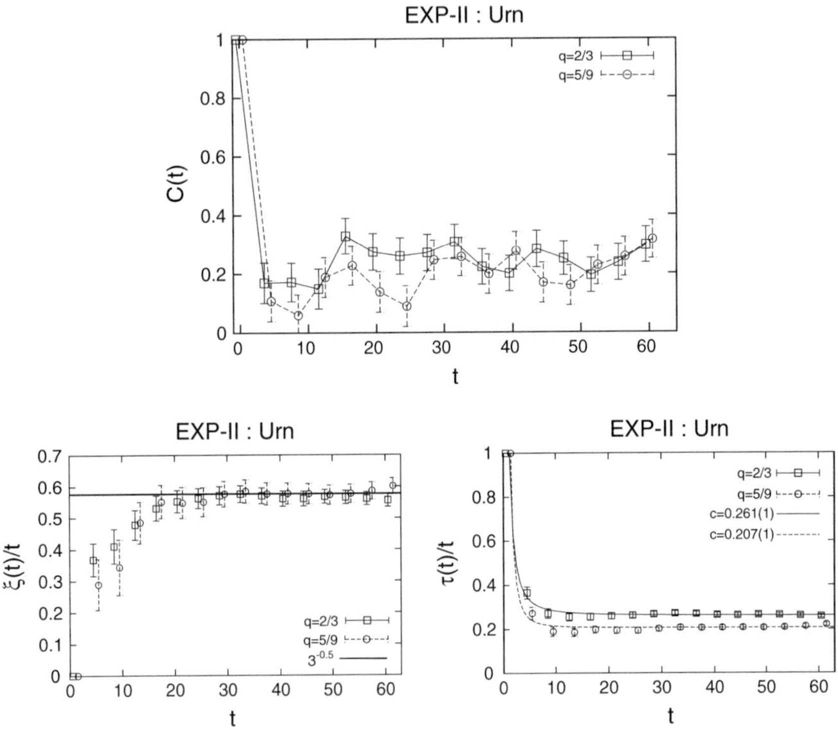

Fig. 1.4 $C(t)$, $\xi(t)/t$, and $\tau(t)/t$ versus t for EXP-II. We use the symbol *opened square* (*opened circle*) for $q = 2/3(5/9)$. We plot only data with the interval $\Delta t = 4$. To see the behavior clearly, we slightly shift the data horizontally

1.4 Conclusion

We studied the self-correcting nature of information cascades. We proposed the use of the normalized correlation function $C(t)$, which shows how the first subject's choice is propagated to later subjects and measures the strength of the domino effect in information cascades. $c \equiv \lim_{t \to \infty} C(t) > 0$ is a sufficient condition for a non-self-correcting information cascade. In this case, the domino effect continues infinitely. The system is in the two-peak phase, and the probability that $z(t)$ converges to $z_- < 1/2$ is positive. We used data from two types of information cascade experiment: EXP-I, which used a general knowledge quiz, and EXP-II, which used urns. The accuracy q of the private signal is $q \in \{2/3, 5/9\}$ in EXP-II. We estimate $C(t)$ and its integrated quantities $\tau(t)$ and $\xi(t)$. In EXP-I, when the questions were difficult, $c > 0$. In EXP-II, $c > 0$ for both cases of q where $q \in \{2/3, 5/9\}$. In these cases, the system is non-self-correcting.

We focus on the study of the non-self-correcting nature of information cascades. Although $c > 0$ is a sufficient condition for a non-self-correcting cascade, $c = 0$ is not a sufficient condition for a self-correcting cascade. To verify this, one should study the response function $q(z)$ and count the number of solutions for $z = q(z)$. Alternatively, it is necessary to study the limit value of the variance of $z(t)$. If there is only one solution, $z_+ > 1/2$, or the limit value is zero, the system is self-correcting. In EXP-I, we studied these points and concluded that the system is self-correcting for $z_0(T) = 70 \pm 5$ and $80 \pm 5\,\%$ [16]. Our experiment for EXP-II and its analysis are under way [17].

Acknowledgments This work was supported by Grant-in-Aid for Challenging Exploratory Research 25610109.

References

1. Catellano, C., Fortunato, S., Loreto, V.: Statistical physics of social dynamics. Rev. Mod. Phys. **81**, 591–646 (2009)
2. Fernández-Gracia, J., Sucheki, K., Ramasco, J.J., Miguel, M.S., Eguíluz, V.M.: Is the Voter Model a model for voters? Phys. Rev. Lett. **112**, 158701–158705 (2014)
3. Bikhchandani, S., Hirshleifer, D., Welch, I.: A theory of fads, fashion, custom, and cultural changes as informational cascades. J. Polit. Econ. **100**, 992–1026 (1992)
4. Devenow, A., Welch, I.: Rational herding in financial economics. Euro. Econ. Rev. **40**, 603–615 (1996)
5. Lee, I.H.: On the convergence of informational cascades. J. Econ. Theory **61**, 395–411 (1993)
6. Anderson, L.R., Holt, C.A.: Information cascades in the laboratory. Am. Econ. Rev. **87**, 847–862 (1997)
7. Kübler, D., Weizsäcker, G.: Limited depth of reasoning and failure of cascade formation in the laboratory. Rev. Econ. Stud. **71**, 425–441 (2004)
8. Goeree, J.K., Palfrey, T.R., Rogers, B.W., McKelvey, R.D.: Self-correcting information cascades. Rev. Econ. Stud. **74**, 733–762 (2007)
9. Mori, S., Hisakado, M., Takahashi, T.: Phase transition to two-peaks phase in an information cascade voting experiment. Phys. Rev. E **86**, 026109–026118 (2012)

10. Hisakado, M., Mori, S.: Digital herders and phase transition in a voting model. J. Phys. A **44**, 275204–275220 (2011)
11. Hisakado, M., Mori, S.: Two kinds of phase transitions in a voting model. J. Phys. A, Math. Theor. **45**, 345002–345016 (2012)
12. Hill, B., Lane, D., Sudderth, W.: A strong law for some generalized urn processes. Ann. Prob. **8**, 214–226 (1980)
13. Hod, S., Keshet, U.: Phase-transition in binary sequences with long-range correlations. Phys. Rev. E **70**, 015104–015109 (2004)
14. Mori, S., Hisakado, M.: Finite-size scaling analysis of binary stochastic processes and universality classes of information cascade phase transition. J. Phys. Soc. Jpn. **84**, 054001–054013 (2015)
15. Mori, S., Hisakado, M.: Correlation function for generalized Pólya urns: Finite-size scaling analysis. Phys. Rev. E. **92**, 052112–052121 (2015)
16. Mori, S., Hisakado, M., Takahashi, T.: Collective adoption of max-min strategy in an information cascade voting experiment. J. Phys. Soc. Jpn. **82**, 084004–084013 (2013)
17. Hino, M., Hisakado, M., Takahashi, T., Mori, S.: Detection of phase transition in generalized Plya urn in Information cascade experiment. J. Phys. Soc. Jpn. **85**, 034002–034013 (2016)

Chapter 2
Fitting Planar Proximity Graphs on Real Street Networks

Dimitris Maniadakis and Dimitris Varoutas

Abstract Due to the rising progress of sustainable urban infrastructures, modeling realistic street networks is a fundamental challenge. This study contributes to this modeling direction, by suggesting the utilization of planar proximity graphs, and specifically the β-skeleton graphs. Their goodness of fit on producing real-like urban street networks is verified by comparison to real data. In particular, the basic topological and geometrical properties derived from synthetic β-skeleton planar graphs are compared to the properties of five urban street network datasets, all represented using the Primal approach. A good agreement with empirical patterns is found and a possible explanation is discussed.

2.1 Introduction

There are broad agreements that the street patterns shape overlay infrastructure deployment since they define a basic template which strongly constrains the further development of other webs (e.g., power grid or communication networks). Due to the rising progress of sustainable urban infrastructures, understanding and modeling the structure of street networks is an elementary challenge. Despite a large number of studies on street networks, the existing modeling methodologies are mostly long, random-based and simulation-based, which require several assumptions for generating a realistic street layout, e.g., [1].

On the other hand, the construction of planar proximity graphs can be straightforward by using analytical or simulation methods. Planar proximity graphs are planar graphs (edges intersect only in the points/nodes) where two points in Euclidean plane are connected by an edge if they are close in some sense. Each pair of points is assigned a certain neighborhood, and the points of the pair are connected by an

D. Maniadakis (✉) · D. Varoutas
Department of Informatics and Telecommunications, University of Athens, Athens, Greece
e-mail: D.Maniadakis@di.uoa.gr

D. Varoutas
e-mail: D.Varoutas@di.uoa.gr

S. Battiston et al. (eds.), *Proceedings of ECCS 2014*, Springer Proceedings in Complexity, DOI 10.1007/978-3-319-29228-1_2

edge if their neighborhood is empty. The Delaunay Triangulation (DT), the Relative Neighborhood Graph (RNG), the Gabriel Graph (GG) and the Minimum Spanning Tree (MST) are well known examples of proximity graphs. These are constructed by parameter-less algorithms, given the nodes positions. Specifically, the DT for a set of points in a plane is a triangulation such that no point is inside the circumcircle of any triangle; the RNG is defined by connecting two points whenever there does not exist a third point closer to both points; the GG is a graph where two points have an edge between them if no other point exists in the circumball containing the two points; last, the MST is a tree consisting of all points while having the minimum total weight (length). Though, the β-skeleton graphs [2] constitute a parameterized family of planar proximity graphs where different β values give rise to different graphs.

This study contributes to the urban street modeling, examining the fitness of planar proximity graphs, particularly the β-skeleton graphs, on real street networks with complex characteristics. Additionally, a possible explanation is discussed concerning the findings of the analysis.

The rest of this paper is structured as follows. Section 2.2, contains some preliminaries on the β-skeleton concept. The datasets and the methodology used are described in Sect. 2.3, while the results of applying the methodology are presented in Sect. 2.4. Section 2.5 discusses a possible explanation of the findings and finally Sect. 2.6 concludes the study.

2.2 The β-Skeleton Graphs

In the lune-based neighborhoods approach [2], given a spatial distribution of points S in two-dimensional space, two points u and v are connected by an edge whenever the intersection of the two disks of radius r, centered at the points $c1$ and $c2$, contains no points of S (see Fig. 2.1).

The case $\beta = 0$ corresponds to the DT, $\beta = 1$ corresponds to the GG and $\beta = 2$ corresponds to the RNG. For $1 \leq \beta < \infty$, the radius and the disk centers are defined as follows:

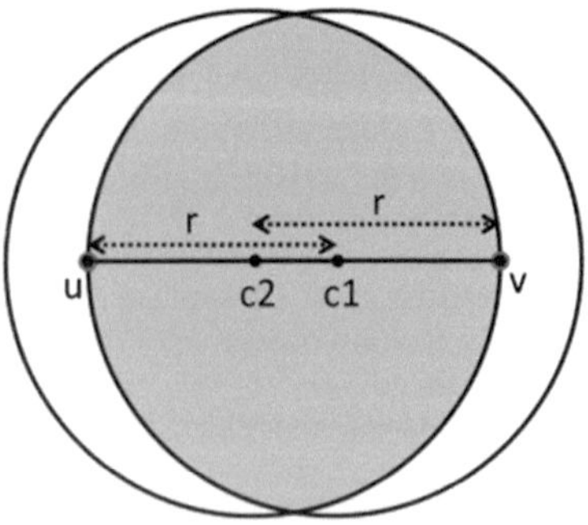

Fig. 2.1 Definition of β-skeleton in the lune-based variant for $1 \leq \beta < \infty$

$$r = \frac{\beta \cdot D(u, v)}{2} \tag{2.1}$$

$$c1 = \left(1 - \frac{\beta}{2}\right) \cdot u + \left(\frac{\beta}{2}\right) \cdot v \tag{2.2}$$

$$c2 = \left(\frac{\beta}{2}\right) \cdot u + \left(1 - \frac{\beta}{2}\right) \cdot v \tag{2.3}$$

while for $0 < \beta < 1$ the two disks pass through both u and v, with radius given by:

$$r = \frac{D(u, v)}{2 \cdot \beta} \tag{2.4}$$

The parameter β determines the size and shape of the lune-based neighbourhood. With the increase of β, the number of edges in the β-skeleton decreases (see Fig. 2.2).

A β-skeleton of a random planar set usually becomes a disconnected graph for $\beta > 2$ and continues losing its edges with further increase of β [3]. On the other hand, as β approaches zero, more and more edges are added to the β-skeleton until it eventually forms the complete geometric graph. For $1 \leq \beta \leq 2$, the following relationships among the different proximity graphs hold for any finite set of points S in the plane:

$$DT(S) \supseteq GG(S) \supseteq \beta - skeleton(S, \beta) \supseteq RNG(S) \supseteq MST(S) \tag{2.5}$$

Since urban street networks are usually connected networks neither DT-like, nor MST-like [4], it is thus of interest to answer to the following questions; (a) is there sufficient accuracy when using β-skeletons with $1 \leq \beta \leq 2$ to reproduce urban street networks? (b) is there a particular β value or subrange of values for which the accuracy is better? (c) what is the possible mechanism that leads real street networks to be associated with particular β values?

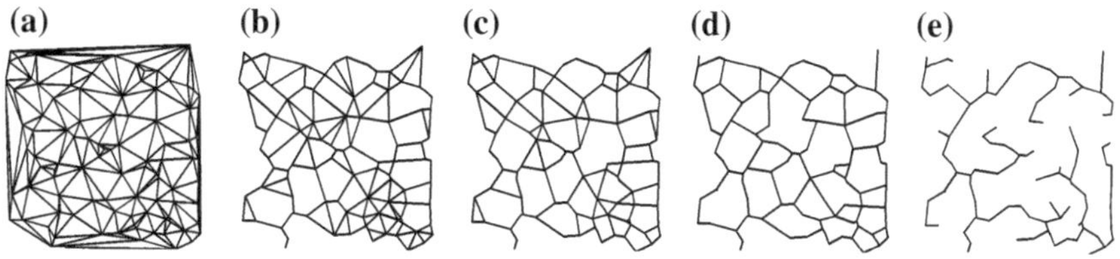

Fig. 2.2 Graph visualizations for the same set of 100 points: **a** delaunay triangulation ($\beta = 0$), **b** Gabriel graph ($\beta = 1$), **c** β-skeleton (here $\beta = 1.4$), **d** relative neighborhood graph ($\beta = 2$), **e** minimum spanning tree

Table 2.1 The datasets used for the goodness of fit verification

Dataset	Number of samples	Area
Cardillo et al. [5]	20	World
Peponis et al. [6]	118	USA
Strano et al. [7]	10	Europe
Chan et al. [8]	21	Germany
Maniadakis et al. [9]	100	Greece

2.3 Dataset and Methodology

2.3.1 Data

Five literature datasets of street networks [5–9] are used in order to compare their properties to those derived from the β-skeleton graphs. The properties of the 269 dataset samples in total (Table 2.1) are here normalized to correspond to 1 km^2 surface.

2.3.2 Methodology

The Primal approach [10] is used in studies [5–9] in order to turn Geographic Information System (GIS) data into spatial, weighted, undirected graphs $G(V, E, L)$ by associating nodes, V, to street intersections and edges, E, to streets (see Fig. 2.3), with length, L, as a weight.

For every sample,[1] obtained using the Primal approach, beyond the number of nodes $|V|$ (graph order), the basic statistical metrics are calculated (see Fig. 2.4); the number of edges $|E|$ (graph size), the density, the average node degree, the diameter, the average shortest path length and the cost (total wiring length).

Then, β-skeleton graphs are produced by simulations, tuning only the number of nodes $(10 \leq |V| \leq 1000)$ and the β parameter $(1 \leq \beta \leq 2)$. The same set of properties is computed for the β-skeleton graphs as well, and the resulting goodness of fit, with respect to the properties of the samples, is evaluated. In particular, the Mean Absolute Percentage Error (MAPE) measure is used for evaluating the comparison of the derived β-skeleton properties with the actual properties.

[1] In the samples where the entire set of these properties is not available, only the available properties are kept.

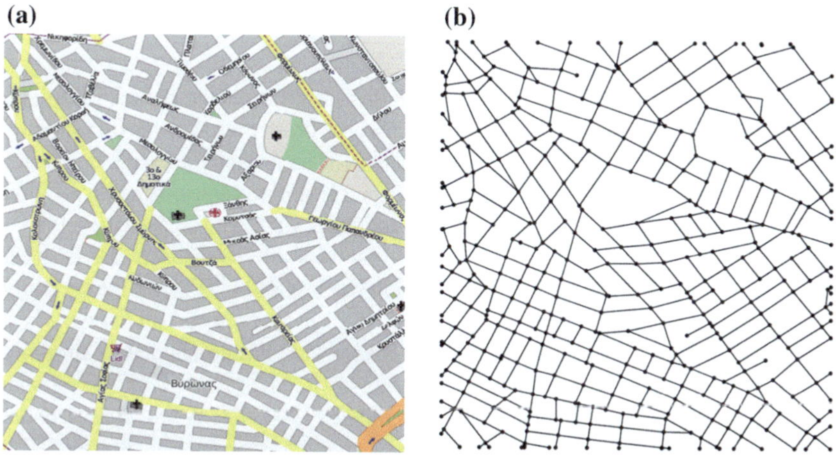

Fig. 2.3 A sample from the dataset of [9]. **a** The conventional street map, **b** the corresponding Primal graph

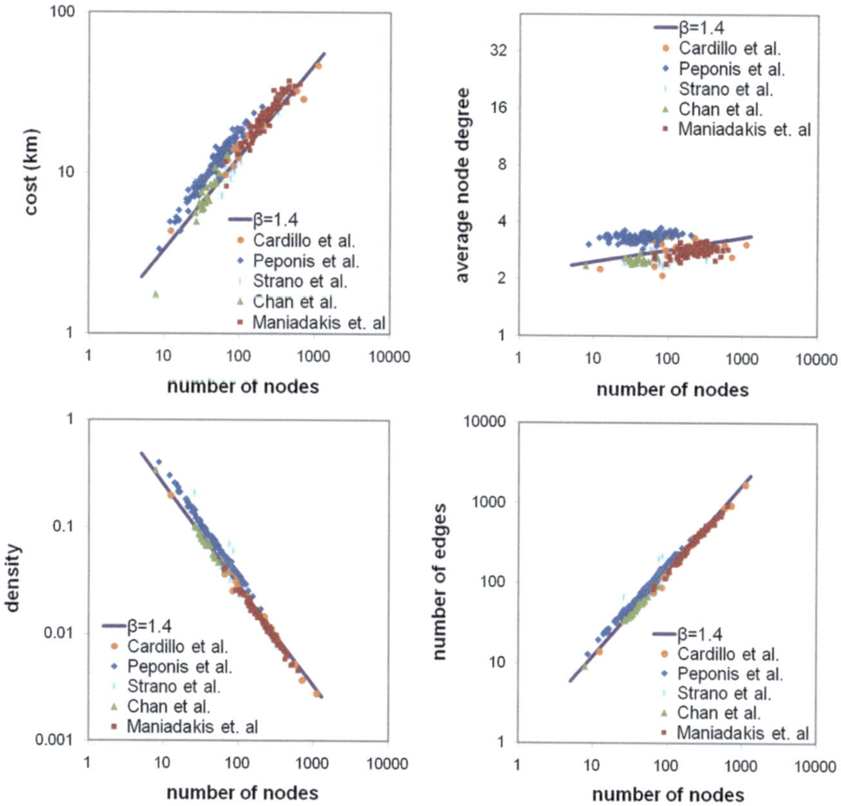

Fig. 2.4 Four of the properties as derived from real street datasets and from synthetic β-skeletons with $\beta = 1.4$ (the rest of the properties are not presented here due to lack of space)

2.4 Results

What the results of this study indicate is the sufficient ability of the β-skeleton graphs on the reproduction of the real urban street networks properties. The MAPE variance is observed in Fig. 2.5 per β value and per dataset. The β-skeletons with $1 \leq \beta \leq 2$ can lead to fitting errors of less than 10 % for certain β values. Specifically, the values of β yielding the highest averaged goodness of fit (lower MAPE), for the entire set of the examined statistical properties and for all five datasets, are found to span between 1.2 and 1.4 with $\beta = 1.4$ having the less errors (Fig. 2.6). Especially, for parameter $\beta = 1.4$ the average of MAPE of all properties and all datasets is as low as 8 %. A recent study by Osaragi and Hiraga [11] has concluded to similar β values too. In particular, they found that β lying between 1.15 and 1.45 corresponds to a maximum "agreement rate" between synthetic β-skeletons and streets of the Tokyo metropolitan region. Even though their study was geographically restricted and limited in the "agreement rate" index without investigating global topological and geometrical properties, however this matching of β values is remarkable.

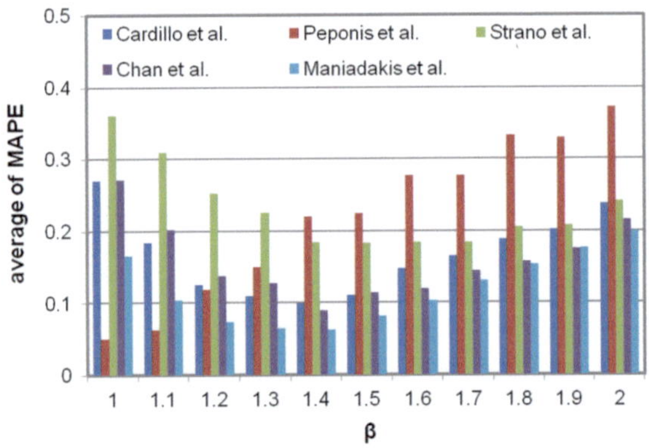

Fig. 2.5 The average of MAPE for each of the five datasets

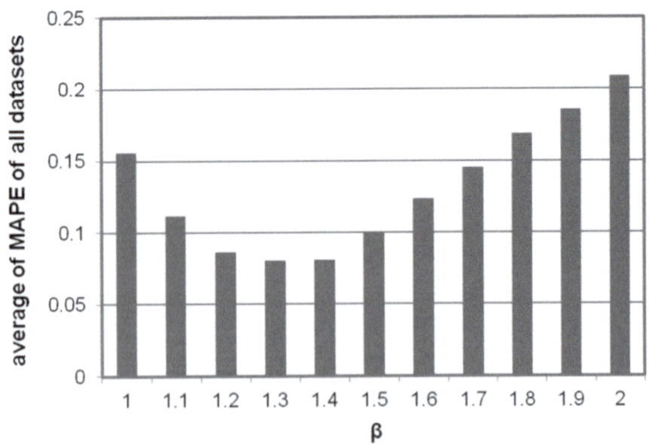

Fig. 2.6 The average of MAPE of all five datasets

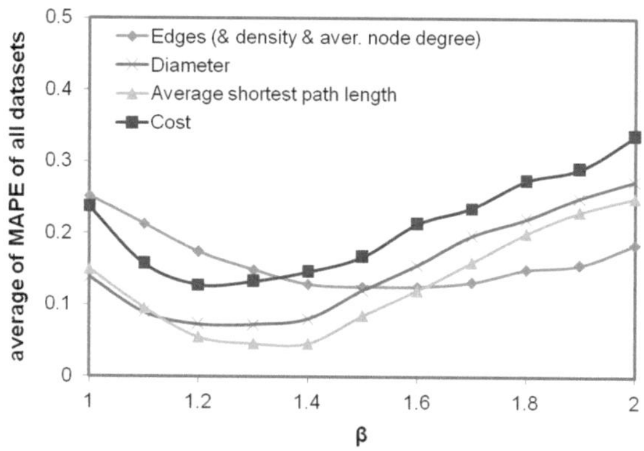

Fig. 2.7 The average of MAPE of all five datasets for the set of the basic statistical properties

The GGs ($\beta = 1$) and the RNGs ($\beta = 2$) can exhibit large errors when fitting real street properties. For instance, for the same planar set of nodes, the GGs produce higher number of edges and the RNGs generate graphs of quite less length compared to actual street networks. More specifically, as observed in Fig. 2.7, the diameter, the average shortest path length and the cost, all have the same behavior in terms of MAPE minimization with corresponding β values varying between 1.2 and 1.4. Regarding the rest of the basic properties, i.e., edges, density and average node degree, the MAPE minimization is shifted to slightly higher β values, approximately between 1.4 and 1.7. This implies that β-skeletons for $1.2 \leq \beta \leq 1.4$ may have almost identical properties with real street networks; however this arises with slightly increased number of edges compared to real street networks of the same order.

2.5 Discussion

Following the findings presented in the previous section, an intriguing question emerges. What is the mechanism that leads the majority of real street networks to have β-skeletons as equivalent graphs with β ranging between 1.2 and 1.4? Since it is natural for one to expect lower β values, as this would imply larger efficiency, is there a mechanism that restricts the network size and structure from going toward the DT characteristics ($\beta = 0$)? In this section, a possible explanation is attempted.

The hypothesis assumed here about the mechanism behind deriving the particular β values is that it may be a consequence of the percentage of land occupied by streets.

In their book [12] Meyer and Gómez-Ibáñez used data from the 1960s to examine the relationship between population density and land area in streets for 15 large cities in the United States. The majority of the investigated cities had a share of land in streets between 20 and 30 %. In addition, the results of a more recent study [13] indicate similar percentages of land for street space. It is likely to believe that this proportion is bounded at a certain level in urban areas, since the remaining land is needed for buildings, parks, plazas, parking, etc., and according to the above-mentioned empirical data this is around 30 %. Starting from this share of land used in streets, i.e., 20–30 %, and then constructing β-skeleton as street network, it is expected that particular β values will derive, as this percentage would limit the street network size and thereafter affect its structure.

More specifically, it is found that specific shares of land in streets correspond to specific values of normalized street network cost. Normalized cost ($cost_{rel}$) is a cost measure defined in [5, 9] with the introduction of two auxiliary graphs for each sample, which serve as two extreme cases; the respective MST (minimum cost) and the respective DT (maximum cost):

$$cost_{rel} = \frac{L_{graph} - L_{MST}}{L_{DT} - L_{MST}} \tag{2.6}$$

Let s be the share of land in streets, e be the surface and w be the width of the street. Then, L_{graph} is defined as:

$$L_{graph} = \frac{s \cdot e + |V| \cdot <k> \cdot w^2}{w} \tag{2.7}$$

where the term $<k>$ stands for the average node degree, therefore $|V| \cdot <k> \cdot w^2$ is added in order to take into account the multiple counting of the land at street intersections. Even though L_{graph} can vary with the number of nodes, the $cost_{rel}$ only slightly varies with the number of nodes, since it is normalized between the MST and the DT values. Thus, the figures that follow depict an average of values, whereas no large deviation is observed when testing 10–1000 nodes/km^2.

The derived normalized cost[2] values are similar to those actually observed in real street networks, e.g., [5]. The street width obviously can vary by city and by country. Here, the indicative values $w = 10$ m, $w = 15$ m, $w = 20$ m are set, with more realistic the case of $w = 15$ m. Actually, a large width w implies less total wiring and thus lower normalized cost. It is observed that 20–30 % share of land in streets corresponds to 20–30 % normalized street network cost (see Fig. 2.8a).

Then, it is possible to associate the normalized cost values with β values, since each β-skeleton corresponds to a cost between the respective MST and the respective DT costs. By running simulations, synthetic β-skeletons are produced and the relationship between normalized cost and β values is found (see Fig. 2.8b). This relationship is used for mapping the normalized cost to specific β values. Indeed, for 20–30 % normalized cost, the corresponding β values belong to the subrange between 1.2 and 1.4 (see Fig. 2.8b). As expected, these are the values of β associated with real urban street networks. This is more clearly seen in Fig. 2.9, where the overall relationship between parameter β and the share of land in streets is drawn, depicting the sensitivity of street width as well. In short, a certain level of structural

[2] It should be noted that the normalized cost is not a measure of construction cost, but only an index of how long the wiring of the graph is, compared to the respective extreme planar graphs (MST and DT).

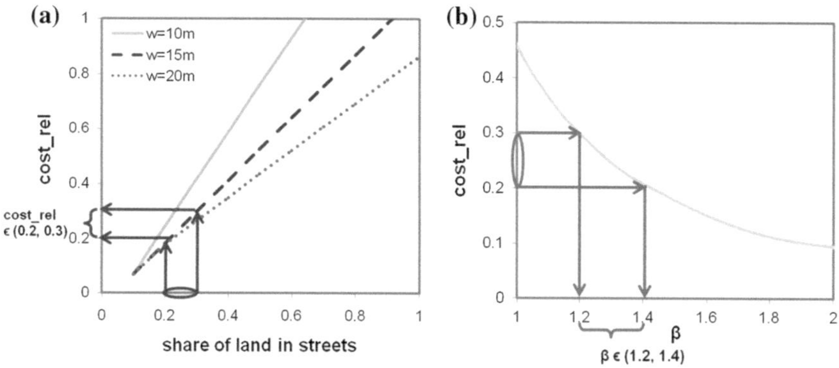

Fig. 2.8 **a** The relationship between normalized cost and share of land in streets, **b** the relationship between normalized cost and parameter β

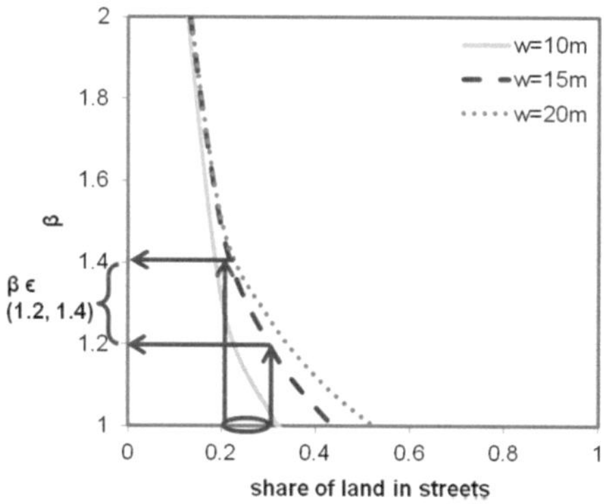

Fig. 2.9 The relationship between parameter β and the share of land in streets

similarities across urban areas as well as some differences may be well captured by the different shares of land in streets, and thus the different β values. Though, for realistic values of street width and reasonable shares of land used in streets, the urban street networks are found to be associated with a specific range of β values; the same range that was empirically observed.

2.6 Conclusion

Planar proximity graphs based on β-skeletons, which change as a response to variations in parameter β, were employed in the present study. Particularly, they were considered from the viewpoint of the topological and geometrical structure and were compared to real street networks.

The good agreement with empirical street data that was found to characterize the β-skeleton graphs induces their utilization—particularly for β values between 1.2 and 1.4—in modeling complex urban street networks and assisting various applications, more essentially those concerning street-constrained processes. Besides, the reasoning behind associating real street networks with the particular β values was discussed and a possible mechanism based on the share of land in streets was sketched.

References

1. Barthélemy, M., Flammini, A.: Modeling urban street patterns. Phys. Rev. Lett. **100**, 138702 (2008)
2. Kirkpatrick, D.G., Radke, J.D.: A framework for computational morphology. Comput. Geom. 217–248 (1985)
3. Adamatzky, A.: On growing connected β-skeletons. Comput. Geom. **46**, 805–816 (2013)
4. Hamaina, R., Leduc, T., Moreau, G.: A structural analysis of the streets network to urban fabric characterization. In: 25th International Cartographic Conference (ICC2011), pp. 1–8. Paris, France (2011)
5. Cardillo, A., Scellato, S., Latora, V., Porta, S.: Structural properties of planar graphs of urban street patterns. Phys. Rev. E. **73**, 066107 (2006)
6. Peponis, J., Allen, D., Haynie, D., Scoppa, M., Zhang, Z.: Measuring the configuration of street networks. In: 6th International Space Syntax Symposium, pp. 1–16. Istanbul, Turkey (2007)
7. Strano, E., Viana, M., Costa, L. da F., Cardillo, A., Porta, S., Latora, V.: Urban street networks, a comparative analysis of ten European cities. Env. Plann. B: Plann. Des. **40**, 1071–1086 (2013)
8. Chan, S.H., Donner, R.V., Lämmer, S.: Urban road networks-spatial networks with universal geometric features? Eur. Phys. J. B. **84**, 563–577 (2011)
9. Maniadakis, D., Varoutas, D.: Structural properties of urban street networks of varying population density. In: 10th European Conference on Complex Systems (ECCS'13), pp. 1–6. Barcelona, Spain (2013)
10. Porta, S., Crucitti, P., Latora, V.: The network analysis of urban streets: a primal approach. Environ. Planning B: Planning Des. **33**, 705–725 (2006)
11. Osaragi, T., Hiraga, Y.: Street network created by proximity graphs: its topological structure and travel efficiency. In: 17th Conference of the Association of Geographic Information Laboratories for Europe on Geographic Information Science (AGILE2014), pp. 1–6. Castellón, Spain (2014)
12. Meyer, J.R., Gómez-Ibáñez, J.A.: Autos, Transit, and Cities: A Twentieth Century Fund Report. Harvard University Press, Cambridge (1981)
13. Manville, M., Shoup, D.: People, parking and cities. J. Urban Planning Dev. **131**, 233–245 (2005)

Chapter 3
Qualitative Methods for the Exploration of Complexity in Human Social Systems: Applications in Family Psychology

Ana Teixeira de Melo and Madalena Alarcão

Abstract The study of Complex Systems has expression in the many fields dedicated to human social systems. However, Family Psychology has been slow in integrating contributions from complexity science, although the interaction could open new research avenues towards a deeper understanding of the processes underlying family emergence, development, change and adaptation. The mathematics of complex systems are appealing and many authors have applied them in psychological studies. However, there are many aspects of the family, as a human social systems, which are not amenable to quantitative analyses, at least without significant loss of the same features of complexity that need to be understood. In this paper, we discuss the use of qualitative methods in complexity research and present an integrative view of a Complexity-Informed, Qualitative, Case-Based Discovery-Oriented Family Psychology, oriented by a General Complexity Thinking approach where Abduction, along with induction and deduction, drives the enlargement of our understanding of complexity.

3.1 Introduction

Family science, in general, and family psychology, in particular, have, at their core, a systemic orientation [1]. Nevertheless, the systemic thinking easily vanishes when the methods chosen or the research designs oversimplify or fail to attend to the complexity of the family as a multidimensional collective entity, with particular properties. As a complex system, the family emerges from multiple and complex interactions among different, also quite complex, elements and by a close and mutually determined relationship with other social systems and ecological conditions.

A. Teixeira de Melo (✉) · M. Alarcão
Faculty of Psychology and Education Sciences of the University of Coimbra, Coimbra, Portugal
e-mail: anamelopsi@gmail.com

M. Alarcão
e-mail: madalena.alarcao@uc.pt

S. Battiston et al. (eds.), *Proceedings of ECCS 2014*, Springer Proceedings in Complexity, DOI 10.1007/978-3-319-29228-1_3

Family Psychology has been slow in incorporating contributions from Complexity science in order to deepen its knowledge about the collective properties of the family, and the processes sustaining it and leading it through transformations in time, which affect and are affected by its individual elements. There are some successful examples of integration of a complex dynamical systems perspectives in both quantitative and qualitative or mixed methods research in the field of psychology and developmental research [2]. Nevertheless, a lot of the research to date presents a very specific, relatively narrow focus or strategic orientations (e.g. aiming to understand the dynamics of a very particular dimension of a family relationship) that are insufficient to address fundamental questions regarding how the families emerge, what sustains them and how they change [3].

Complexity may inspire the raise of (re)new(ed) research questions and open new fields for exploration. It also offers a great variety of methodological and analytic tools that can be used, adapted and combined with other methods and even inspire the development of new ones [4], more suited to address core questions [3]. Complex systems perspectives have informed a revision of methods in psychology, particularly concerning quantitative methods and the number of [5]. In the absence of an integrative complex thinking perspective, research outcomes in psychology, in general, and family psychology in particular, will continue to appear too limited for a true expansion of our understanding.

Family psychology suffers from many of the fragilities affecting psychology in general. Among them is an excessive reliance on quantitative methods, research programs which are mostly variable-oriented and inspired by the search for nomothetic laws, the reliance on the study of groups, at the cost of learning from what is unique for individual cases [6, 7]. In fact, as other authors have noticed for psychology in general, family psychology has made its course more focused on building isolated pieces of evidence than in developing integrated theories and on studying isolated parts (even if they are sub-systems) more than the whole [7]. Although there are exceptions, it is more frequent for family studies to consider sub-systems or dyads (e.g. parent-child subsystem) rather triads or the family functioning as a whole, and, oftentimes those sub-system are studied in isolation from the larger one [8]. On the other hand, albeit exemplary exceptions [9], the investigation of the dynamics of families as complex systems does not take a leading role in fundamental and applied/intervention family psychology research.

Under the influence of Complexity Science, Family Psychology could re-connect with its systemic heritage in order to expand its outlook and find novel and potentially more useful ways to understand how families emerge and develop into such diverse forms. Under the scope of Complexity, Family Psychology could better understand the diversity of experiences among families and learn more about what connects the individual and the whole, as well as their mutual influence and change processes. It could also, more thoroughly, explore family formation, development and change as well as the multiple forms and pathways of causality associated with different developmental outcomes.

However, the field of Complex Systems is quite diverse, holding different "Complexities". A great deal of research is "simplistic" and "restricted" [10], aimed at universal, one-size-fits-all laws, driven by an urge to predict and control and to portrait a very "clean", deterministic and straightforward view of a world that seems sometimes pretty clean, sometimes not so clean and, sometimes, really messy [11, 12]. It is a Complexity still grounded in split metatheories of the world [7], which neglect, or even reject, many important aspects of how we participate, both individually and collectively, in the world. It is, therefore, a complexity science that fails to fully embrace the complexity thinking that needs to be present in a general complexity [10].

It is probably in the realm of human existence that we may find the most intriguing and fascinating features of complex systems, and the most complex forms of causality.

Many aspects of family systems are not amenable to quantitative analysis, nor suited for a strict application of mathematical or computational approaches, at least if one aims at embracing and understanding family complexity without forcing the results or losing sight of the uniqueness of these systems [13]. On the other hand, exploration of family complexity should be respectful of the time-bound and context-specific nature of the complex configurations of features and processes determining a family's trajectory [14–16].

We join others who advocate for the development of Complexity-Informed Social Sciences [17] to defend a Complexity-Informed Family Psychology, capable of accepting and respecting the uniqueness of the systems under attention, and of investigating human dynamics with methods whose features are congruent with those systems. We defend a Science that is not afraid to pay tribute to the rich heritages and traditions of qualitative and case-based research coming from the Social Sciences. We believe this integrative Family Psychology should be able to use traditional methods creatively, with a "complexity twist", both alone or in combination with other methods and analytic tools.

In this paper, we present an integrative view of a Complexity-Informed, Qualitative and Discovery-Oriented Family Psychology that, without denying the value of quantitative research calls for a complementary qualitative focus. We reflect on some of the synergies emerging from the interactions of Complexity Sciences, Complexity Thinking, Qualitative and Case-based methods and Abduction for the field of family psychology. We briefly discuss how these contributions may enrich Family Psychology and what emphasis and caution should be used in their integration.

3.2 Investigations of Family (Through) Complexity

3.2.1 Complexity Science and Family Psychology

The American Psychological Association defines Family Psychology as "a broad and general specialty in professional psychology founded on principles of systems theory with the interpersonal system of the family the focus of assessment, intervention and research" [18]. Family psychologists are expected to be able to understand human behavior through the application of systemic concepts [18].

Traditional themes of family psychology research include the investigation of family functioning and interpersonal relationships within the family, but also the relationship between the individual and the family relational context, with consideration of a diversity of factors from the biological to intra or interpersonal as well as social and cultural ones [19]. Family psychology researchers have sought to understand processes implicated in the formation and transformation of relational bonds exploring their relation to individual and family development, positive, adaptation and resilience [20].

However, the dominant research paradigms are mostly hypothetical-deductive and, therefore, limited to testing existing theories. The research process tends to focus on variables and on measuring the extent their variation is associated with variations in some outcome of interest, mostly relying on assumptions of linear causality [7]. Even when interactions are considered, the analyses remain quite limited in capturing that which is truly complex about the family and in seeing the family as a whole, or its unique collective emergent properties beyond isolated variables.

Dominant research programs tend to overlook more wholesome configurations of traces [21] characterizing the family as a complex system, and both distinguishing and classifying the families as belonging to the same type of system.

Developmental dynamic systems theorists brought complexity closer to family psychology, by importing, adapting and even creating new methods inspired in Complex and Dynamical Systems Sciences to study different aspects of development [22]. However, they are not the majority.

Complexity science and its core concepts [23] invite family psychology to look at the family in new ways and with new methods. New research questions or different research focus may be established around topics such as:

- What are the specific properties that allow the identification of a system as a family?
- What processes constitute a family and underlie its systemic emergence and collective properties?
- What is the nature and properties of the information associated with coupling and (similar and different) family bonds?
- What changes and how does it change in family through time? What processes of change are associated with positive development, adaptation and resilience?
- What bottom-up and top-down processes are implicated in the relationship between the individual and the whole and what role do they play?
- What are relevant order parameters or coordination variables to understand different phase transitions? What relevant states and transitions can be investigated regarding relevant developmental and adaptation outcomes?
- What different type of dynamics and (non linear) processes are associated with family functioning, change and development? How can they be described?
- What features and properties define the networks or configurations of family bonds and how do they relate to developmental outcomes?

While inspiration from Complexity Science is welcomed, importation should be cautious and empirically grounded and influenced by a General Complexity approach.

3.2.2 Family Psychology and General Complexity Thinking

Traditional research often fails to incorporate a general complexity thinking, reducing the family to simple elements or testing some interactions, without being able to capture and articulate the special emergent properties of the whole and the relation between levels or boundary conditions. In the lack of a general complexity orientation, even studies assuming a complex or dynamical systems approach may be narrow in focus. In principle, any given aspect of family functioning or its sub-systems can be assessed through time. However, this alone, even if relying of dynamical systems mathematics gives us limited information of the properties of the family as a singular complex eco-self-organized [10] entity or its special features such as its autonomy or potential for adaptation and change.

As Morin [10] suggested, complexity thinking needs to be able to separate and distinguish but also to relate and connect. Zooming in may be necessary but it is fundamental to reflexively integrate the research results back into a whole.

The majority of studies assume, a priori, some properties of the system and test hypotheses deducted from theories. This has prevented the field from moving forward into understanding that which is specific and unique in the organization and dynamics of the system we call a family [3].

When approaching the family or other human social systems, it is important to make use of complex tools in order to embrace the naturally occurring complexity. Our methods often oversimplify and we lose touch with what really constitutes a foundation of our humanity: relationships embedded in other relationships. We must consider that relationships involve contradictions and dualities or complementary aspects. Family complexity research should attend to the complementary nature of many complex systems in order to systematically explore the utility of some complementary pairs and their dynamics [23] in understanding family bonds and inter-level influences. Exploratory research should look into identifying relevant complementary aspects of family functioning some, which may include aggregation ~ segregation, synchronization ~ uncoordination, closeness ~ distance; independence ~ influence; conflict ~ agreement; order ~ disorder; health ~ pathology [23]. If information emerges from differences, then it is likely that the analyses of differences, between families and within families, within and between studies, regarding the polarities of family relations and their complementary dynamics gives us new insight into the processes or information that matter the most to understand family functioning and inform family interventions. A general and complementary view of family complexity should also include an integrated complementary view of the biological, the psychological and the social and cultural aspects of family life. Biology must be understood in the context of the psychological and the psychological in the context of the social, and vice-versa [24]. Sub-systems must be understood considering lower and upper-levels of family functioning [23] and each level in relation to between level processes. The study of the relationship and the processes operating between levels may be of special relevance to understand natural occurring and therapeutic change.

There is a considerable terrain for exploration pertaining the relationship between the family boundary conditions and external coupling with the properties of internal coupling in the family.

Coupling is a central concept to understand a complex system, particularly living systems, but the study of information and communication in the family is still quite restrictive and focused on simple feedback cycles. We believe there are many domains and sources of information implied in the coupling holding family members together, and contributing to particular types of bonds, that are still not recognized or duly understood.

A Complexity-Informed Family Psychology must embrace the principles of general complexity. The acts of making/analyzing distinctions must join an ability to accommodate apparent contradiction and explore their complementarity. Differences and distinctions must be integrated in higher levels of analyses, where their relationships are easier to describe, comprehend and explain.

Family Psychology must be able to understand how the family exists in an integrated way in a multidimensional space and how it is able to simultaneously construct and be constructed by that space and the ones surrounding it. It must build, from the ground-up, specific and tailored concepts models, collected from and close to "real" family life. On the other hand, it must use to a top-down approach to explore family life through the lenses of the concepts applied to other systems, test their fitness and need for adaptations. The same families should be studied under diverse conditions. Likewise, similar conditions should be explored for different types of families in order to identify configurations of processes and features implicated in different family forms and developmental outcomes.

The methods chosen should also be able to capture or respect the core of the experiences associated to being a family or participating in a family. Researchers should ask: Do the descriptions allowed/informed by a complex systems approach respects or retain a sense of what we, in common sense, "feel" families to be like? Do they allow us to stay close to what is like for people to be a part of a family? Is the knowledge we are producing useful? How may it be used to benefit families?

Complexity thinking reminds us that the study of parts is just as important as the study of whole, and the relationship between the two. Nevertheless, whenever the researcher zooms in on the family, he/she should attempt to frame and make sense of the results in the context of the immediate/adjacent parts and the whole. During this process, the researcher must explore, explain and hypothesize about the results and possible variations regarding their implications for understanding family structural, functional and dynamical complexity. Research programs should mimic the nested and integrative nature of complex system by conducting related series of single and multiple intensive case studies and integrating the results.

3.3 Discovery-Oriented Research and Abduction

Family psychology should not over rely on hypothetical-deductive research designs. Under the scope of complexity, it should embrace the modes of thinking and the kind of research where novelty is more likely to emerge and be reasoned in depth. Family psychology needs to explore new insights in a data-driven but reflexive, flexible and creative way. This is more likely to happen through methods where the researchers' control is only partial (and therefore not limited to what he/she already knows), and where the families can guide him/her into discovering new features of family life. A discovery-oriented approach is essential for new hypothesis, new concepts, contents, forms and descriptions as well as explanations to emerge and enrich the field with greater comprehensiveness. Complexity Science and Complexity Thinking play an important role in guiding and inspiring the family researcher. Family psychology needs to fully embrace research programs where both induction and deduction have a place but also where abduction is likely to be needed and propel the research process. As a form of reasoning aimed at finding the best possible explanations for otherwise, poorly understood empirical facts, abduction is a form of reasoning many authors have considered to be at the heart of true science [25]. We will not explore the specificities of abduction here, inviting the reader to consult other sources [26]. Nevertheless, we would like to stress that abduction is about taking a cognitive leap beyond the data to explore novel explanations. It is about making comparisons and exploring differences and similarities in observations in order to find alternative explanations which can then be further tested in research cycles involving abduction, induction and deduction [26].

Complexity-Informed Family Psychology may confront the researcher with new ways of looking into existing empirical data or new facts or experiences and invite new explanations. For this, the researcher needs to have access to different types of data but also make a reflective use of the resonances emerging from the interaction between his/hers experiential and academic knowledge, and the information produced during research and when contacting with the families. He needs to be able to use the theory and the techniques as much as him/herself, as a complex tool, into building "informed" grounded theories [27]. The researcher's knowledge about other systems may be helpful if he/she can use that information to raise relevant questions to the data and construct new pathways for empirical and reflexive cognitive exploration [27]. He/she must be able to assume its own complexity by recognizing and combining multiple reasoning abilities, exercising an open mind and maintaining a reflexive stance. In sum, he/she should be able to recognize, experience, "feel" and purposively deal with the complexities under and of the investigations.

3.4 Complexity-Informed Qualitative Based Research

The research methods used to investigate complexity should be congruent with the nature of the system under investigation. Human systems are immensely different from physical or other biological systems. Time and context, meaning, history and culture are also of extreme importance to understand human systems and the research methods used must attend to them. The methods should be flexible enough to allow the researcher to revise his/hers assumptions while collecting or analyzing the data in order to adjust them to emergent results [28], particularly in early stages.

As other social systems, families can be studied as cases, or complex configurations of features [11, 29, 30], to systematically explore the multiple, complex and contingent forms of causality [21, 31] implied in family functioning while guaranteeing the necessary "thickness" to ground their interpretations and guard against oversimplification [32]. Case-based research is suited to capture the rich qualities of the family.

Qualitative methods have long proven their value. Other authors have used and proposed qualitative analysis for the study of complex social systems [32, 33].

For our current purpose, we will assume a pragmatic stance and avoid discussing the epistemological or ontological implications of the choice of the research methods, in order to focus their pragmatic possibilities.

In many ways, both isolated and in combination with quantitative methods, qualitative research fulfills the needs of a Complexity-Informed Family Psychology, as discussed in this paper. It serves both a logic of justification and discovery/exploration [26] and can accommodate different epistemological and ontological views. They can stand alone or in combination with quantitative methods and be applied both to qualities/texts/descriptions and quantities/numbers/measurements.

Qualitative research affords multiple and flexible ways of sampling [32] and may retrieve contextualized and contingent information regarding many aspects of the family functioning. Its methods give the researcher the possibility of collecting time-bounded, historical, sequential (e.g. in life narratives) and multi-level information.

Qualitative analyses can deal with "messy" and multiple types of data (oral/visual; self/hetero-reported; retrospective/prospective; in vivo vs processed data, etc.) and methods for data visualization. They can be structured and organized and still guarantee room for the researcher's intuitions and subjective perceptions to propel the analyses. They can attend to the what's, the how's and the why's of family functioning and look into both spatial/contextual and dynamic/time information.

Qualitative methods allows the researcher to be immersed in the data and put information in context, attending to multiple and concurrent potential causes [34]. They have the potential to integrate multiple aspects of reality as experienced, perceived and constructed by family members and can accommodate and integrate the contradictions and complementary/apparently aspects associated with family life. They also help the researcher to stay close to "real life complexity", namely how people experience their lives and the complex features of being a part of a family.

Different categorizing and connecting strategies may contribute to explore different aspects of the family system at both a descriptive and explanatory level [34]. Qualitative descriptions may be constructed regarding general or specific properties of the family [7]. Exploratory, discovery-oriented inspections of the qualities of the families and its bonds may lead the researcher to the identification and description of particular features of the information associated to different forms and consequences of internal and external coupling. Specific dynamics can also be portrayed in qualitative terms [35] with the advantage that the researcher may not only use imported concepts but also adapt them as suited and build new ones [27]. Qualitative methods support the identification of concepts, categories, themes, contents, patterns and profiles that may compose a map of the family's structural and dynamical complexity providing conditions for the emergence of a tailored language for the emergent theoretical constructions.

The researcher may develop qualitative coding schemes to explore specific structural and dynamical aspects of family functioning, from a complex systems perspective. The codes may be built after emergent conceptual categories and properties, as theories are constructed from the ground up [36, 37], or after established theories. This sort of coding schemes can be converted into valid input for other sort of analyses, including mathematical.

The exploration of complexity within human systems can rely on qualitative comparative methods [30, 38], including case-based comparative studies [30] in order to elucidate how different and similar dimensions and variables, processes and mechanism lead to similar or different outcomes, for example, regarding the family's positive adaptation. These methods can help the researcher discern patterns and make sense of such differences and similarities from which to construct theoretical building blocks.

Comparative methods have been at the core of approaches such as Grounded Theory aimed at unraveling the complexity of social phenomena [36, 37]. The development of grounded theory implies the intensive elaboration of the properties of categories emerging from the data, describing and explaining the phenomena under study, after thorough, initial coding. Grounded theory is an example of a qualitative approach suited to answers some of the core questions of a Complexity-Informed Family Psychology.

Comparative practices help us construct more complex pictures of our human world, and may assist us in the exploration of the innumerous dimensions and properties of family life. They may provide us with concrete clues about the specific type of structural and dynamical traces associated with family complexity and different types of outcomes and construct a more solid theoretical base to choose appropriate mathematical methods for further explorations.

Qualitative methods can inspire researchers to respect the human nature of family systems by including information from particular viewpoints internal to the family. Information collected from various sources may be rich enough for the researcher to develop an understanding the multidimensionality of family functioning and the differences and similarities of the organization at different levels.

Finally, qualitative analyses afford unique explorations of causality, through contiguity analyses [34] in ways that are flexible and not so much dependent on the controllability of the sampling conditions and data collection. Systematic inspections of the data can aim at the identification of connections between its building blocks. The qualitative community has explored methods such as matrixes and displays to explore causal relationships, often supported by intensive memoing [39]. Qualitative Comparative Analysis is also a powerful tool to analyze matrixes of variables, and compare configurations of features, while considering the fuzziness of the social world [30]. This potential of this sort of method has been underexplored in family psychology but may be useful in a Complexity-Informed Family Psychology.

The meanings of the parts in the context of the wholes and vice-versa have long been the subject of hermeneutics. Juarrero [15] pinpointed hermeneutics as a privileged interpretative strategy to work on the reconstitution and the elaboration of narrative and historical explanations for the course of human action. The author highlights how hermeneutics resembles the dynamics of self-organization. She identifies its comprehensive potential in the extent that life events gain unique meanings and may illuminate points of decision and transformation in human trajectories as well as the conditions contributing to them. This sort of reasoning is not very strange to family therapists used to elaborate clinical hypotheses from perspectives that recognize the importance of time and place. Rennie presents hermeneutic analysis as an essential component of a general framework of qualitative analysis in psychology [40]. But family research has still to fully explore the contributions of case-based, hermeneutical, interpretative narrative and historical hypotheses, for the explanation of the family outcomes of interest. These explorations of part-whole are facilitated by the diversity of data and data collection methods that can be used.

3.5 Conclusion

In this paper, we have discussed how Family Psychology may integrate Complexity Science to deepen the study of the family as a complex system. We have also stated that to shed light into that which is truly complex in a family, as a system, and that, which is truly unique in the system we call a family, the influence of Complexity Science should occur under the scope of General Complexity. We have highlighted (re)new(ed) research questions emerging from these interaction and stated that the methods used to investigate them should be congruent and suited to the specific features of families as particular forms of human social systems. These are probably among the most complex of complex systems. They are not only multidetermined and multidimensional as they exist in a complex network of influence effects between individual and collective realms of existence. We considered how a Complexity-Informed Family Psychology should build its own pathways of research and its own body of knowledge by combining the influence of Complexity Science and Complexity Thinking, and the specific research questions emerging from these interactions, with a rich tradition of qualitative research methods in the social sciences. We have proposed that the field should be driven by exploratory intentions and methods that are somehow congruent with the hypothesized nature and properties of human social systems, such as the family. We, therefore, advocated for a creative use of qualitative methods in flexible research designs, encompassed by integrative and multi-methods research programs. In order to progress and explore new research questions, studies must adopt an exploratory, discovery-oriented stance, where new types, new arrangements, or interpretations of data can call for novel explanations and new ways of thinking about the family as a complex system. Abduction, and the researchers' own experiences, knowledge and intuitions are, therefore, prone to play a pivotal role in the unfolding of a Complexity-Informed Family Psychology.

Acknowledgments This work was supported by a post-doctoral research fellowship (SFRH/BPD/77781/2011) attributed to the first author by the Portuguese Foundation for Science and Technology.

References

1. Cox, M.J., Paley, B.: Families as systems. Annu. Rev. Psychol. **48**, 243–267 (1997)
2. Hollenstein, T.: Twenty years of dynamic systems approaches to development: significant contributions, challenges, and future directions. Child Dev. Perspect. **5**(4), 256–259 (2011)
3. Melo, A.T., Alarcão, M.: Beyond the family life cycle: understanding family development in the twenty-first century through complexity theories. Fam. Sci. **5**(1), 52–59 (2014)
4. Van Geert, P.: Dynamic systems. In: Laursen, B., Little, T.D., Card, N.A. (eds.) Handbook of Developmental Research Methods, pp. 725–741. The Guilford Press, New York (2012)
5. Guastello, S.J., Koopmans, M., Pincus, D. (eds.): Chaos and complexity in psychology. The theory of nonlinear dynamical systems. Cambridge University Press, New York (2009)
6. Valsiner, J., Molenaar, P.C.M., Lyra, M.C.D.P., Chaudhary N.: Preface. In: Valsiner, J., Molenaar, P.C.M., Lyra, M.C.D.P., Chaudhary, N. (eds.), Dynamic process methodology in the social and developmental sciences, pp. v–xi. Springer, Dordrecht (2009)
7. Toomela, A.: How methodology became a toolbox-and how it escapes from that box. In: Valsiner, J., Molenaar, P.C.M., Lyra, M.C.D.P., Chaudhary, N. (eds.) Dynamic Process Methodology in the Social and Developmental Sciences, pp. 45–66. Springer, Dordrecht (2009)
8. Cox, M.J., Paley, B.: Understanding families as systems. Curr. Dir. Psychol. **12**(5), 193–196 (2003)
9. Gottman, J.M., Murray, J.D., Swanson, C.C., Tyson, R., Swanson, K.R.: The Mathematics of Marriage: Dynamic Nonlinear Models. The MIT Press, Cambridge, MA (2002)
10. Morin, E.: Restricted complexity, general complexity. In C., Gershenson, D. Aerts, and B. Edmonds (eds.), Worldviews, Science, and Us: Philosophy and Complexity, pp. 5–29. Singapore, World Scientific (2007)
11. Byrne, D.: Complexity, configurations and cases. Theory Cult. Soc. **22**(5), 96–111 (2005)
12. Morin, E.: Introduction à la pensée complexe (Introduction to complex thinking). Éditions Points, Paris (2002) (originally published in 1990)
13. Byrne, D.: Complexity theory and social research. Soc. Res. Upd. **18**, Autumn. (1997)
14. Castellani, B., Hafferty, F.: Assemblage: Aa method for doing qualitative and historical complexity science. Available at: https://www.academia.edu/2890671/Assemblage_A_Method_for_Doing_Qualitative_and_Historical_Complexity_Science_Brian_Castellani_and_Fred_Hafferty. (n.d.)
15. Juarrero, A.: Dynamics in Action: Intentional Behavior as a Complex System. MIT Press Cambridge, MA (1999)

16. Fogel, A.: Interindividual communication: the transformation of meaning-making. J. Dev. Proc. **1**(1), 7–30 (2009)
17. Byrne, D., Gallaghan, G.: Complexity Theory and the Social Sciences. The State of the Art. Routledge, New York (2014)
18. http://www.apa.org/ed/graduate/specialize/family.aspx
19. Liddle, H.A., Santisteban, D.A., Levant, R.F., Bray, J.H. (eds.): Family Psychology. Science-Based Interventions. American Psychological Association, Wahsington, DC (2006)
20. Walsh, F.: Normal Family Processes: Growing Diversity and Complexity. The Guildford Press, New York (2012)
21. Byrne, D., Uprichard, E.: Useful complex causality. In: Kinkard, H. (ed.) Oxford Handbook of Philosophy of Social Science, pp. 109–129. Oxford University Press, Oxford (2012)
22. Hollenstein, T.: Twenty-years of dynamic systems approaches to development: significant contributions, challenges and future directions. Child Dev. Perspect **5**(4), 256–259 (2011)
23. Érdi, P.: Complexity Explained. Springer, Berlin (2008)
24. Overton, W. F.: Developmental psychology: philosophy, concepts, methodology. In: Damon, W., Lerner, R.M. (eds.): Handbook of Child Psychology, vol. 1, Theoretical models of human development, pp. 18–88. Wiley, New Jersey (2006)
25. Rozenboom, W.W.: Good science is abductive, not hypothetico-deductive. In: Harlow, L.L, Mulaik, S.A., Steiger, J.H. (eds.), What If There Were No Significance Tests?, pp. 366–391. Erlbaum, New Jersey (1997)
26. Reichertz, J.: Induction, deduction, abduction. In: Flick, U. (ed.) The Sage handbook of qualitative data analysis, pp. 121–135. Sage, London (2014)
27. Thornberg, R.: Informed grounded theory. Scand. J. Educ. Res. **56**(3), 243–259 (2012)
28. Gubrium, J.F., Holstein, J.A.: Analytic inspiration in ethnographic fieldwork. In: Flick, U. (ed.) The Sage handbook of qualitative data analysis, pp. 35–48. Sage, London (2014)
29. Byrne, D., Ragin, C.C. (eds.): The Sage Handbook of Case-Based Methods. Thousand Oaks, Los Angeles (2009)
30. Ragin, C.C.: Fuzzy-Set Social Science. The University of Chicago Press, Chicago (2000)
31. Given, L. M.: The Sage Encyclopedia of Qualitative Research Methods, vols. 1, 2. Sage Publications, Thousand Oaks (2008)
32. Castellani, B.: Qualitative/Narrative Complexity Science. Blog post. http://sacswebsite.blogspot.pt/2009/03/qualitativenarrative-complexity-science.html (2009)
33. Castellani, B., Hafferty, F.: Assemblage: A method for doing qualitative and historical complexity science. https://www.academia.edu/2890671/Assemblage_A_Method_for_Doing_Qualitative_and_Historical_Complexity_Science_Brian_Castellani_and_Fred_Hafferty (n.d.)
34. Maxwell, J.A., Chmiel, M.: Notes toward a theory of qualitative data analysis. In: Flick, U. (ed.) The Sage Handbook of Qualitative Data Analysis, pp. 21–34. Sage, London (2014)
35. Vedeler, D., Garvey, A.P.: Dynamic methodology in infancy research. In: Valsiner, J., Molenaar, P.C.M., Lyra, M.C.D.P., Chaudhary, N. (eds.) Dynamic Process Methodology in the Social and Developmental Science, pp. 431–453. Springer, Dordrecht (2009)
36. Glaser, B.G., Strauss, A.L.: The discovery of grounded theory: strategies for qualitative research. Aldine Transactions, New Brunswick (2008) (Originally published in 1967)
37. Charmaz, K.: Constructing Grounded Theory. A Practical Guide Through Qualitative Analysis. Sage, London (2006)
38. Palmberger, M., Gingrich, A: Qualitative comparative strategies. In: Flick, U. (ed.) The Sage Handbook of Qualitative Data Analysis, pp. 94–108. Sage. London (2014)
39. Miles, M.B., Huberman, A.M.: Qualitative Data Analysis. An Expanded Sourcebook, 2nd edn. Sage Publications, Thousand Oaks, CA (1994)
40. Rennie, D.L.: Qualitative research as methodical hermeneutics. Psychol. Methods **17**(3), 385–398 (2012)

Chapter 4
Tangible Networks: A Toolkit for Exploring Network Science

Espen Knoop, Edmund Barter, Alonso Espinosa Mireles de Villafranca, Antoni Matyjaszkiewicz, Christopher McWilliams and Lewis Roberts

Abstract We present Tangible Networks (TN), a novel electronic toolkit for communicating and explaining concepts and models in complexity sciences to a variety of audiences. TN is an interactive hands-on platform for visualising the real-time behaviour of mathematical and computational models on complex networks. Compared to models running on a computer, the physical interface encourages playful exploration. We discuss the design of the toolkit, the implementation of different mathematical models and how TN has been received to date.

4.1 Introduction

Our work focuses on communicating ideas from complexity science to a non-specialist audience. Academics are frequently required to communicate their research to funding bodies. Since the Wolfendale committee [26], there has been an increasing drive to communicate research to the public and promote dialogue [5, 27]. We believe that public engagement is of particular importance to complex systems research because of its relevance to a wide range of systems in nature, engineering and social sciences. Drawing on our experience of public engagement [3], we have developed Tangible Networks (TN) for communicating key concepts from complexity science, and more generally facilitating learning.

E. Knoop (✉) · E. Barter · A.E.M. de Villafranca · A. Matyjaszkiewicz ·
C. McWilliams · L. Roberts
Department of Engineering Mathematics, University of Bristol,
University Walk, Bristol BS8 1TR, UK
e-mail: espen.knoop@bristol.ac.uk

E. Knoop
Bristol Robotics Laboratory, T Block, Frenchay Campus, Coldharbour Lane,
Bristol BS16 1QY, UK

S. Battiston et al. (eds.), *Proceedings of ECCS 2014*, Springer Proceedings
in Complexity, DOI 10.1007/978-3-319-29228-1_4

Complexity scientists make extensive use of tools which can be unfamiliar to non-specialists, including advanced mathematical techniques and computer algorithms. From our experience it can be a challenge to effectively communicate concepts in complex systems without referring to these tools. Furthermore, mathematical and algorithmic models are becoming increasingly pervasive in today's society. Therefore, making these models more accessible and engaging is an important task.

One effective way of communicating such concepts is by using interactive models, where users can perturb the simulation and observe how the dynamics are affected. For example, *NetLogo*[1] is an interactive visualisation software package developed for exploring agent-based systems in real time. *Danceroom Spectroscopy* [12, 19] is an interactive model of molecular dynamics where the visualisation of the moving molecules is projected on a screen and the bodies of users become "energy landscapes" that directly affect the forces on the molecules. Danceroom Spectroscopy has been used as an art installation, in dance performances, for education and also for research. As well as teaching, these platforms are useful for raising awareness of mathematical modelling.

Mathematical models are generally presented on a computer. The idea of Tangible Interfaces [15] is to interact with the digital world by manipulating physical objects. It has been argued that Tangible Interfaces encourage playful learning and creative exploration, are more accessible and are well suited for collaboration [18]. There are a number of educational toolkits for teaching electronics (e.g. *LittleBits* [4]) and robotics (e.g. *cubelets*, previously *roBlocks* [24]). A tangible interface for teaching mathematics is *Smart Blocks* [11] where shapes can be constructed by snapping blocks together, and the volume and surface area of the shape is computed. Horn [14] presents a tangible tool-kit for teaching programming through interaction with physical blocks.

Taking inspiration from interactive models and tangible interfaces, we have created a tangible interactive network model. Tangible Networks makes the exploration of science and complex systems more approachable and inviting to a wider range of audiences. TN is a physical platform for network simulations that makes the key components (nodes and links, [21]) physical building blocks that can be manipulated while simulations are running, as illustrated in Fig. 4.1. The platform has been designed to make the network topology clearly visible and reconfigurable, so that users can get an understanding of how network topology affects behaviour.

There are several examples of robotic swarms being used to demonstrate complex network behaviour (e.g. [23]), where typically robots will communicate wirelessly with other neighbouring robots and the swarm exhibits a global behaviour from the local interactions. Although these platforms are excellent for demonstrating concepts such as swarming and emergence, it is more difficult to build a fundamental understanding of ideas such as network topology and how topology affects behaviour. We believe TN is well suited for teaching such concepts.

---

[1] Available online at https://ccl.northwestern.edu/netlogo/.

Fig. 4.1 We want to create a hands-on interactive visualisation of complex network models

4.2 The Toolkit

Many complex systems are modelled as interactions between simple agents connected in a network where collective behaviour emerges from localised interactions. With TN, we have created a physical network, where nodes and links are represented by electronic units and connecting wires respectively. Each node runs a mathematical model of the local dynamics. Users can interact with the network and observe how the behaviour changes. The TN toolkit is shown in Fig. 4.2, with key features labelled.

A central objective in the development of TN was to create a simple and robust platform allowing for playful interaction. We wanted users to be able to change the network topology and add or remove nodes from the network with the simulation running.

In many real-world systems, agents only have local information. In keeping with this, we have implemented a distributed simulation where each node runs a model of the local dynamics. This implementation also improves robustness. Two nodes can only exchange information if there is a link between them. This makes the behaviour of the model more transparent—there is a very close link between the physical and mathematical structures. There are some disadvantages of distributed simulations, e.g. computing a global state of the system is difficult. However, for many systems a distributed model is highly appropriate.

In our implementation, links are directed (Fig. 4.2d). This is a more general case, as undirected links can be made by combining two directed links (Fig. 4.2e). The maximum degree of each node is determined by the number of physical connectors

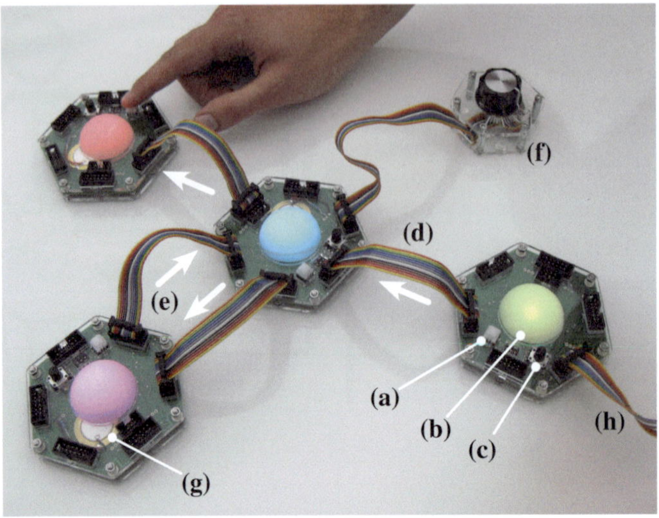

Fig. 4.2 The Tangible Networks toolkit. Labelled are **a** pushbutton switch; **b** glowing dome; **c** potentiometer; **d** directed link; **e** undirected link; **f** master controller; **g** piezo speaker; **h** power supply. Online version in colour

it has. In TN, nodes have a maximal in- and out-degree of three. This is the simplest case where non-trivial undirected networks can be built—with a degree of two only lines and rings would be possible.

Users can interact with the running network simulation. The topology of the network can be changed by reconfiguring the wires, and nodes can be added or removed from the network. Each node can be perturbed with a pushbutton switch (Fig. 4.2a), and local parameters can be changed with a potentiometer (Fig. 4.2c). The pushbutton and potentiometer can be programmed to have any function.

As well as local control, users can control global properties by connecting a master controller to any one node. A master dial (Fig. 4.2f) can be used to adjust a global parameter, such as the coupling parameter in models of coupled dynamical systems. This sends a continuous value to all of the nodes in the network that can be read by each node. Alternatively, a master pushbutton switch can be used for digital input such as to reset the simulation.

We require nodes to output their current state. The output must be in a form such that the collective behaviour of a large network can be easily observed. For this reason, nodes produce visual output by means of a glowing dome (Fig. 4.2b). The state of each node can be visualised with the brightness and colour of the dome or by changing the frequency of brightness oscillations. Nodes can also produce sound by means of a piezo speaker (Fig. 4.2g).

4.2.1 Technical Description

4.2.1.1 Processor

An Atmel ATMega 328p microcontroller, as used in the widespread open-source Arduino platform [1], runs the local model on each node. The Arduino programming language is essentially C++, with low-level hardware control hidden in wrapper functions but still being available if required. Arduino is designed to be easy to use and is very well supported on-line, requiring no experience in microcontroller programming. We have written a library for interfacing with the TN hardware. Programs are uploaded to the TN nodes with an In-System Programming (ISP) hardware programmer.

4.2.1.2 Electronics

The state of each node is shown with a glowing dome, lit up using a high-brightness Red-Green-Blue (RGB) Light Emitting Diode (LED) behind a diffuser. LEDs are controlled with Pulse Width Modulation (PWM). Power is distributed through the network, and the power supply can be connected to any node (Fig. 4.2h). Neighbouring nodes communicate via analogue signals, so each connection transmits a single real number in a limited range. This is very robust, and forces models to be simpler and more intuitive. The master controller sends an analogue voltage to all the nodes in the network. The nodes have a tactile momentary pushbutton switch; a potentiometer connected to an analogue input pin and three DIP configuration switches. Insulation Displacement Connectors (IDCs) and ribbon cables are used for the links.

4.2.1.3 Mechanical

The nodes are designed to be simple to fabricate with limited facilities. Each unit has a single Printed Circuit Board (PCB); a diffusor fabricated from a bisected table tennis ball and an enclosure comprised of two pieces of laser cut acrylic. A CAD drawing of a TN node is shown in Fig. 4.3.

The nodes are hexagonal, with one connector along each edge of the hexagon and a centred glowing dome. The hexagonal shape gives the units a visual identity and the symmetry lends itself well to different network topologies. Input and output connectors are alternated, so a bidirectional connection is achieved by having links to two adjacent connectors. Each node is 90 mm across, which is sufficiently small so that it can be picked up by a child and yet large enough that the unit can be supported with the other hand when plugging in cables.

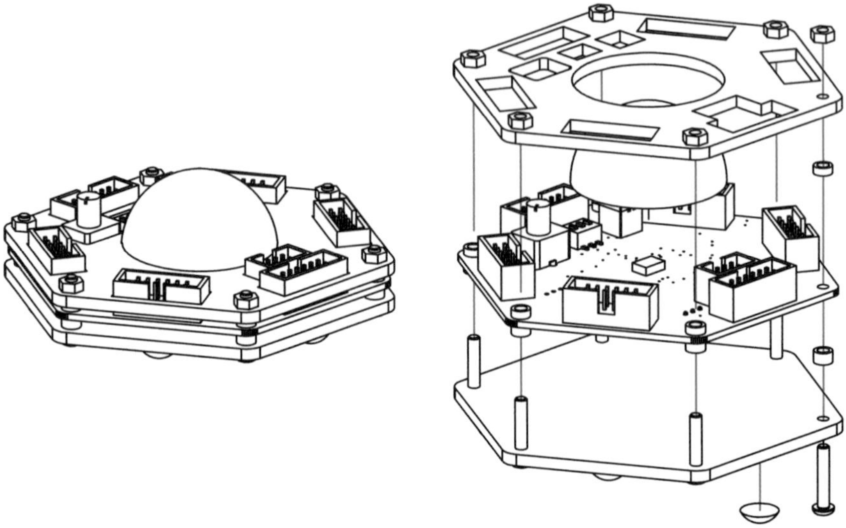

Fig. 4.3 CAD drawing of a TN node, in assembled and exploded states

4.2.1.4 Additional Functionality

For future expansion, each node also features an auxiliary input and auxiliary output as well as a serial connection. The serial connection could be interfaced to other hardware, or to a computer for plotting real-time graphs of the network dynamics. A further use of the serial port would be to combine Tangible Networks with a computer-simulated network so that the physical nodes could be connected to a large virtual network running on the computer. This would allow for much larger network simulations, where a small part of the network can be interacted with. The aux out port is connected to a general purpose pin on the microcontroller that can generate PWM signals. We have used the aux out to drive a RC servo for mechanical output, and also a piezo speaker. The auxiliary in port is connected to a general purpose pin on the microcontroller that can function as an analogue input. We have used the auxiliary input to connect a light sensor and also a single-axis accelerometer, but these are only examples of what is possible. There are a large number of examples and code snippets online for interfacing the Arduino with a range of sensors and other hardware.[2]

4.3 Implemented Models

Here we present some of the network models we have implemented on Tangible Networks, based on our research in complexity science. These demonstrate some

[2] See for example https://forum.arduino.cc.

key concepts from ongoing research in our group in an approachable way. We have also developed a network based game, which demonstrates the potential for TN as a tool for teaching pure mathematics through problem solving. The following model descriptions are brief, but are intended as examples of what is possible with the TN toolkit. They include discrete-time models and dynamical systems, different types of local and global interaction, and possible ways of presenting the model output.

4.3.1 Excitable Neurons

The Fitzhugh-Nagumo (FHN) model [10, 20] is widely used to describe the spiking patterns of excitable cells such as neurons or muscle cells. Excitable cells generate a spiking electric current (an action potential) when stimulated. In our implementation, each TN node is an excitable cell. The action potential is visualised with the colour and brightness of the dome, and nodes emit sound when they spike. The pushbutton gives the cell an instantaneous stimulus, and the potentiometer sets a level of continuous stimulation. Cells also receive stimulus from their neighbours, and the global coupling parameter is set with the master dial. This introduces the idea that the specific environment that cells are in, such as the presence or absence of different substances can affect the overall dynamics. The pattern of spiking is dependent on the topology, and on the type and strength of coupling. Excitatory (positive) coupling can lead to travelling waves or synchronous oscillations, with increased coupling increasing the wave speed of propagating spiking patterns. Inhibitory (negative) coupling leads to a range of asynchronous oscillations due to post-inhibitory rebound spiking, including sustained oscillations with neighbouring nodes in antiphase for some topologies.

4.3.2 Synchronising Oscillators

The Kuramoto model [17] describes a wide variety of synchronisation phenomena [2]. Examples include flashing fireflies [6], power grid systems [8, 9] or a conductor keeping an orchestra in time. Each TN node is an independent first order oscillator with a natural frequency that is adjusted with the potentiometer. Neighbouring nodes are coupled, and the global coupling parameter is controlled with the master dial. Stronger positive coupling increases the level of synchronisation, while negative coupling causes neighbouring nodes to oscillate in antiphase. The phase of each oscillator is shown with variations in brightness. The colour represents the local degree of synchronisation, computed as the mean phase difference with its neighbours. Oscillators can be stopped and held at a constant phase by holding down the pushbutton. Users can explore how the level of synchronisation is affected by the topology and natural frequencies.

4.3.3 Opinion Dynamics

The majority-vote model produces qualitative results similar to the patterns of opinions in networks [7]. Each node is a person who holds one of two opinions and is influenced by their neighbours. Stubborn nodes change their opinion if more than half of their neighbours disagree with them. Fickle nodes change their opinion if at least half of the neighbours disagree with them. Nodes update their opinion in discrete timesteps. Users can set the initial opinions and stubbornness of each node, with the potentiometer. Opinions are shown as green and blue, with fickle nodes being more pale. Pressing the pushbutton toggles the opinion of that node, and the user can then observe whether this causes any further nodes to change their opinion. The simulation is reset with the master switch. The network simulation can demonstrate consensus and clustering of opinions, along with ideas such as group influence. It can also show how some nodes are more influential than others, and that the most influential nodes are not necessarily the most central or most connected nodes.

4.3.4 Predator-Prey Dynamics

The Lotka-Volterra equations describe predator-prey dynamics in ecological systems of two or more species. In our implementation, each node is a species. The brightness indicates the current population, and the colour indicates the trophic level (red: top predator, yellow: intermediate predator, green: primary producer). The trophic level is set with the DIP switches. The potentiometer sets the intrinsic growth rate of the species. Pressing the pushbutton increases the population of that species. We can demonstrate simple interactions between a single predator and a single prey that lead to oscillating populations, as well as more complex food webs. The model can be used to demonstrate meaningful ecological concepts such as competitive exclusion, apparent competition and biological pest control.

4.3.5 Hamiltonian Paths

A Hamiltonian path is a route through the network that visits each node exactly once. In our implementation, we introduce this concept through a problem-solving exercise: Pressing the master switch resets the game, making all nodes cyan. Pressing the pushbutton on one node selects the starting node and starts the game. The current node is green, visited nodes are yellow and unvisited nodes are blue. Pressing the pushbutton on an unvisited node adjacent to the current node moves the player to that node. The game ends when there are no more possible moves, at which point visited nodes turn green and unvisited nodes turn red.

The Hamiltonian path problem is simple to solve heuristically on a small network, however the exercise can be scaffolded to ask for deeper understanding and problem solving. The game has been developed in tandem with a structured worksheet, and has been given a storyline to motivate younger children to solve the problem.

4.4 Discussion

We have presented Tangible Networks, a toolkit for interacting with mathematical models and communicating ideas from complex systems. A TN user can explore and learn about these systems and models even if they are unfamiliar with the underlying mathematics. TN presents the models in a more engaging and approachable way by allowing for direct hands-on interaction. From interacting with TN, users can explore how local interactions lead to global phenomena; one of the fundamental concepts of complex systems.

TN has been demonstrated at a range of events including science festivals, university open days, school lessons, summer schools, undergraduate courses and academic conferences (Fig. 4.4). The attractive visualisation and interactive aspect has

Fig. 4.4 We have demonstrated tangible networks at a number of events, and reception has been overwhelmingly positive

captured the interest of a wide range of audiences ranging from young children to senior academics. Through displaying multiple models at events, we demonstrated the adaptability of network science to explain a variety of different systems, and people have been impressed by the breadth of systems studied with networks. We have used TN for demonstrating particular system behaviours, as well as letting users freely explore and interact with the models and facilitating further discussion.

We believe that TN is well suited for educational use. Alternative learning activities are used to vary teaching styles in order to address different ways of learning [13, 16, 22]. TN offers a way to open the 'black box' of in-silico simulations in order to facilitate the understanding of concepts in mathematical modelling, network science and graph theory. These models can be discussed in as much technical detail as required, making it suitable from primary schools to universities. The TN toolkit could also be used to teach programming and electronics; making a device or program that 'does something useful' is motivating and rewarding.

We would like to encourage others to use the platform as it is, or develop it further for their own work. The necessary files to make the TN hardware and software are all open source. Designs are available online, along with information about the project [25]. We have written an Arduino library for the TN hardware which facilitates further software development.

Communication of scientific ideas is of utmost importance, be it to funding bodies, schoolchildren, undergraduate students, fellow academics and members of the general public. We hope that TN will be useful in this regard.

Acknowledgments This work has been funded by the Bristol Centre for Complexity Sciences, through EPSRC grant EP/I013717/1. EK wishes to acknowledge funding from the James Dyson Foundation.

References

1. Arduino Project: Arduino. www.arduino.cc
2. Arenas, A., Díaz-Guilera, A., Kurths, J., Moreno, Y., Zhou, C.: Synchronization in complex networks. Phys. Rep. **469**(3), 93–153 (2008)
3. BCCS: BCCS Outreach (2014). http://www.bristol.ac.uk/bccs/public-engagement/
4. Bdeir, A.: Electronics as material: littleBits. In: Proceedings of the ACM TEI '09, pp. 397–400 (2009). http://dl.acm.org/citation.cfm?id=1517743
5. Bubela, T., Nisbet, M.C., Borchelt, R., Brunger, F., Critchley, C., Einsiedel, E., Geller, G., Gupta, A., Hampel, J., Hyde-Lay, R., et al.: Science communication reconsidered. Nat. Biotech. **27**(6), 514–518 (2009)
6. Buck, J.: Synchronous rhythmic flashing of fireflies. ii. Quarterly review of biology, pp. 265–289 (1988)
7. Campos, P.R.A., de Oliveira, V.M., Moreira, F.G.B.: Small-world effects in the majority-vote model. Phys. Rev. **E 67**, 026104 (Feb 2003). http://link.aps.org/doi/10.1103/PhysRevE.67.026104
8. Dorfler, F., Bullo, F.: Synchronization and transient stability in power networks and nonuniform kuramoto oscillators. SIAM J. Control Optim. **50**(3), 1616–1642 (2012)
9. Filatrella, G., Nielsen, A.H., Pedersen, N.F.: Analysis of a power grid using a kuramoto-like model. Eur. Phys. J. B-Condens. Matter Complex Syst. **61**(4), 485–491 (2008)

10. FitzHugh, R.: Impulses and physiological states in theoretical models of nerve membrane. Biophys. J. **1**(6), 445–466 (1961)
11. Girouard, A., Solovey, E., Hirshfield, L.: Smart Blocks: a tangible mathematical manipulative. In: Proceedings of ACM TEI '07, pp. 183–186 (2007). http://dl.acm.org/citation.cfm?id=1227007
12. Glowacki, D.R., O'Connor, M., Calabro, G., Price, J., Tew, P., Mitchell, T., Hyde, J., Tew, D., Coughtrie, D.J., McIntosh-Smith, S.: a GPU-accelerated immersive audiovisual framework for interaction with molecular dynamics using consumer depth sensors. Faraday Discussions (2014). http://pubs.rsc.org/en/Content/ArticleLanding/2014/FD/c4fd00008k
13. Grasha, A.: Teaching with Style: A Practical Guide to Enhancing Learning by Understanding Teaching and Learning Styles. Alliance Publishers, Curriculum for change series (1996)
14. Horn, M.S., Jacob, R.K.J.: Designing tangible programming languages for classroom use. In: Proceedings of ACM TEI '07, pp. 159–162 (2007). http://dl.acm.org/citation.cfm?id=1227003
15. Ishii, H., Ullmer, B.: Tangible bits: towards seamless interfaces between people, bits and atoms. In: Proceedings of ACM CHI '97, pp. 234–241 (1997). http://dl.acm.org/citation.cfm?id=258715
16. Kolb, D.A., et al.: Experiential learning: Experience as the Source of Learning and Development, vol. 1. Prentice-Hall Englewood Cliffs, NJ (1984)
17. Kuramoto, Y.: Chemical oscillations, waves, and turbulence. Courier Dover Publications (2003)
18. Marshall, P.: Do tangible interfaces enhance learning? In: Proceedings of ACM TEI '07, pp. 163–170 (2007). http://dl.acm.org/citation.cfm?id=1227004
19. Mitchell, T., Hyde, J., Tew, P., Glowacki, D.: danceroom Spectroscopy: at the frontiers of physics, performance, interactive art and technology. Leonardo, p. 140826115909005 (Aug 2014). http://www.mitpressjournals.org/doi/abs/10.1162/LEON_a_00924
20. Nagumo, J., Arimoto, S., Yoshizawa, S.: An active pulse transmission line simulating nerve axon. Proc. IRE **50**(10), 2061–2070 (1962)
21. Newman, M.: Networks: an introduction. Oxford University Press (2010)
22. Riechmann, S.W., Grasha, A.F.: A rational approach to developing and assessing the construct validity of a student learning style scales instrument. J. Psychol. **87**(2), 213–223 (1974)
23. Rubenstein, M., Cornejo, a., Nagpal, R.: Programmable self-assembly in a thousand-robot swarm. Science **345**(6198), 795–799 (Aug 2014). http://www.sciencemag.org/cgi/doi/10.1126/science.1254295
24. Schweikardt, E., Gross, M.: roBlocks: a robotic construction kit for mathematics and science education. In: Proceedings of ACM ICMI '06 (2006). http://dl.acm.org/citation.cfm?id=1181010
25. Tangible Networks. www.tangiblenetworks.net
26. Wolfendale committee and others: Wolfendale committee final report (1995)
27. Wynne, B.: Public engagement as a means of restoring public trust in science—hitting the notes, but missing the music? Public Health Genomics **9**(3), 211–220 (2006)

Chapter 5
The Geometric Origins of Complex Cities

Ruiqi Li, Lei Dong, Xinran Wang and Jiang Zhang

Abstract Due to the rapid urbanization, cities have become a hot topic. Extensive complex phenomena, such as scaling laws with respect to population, morphology, spatial distribution within cities have been revealed and validated by the empirical studies. Yet there's still no clear answer to the question that what's the underlying mechanism responsible for these observed complex phenomena. Most of previous studies only focus on one aspect of the city. However, focusing on only one aspect may lose the whole picture of it. Based on a very simple "matching growth" rule and two more simple assumptions, which are all performed locally, we propose a simple model which can derive most of observed macro scaling relations and spatial distribution. All these theoretical deductions can be well supported by empirical data. And the consistency between the exponents of different cumulative spatial distribution may indicates that the city really follows the rules we assumed.

5.1 Introduction

The urbanization rate of the world was more than 50 % at the end of 2008, and for developed countries is about 80 %. And according to urbanization experience in developed countries, urban population of the developing countries will keep accumulating rapidly for certain decades [1]. This rapid pace of urbanization is one of the crucial problems with which the world is faced now and will be in the future. Urbanization seems to be a double-edged sword. The bright side is that economic variables like total income, GDP and innovation related variables such as the total number of new patents, R&D, etc. have a super-linear scaling relation with population (that

R. Li · J. Zhang (✉)
School of Systems Science, Beijing Normal University, Beijing 100875, China
e-mail: zhangjiang@bnu.edu.cn

L. Dong
School of Architecture, Tsinghua University, Beijing 100084, China

X. Wang
College of Resources Science and Technology, Beijing Normal University,
Beijing 100875, China

S. Battiston et al. (eds.), *Proceedings of ECCS 2014*, Springer Proceedings in Complexity, DOI 10.1007/978-3-319-29228-1_5

means GDP grows faster as population agglomerates); meanwhile, some variables related with infrastructure like total road (and cable) length and urban area are sub-linear. These phenomena mean that in larger cities, we can get more outcome with a relatively low input at a relatively higher efficiency. However, the dark side is that rate of serious crimes, disease (such as new AIDS cases) [2–5] are also super-linear. Besides, there are also many other issues such as heavy congestion, social conflicts, environmental degradation in large cities (especially for some Chinese cities).

So, quite different from agglomeration of particles, during the agglomeration of people, there are bunch of nonlinear interactions, which might be the origin of scaling laws. This agglomeration not only brings more innovations, wealth and economic growth, but also more pollution, traffic congestions, diseases, crimes and social issues [6]. But what's the underlying mechanism(s) responsible for these complex phenomena? Furthermore, is the city can be simple enough to be understood?

People argued for a long time about the evolution mechanism of cities [7–9]. More and more researchers tend to believe that cities are self-organized complex systems whose infrastructural, economic and social components are strongly interrelated [4]. In its wake, some remarkable ubiquitous empirical laws have been observed when we focus on the global properties of city systems [10]. For example, early studies pointed out that Zipf law is a significant feature for city systems, which states that the population distribution of different cities are long tails [11, 12]. Furthermore, the growth rate of a city is proportional to its size which is known as Gibrat's law [13]. This empirical law is tightly connected to the Zipf law [14].

Zipf law and Gibrat law only describe one variable, the size of cities which is usually measured by population, other studies focused on the relationships between two variables of cities. For example, the earliest studied relationship is between urban area and population. Nordback [15] found that area scales with population, and the exponent is 2/3. However, more empirical studies found that the exponents are depended on the definition of city areas [16, 17].

Compare to the fast development of empirical work of city scaling, the development of theoretical understanding of cities is slow. Batty et al. found the fractal nature of cities, and introduced the diffusion limited aggregation model (DLA) to simulate the growth of a city [18]. Makse et al. pointed out that this model can only generate one cluster and is less compact than the real cities. So they built a new model by correlated percolation [19]. This model can only generate population density and Zipf law but the scaling relationships were never considered. In [13], the authors devised an economic geographic model to reproduce the patterns of Zipf law and allometric scaling between area and population.

After the empirical super-linear scaling law between socioeconomic output and population was discovered and highlighted, Pan et al. gave another network model to show the relationship between super-linear scaling with population density in cities, however, their work didn't focus on reproducing the spatial distribution of variables within the city [20]. Recently, Bettencourt proposed a model to reproduce various scaling relationships successfully by taking several assumptions [21]. But this model is a little complex and the spatial distribution of population and wealth is not considered.

So far, there's still no integrated theory to tie those important factors (spatial distributions, morphology, scaling laws) mentioned above together to give a more comprehensive insight. Most researchers only focus on one or two aspects of the city. To overcome the known shortcomings of the previous models and make up the gap between the micro-spatial distribution and macro-scaling laws, this paper, inspired by the realistic evolution of the city [22] presents a new model based on a simple evolving mechanism which is called "matching growth" (new node can only grow if it is close enough to the existing nodes).

5.2 The Model

In our model, a planar network will be grown in a $L \times L$ 2-d Euclidean space. The planar network can be used to simulate the minor roads and main streets. So each node on the network stands for a community with a constant population, then it's natural to regard the links between communities as roads (actually the word connection is more appropriate and we will explain it later). The network is grown as the nodes (communities) added sequentially based on an important rule called "matching growth". That is, the new coming node can be added successfully only if it is matched with existing nodes.

Concretely, suppose our network is grown from an initial seed node locating at the center of the free space at the beginning. At each time step, one node is generated with a random position evenly distributed within the space. However, a new node can be added into the network only if it is close enough to any existing node in the network (the Euclidean distance is smaller than a given threshold r), otherwise it will be removed. In this way, new nodes join the network unceasingly only if they matched with existing nodes, the network is growing. This matching growth process is illustrated in Fig. 5.1. The process of new nodes joining the network models the building of new community. Although it is hard to imagine that the new nodes randomly generating mechanism is true, our model generate an accelerating growth phenomenon, which means the time interval between any two nodes joining the network is shorter and shorter. This is comparable with our observations. We can define P_t as the total number of nodes in the current network and A_t as the total area that covered by the disks. They represent the population and urban area of the modeled city.

Next, we will show how the road network is built. The simplest way is minimum spanning tree(MST): i.e. roads are built by linking several nearest nodes. Because the road links are always space filling, the total length of road links, $L_t = \sum l_i$ (l_i is the total length of road links built by i), can be understood as the total road volume in a city, since in real cities, the variance of road width is not very big).

However, for a closer look, especially on urban scale, the roads are not directly linking building blocks but surrounding them and there are roads pointing outside at the edge of city, which means the method above just mimics the connection between community but not the real situation of roads between them. We assume a better way

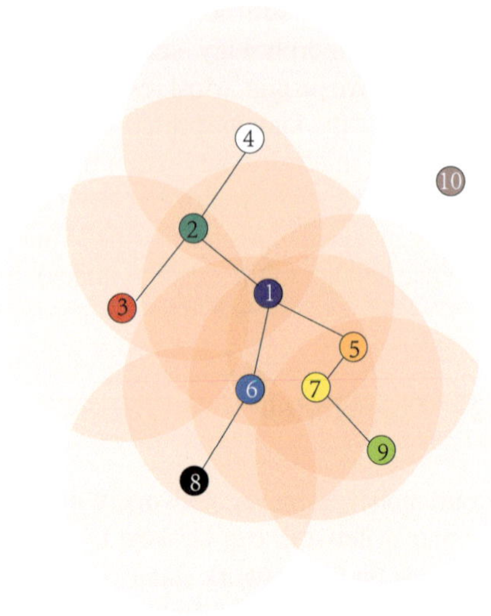

Fig. 5.1 A sketch of "matching growth" process and "road network". The numbers in nodes are the orders they joining the network. The *large disks* surrounding nodes are their range of interaction with radius r. The nodes can only survive if it locates in the shaded area. Therefore node 10 cannot exist. The *links* between nodes represent the roads. In this sketch, each node can build only $k = 1$ link to the nearest existing node.

to generate the ideal road network within the city is the idea of Voronoi diagram, which can guarantee the fairness for two building blocks besides a road, i.e. the distance for both of them to get access to the road are the same, as roads are public infrastructure. And the corresponding Delaunay triangulations can mimic the minor roads which allow the community to reach the road networks as quick as possible, which is exactly the dual graph of Voronoi diagram.

Furthermore, we assume that all interactions among people are taking place surrounding the road areas. This assumption is in accordance with our observation that all the markets, companies and shops are built around road sides. For people in node i, it is able to interact with all the people within its interaction range along the road links. Therefore, the total interactions generated by i can be approximated as $g_i \propto l_i|N_r(i)|$, which is the number of neighbors, more specifically, g_i is proportional to the product of l_i and the node density. Thus, the total interactions of the whole system is $G_t = \sum_i g_i$. Furthermore, according to Bettencourt [21], the total output, incomes, crimes are all proportional to the total interactions G_t. We will show that all the variables, A_t, L_t and G_t, scale with P_t.

5.3 Theoretical Analysis of Scaling Laws

According to the two rules introduced in the previous section, we can simulate the growth of the network. However, instead of reporting the numeric results, we derive the theoretical computation of scaling exponents asymptotically (namely, let $t, L \to \infty$).

The shape of the network is irregular and anisotropic when simulation time steps are limited. However, it will transit to a symmetric disk with a rough perimeter when t is very large. We will show, in the mean field approximation, the radius of the whole disk denoted as R_t grows with time linearly,

$$R_t \sim t. \tag{5.1}$$

Actually, each lattice of the 2-d space has an equivalent probability to accept a new generated node at each time step. Therefore, the probability that a new coming node locates on the boundary of the disk is proportional to its perimeter and independent of time. Then, the average time span between two nodes are added on the perimeter of the disk is proportional to $1/R_t$. However, to increase the radius of the disk one unit, we need more and more nodes (the number of nodes increase with $\sim R_t$) to fill its perimeter. Therefore, the average time that the radius of the disk increase one unit is almost a constant ($\sim R_t(1/R_t)$). That means, the speed of radius growth is a constant which leads to (5.1).

Following this equation, we know the total area of the disk increases with time square,

$$A_t \sim t^2 \sim R_t^2. \tag{5.2}$$

To derive the total number of nodes, we need to calculate the node density $\rho_t(R,\theta)$ at any spatial location with radial coordinate (R,θ) and time t as:

$$\rho_t(R,\theta) = \int_{\tau_R}^{t} \frac{1}{L^2} ds \sim (t - \tau_R) \sim (R_t - R) \tag{5.3}$$

where ds is the infinitesimal time, τ_R is the time when the disk's radius is R. Because the probability that the infinitesimal area $d\sigma$ accepts a new node is a constant ($1/L^2$), the average density of population at this infinitesimal area is the accumulation of nodes born in between time step τ_R and t. Therefore, the total population can be computed by integrating (5.3):

$$P_t = \int_0^{R_t} \int_0^{2\pi} \rho_t(r,\theta) r dr d\theta \sim R_t^3 \tag{5.4}$$

According to (5.4) and (5.2), we obtain the scaling relationship between area and population:

$$A_t \sim P_t^{2/3} \tag{5.5}$$

Next, we analyze the total road volume. Suppose the road volume density per capita is $l_t(R,\theta)$ at time t. We know that this density will decrease with time as the population density increases according to the rules that all roads segments are linked to the nearest nodes. In a 2-d space, the minimum distance between any two points evenly distributed in the infinitesimal $d\sigma$ decays in a square root of the density. Therefore, $l_t(R,\theta) \sim \rho_t^{-1/2}(R,\theta)$

Thus, the road volume in an infinitesimal area $d\sigma$ is $\rho_t(R,\theta)l_t \sim \rho_t^{1/2}$. Therefore, the total road volume of the whole network is:

$$L_t = \int \rho_t l_t d\sigma \sim \int_0^{R_t}\int_0^{2\pi} (r-R)^{1/2} r dr d\theta \sim R_t^{5/2} \tag{5.6}$$

Insert (5.4) into (5.6), we obtain the scaling relationship:

$$L_t \sim P_t^{5/6} \tag{5.7}$$

Finally, suppose the density of local interaction per capita is $g_t(R,\theta)$, it is proportional to the local population density times the local road volume according to the assumption that local interaction takes place along the road, therefore, $g_t(R,\theta) = \rho_t(R,\theta)l_t(R,\theta) \sim \rho_t^{1/2}$. Then, the total number of interactions happened in the whole system is:

$$G_t = \int \rho_t g_t d\sigma = \int_0^{R_t}\int_0^{2\pi} (r-R)^{3/2} r dr d\theta \sim R_t^{7/2}. \tag{5.8}$$

So, G_t and P_t has the following scaling relationship:

$$G_t \sim P_t^{7/6}. \tag{5.9}$$

We know the total output in the system is proportional to the total number of interactions, therefore, the socioeconomic output in the system scales with population in a $7/6$ power.

5.3.1 Model Extension and Spatial Distributions

Our model cannot only reproduce the scaling laws of any variable with respect to population, but also generate the spatial distribution of population, roads, and wealth. To make comparable results with the empirical data, we need to extend our basic model and introduce a new parameter. In the real life, people always prefer some particular places due to the heterogeneity of resource distribution on the geographic space. In our extended model, we assume that the spatial preference of a new node follows a power law:

$$P(R,\theta) = \frac{R^{-\beta}}{Z}, \tag{5.10}$$

where R is the distance to city center and $\beta \in [0, 1)$ is a parameter called heterogeneity exponent. $\beta < 1$ is required to make sure that the total road volume and GDP be finite. R is the distance of any point from the center of the city with a range $[R_0, \infty)$, where R_0 is the minimum radius that guarantees (5.10) being a well defined probability distribution. In the following calculations we always let $R_0 \to 0$. Z is the normalization coefficient, $Z = \int_0^{2\pi} \int_{R_0}^{L} R \cdot R^{-\beta} dR d\theta$. With this modified rule, we can get $R_t \sim t^{\frac{1}{1+\beta}}$. Then, we can write down the population density of any given spatial point as,

$$\rho_t(R, \theta) = \frac{1}{Z} \int_{\tau_R}^{t} R^{-\beta} ds \sim R^{-\beta}(R_t^{1+\beta} - R^{1+\beta}), \tag{5.11}$$

After that, the same method as in Sect. 5.5 can be applied to calculate the relations between A_t, L_t, G_t and R_t. Due to the following identity:

$$Y_t = \int_0^{R_t} \int_0^{2\pi} \rho_t^s(R, \theta) r d\theta dRr \sim R_t^{2+s},$$

where s is a positive exponent. Let Y_t stands for A_t, P_t, L_t and G_t, and set $s = 0, 1, 1/2$ and $3/2$ respectively, the exactly same relations as (5.2), (5.4), (5.6), (5.8) proposed can be derived. Then, all the scaling relations (5.5), (5.7), (5.9) are invariant and independent of the exponent β.

Besides the scaling relations, the improved model allows us to analyze the spatial distributions of population, roads and wealth. First, we have derived how the population density decays as the distance from the city center in (5.11). In the downtown area, which means $R/R_t \to 0$, (5.11) is approximated by a power law: $\rho \sim R^{-\beta}$.

This equation is consistent with the empirical observation by Smeed [23] and Batty [24]. However, because population density measurement is inaccurate, we always use the cumulative population along the distance from the city center to validate our model by the empirical data. By integrating (5.11) on the radius, we obtain how population accumulate along the radius:

$$P_t(R) = \int_0^{R} \int_0^{2\pi} \rho(r, \theta) r dr d\theta \sim R^{2-\beta} \left(\frac{R_t^{1+\beta}}{2-\beta} - \frac{R^{1+\beta}}{3} \right), \tag{5.12}$$

where $P_t(R)$ stands for the cumulative population until the distance R. It turns out to be a power law if we only focus on the downtown area ($R/R_t \to 0$):

$$P(R) \sim R^{2-\beta}. \tag{5.13}$$

Equation (5.13) is consistent with the fractal city hypothesis [18]. According to the ball covering method, the fractal dimension of population distribution is $2-\beta$. If we assume that the distribution of buildings is similar with the distribution of population, then the fractal dimension of our city model is also $2-\beta$.

We then can calculate the cumulative road volume along the radius from the city center in a similar way,

$$L_t(R) = \int_0^R \int_0^{2\pi} \rho(r,\theta) l(r,\theta) r dr d\theta \sim R^{2-\beta/2}. \tag{5.14}$$

Therefore, the fractal dimension of road network is predicted as $2-\beta/2$. Compare this equation with (5.13), we suppose that the fractal dimension of road network is slightly larger than the one for population. This is a testable prediction.

Additionally, we can also derive how wealth accumulates along the radius from the city center if we assume that the personal wealth is proportional to the individual output which is proportional to the interactions per capita,

$$G_t(R) = \int_0^R \int_0^{2\pi} \rho(r,\theta) g(r,\theta) r dr d\theta \sim R^{2-3\beta/2}. \tag{5.15}$$

Therefore, wealth accumulate slower than population. This is another testable prediction of our model.

Finally, if we assume that the rent price of a place is also proportional to the local output per capita [21], then we can derive how the rent price decreases with the distance:

$$p(R,\theta,t) \propto g(R,\theta,t) \sim R^{-\beta/2}(R_t^{1+\beta} - R^{1+\beta})^{1/2} \sim R^{-\beta/2} \tag{5.16}$$

Hence, the rent price decays at a slower rate than the population. This is also a prediction.

5.4 Results

In this section, we will show the computer simulation results and compare it with both the theoretical results and empirical works. According to the rules introduced in Sect. 5.2, we can grow a spatial network which resembles the road network in city (Fig. 5.4).

From Fig. 5.2a, b, we observe that the density of population and road links are denser in the center area than the peripheries. Several important roads with large traffic flux (betweeness) naturally emerge. All these results are coherent with empirical

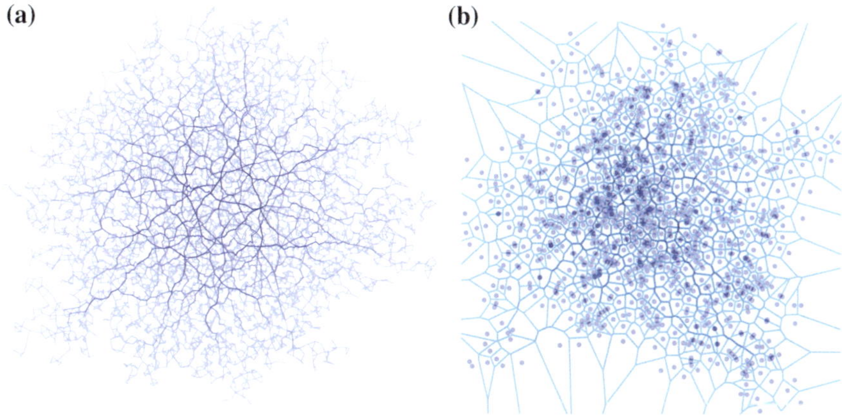

Fig. 5.2 **a** A connection network grown by the rules. In this simulation, we adopt the basic rules (i.e., $\beta = 0$), the number of newly added links is $k = 2$, and the total number of nodes is 10^5. Colors correspond to betweeness of links. **b** A road network generated by the rules. In this simulation, we adopt the basic rules (i.e., $\beta = 0$), the total number of nodes is 10^3

Table 5.1 Exponents of scaling relationships with respect to population for theoretical predictions and observations

Variables	Obs, mean	Obs. range	Theoretical
Area	0.67	[0.56, 1.04]	$2/3 \approx 0.667$
Network volume	0.75	[0.74, 0.92]	$5/6 \approx 0.833$
Socioeconomic rates	1.17	[1.01, 1.33]	$7/6 \approx 1.167$

observations. Besides, our model can generate minor roads which can not be depicted by previous models.

Next, we test the scaling relations in simulation. From previous researches [21], we know the relationships between the total output, total road volume, area and population are power laws. the lower boundary of population for a city varies from hundreds to thousands in different countries. For completeness, we didn't set a threshold. And the exponents are approaching the theoretical predictions.

These exponents are also consistent with our empirical observations in the real cities. Table 5.1 [21] shows the exponents both for theoretical predictions and empirical observations.

In Fig. 5.3a, we show how the normalized population density (normalized by the total population) decays with the distance from the city center. The curves change with parameter β, and it can be compared with the empirical observations. Clark [25] suggested that population density decay with the distance from city center exponentially. However, Smeed [23] proposed that this relationship is a power law if we consider longer distance. In paper [24], the authors claimed that this relationship should be a power law so that the allometric scaling relationship between area and

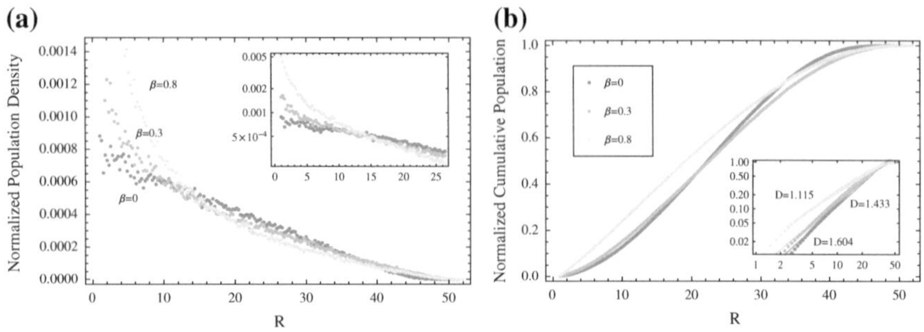

Fig. 5.3 **a** Normalized population density decays with distance from the center. In the simulation, we compare the results of different settings ($\beta = 0, 0.3, 0.8$). The inset shows the semi-log (log y) plot of the same relation. **b** Normalized cumulative population along the radial distance. The inset shows the same curve with log-log plot

population and the fractal city hypothesis are satisfied. As claimed in [24], since the power law exponent is very small in the real observations, it is very hard to discriminate with an exponential function. In our model, the power law relationship is derived. Besides, we also plot the population density with distance in a semi-log plot and take a short range along x-axis shown in the inset of Fig. 5.3a. It looks like a straight line when β is small. Thus, our model can generate similar results as observations.

Another way to study how population density decays with radius is to show the cumulative population along the distance so that fluctuations in population density can be reduced. The empirical studies shows that the cumulative population curve is S-shaped [24]. If we plot our modeled cumulative population curve, an S-shaped curve can be also obtained as shown in Fig. 5.3b.

Due to the cut-off term in (5.12), our model can generate a similar S-shaped curve as observed in real city. If we plot the curve on a log-log coordinate (shown in the inset of Fig. 5.3b, we find that the head part of the figure becomes a straight line which means it can be approximated by a power law, the exponent can be understood as the fractal dimension of population.

Besides the population, we can also study how the other variables like road volume, interaction cumulated with the distance from the center. All of these cumulative variables can be fitted by power laws with different exponents as (5.13), (5.14) and (5.15) predicts shown in Fig. 5.4a.

Our theory in Sect. 5.3 predicts the dimensions are $2 - \beta = 1.7, 2 - \beta/2 = 1.85$ and $2 - 3\beta/2 = 1.55$ respectively which are close to the simulation dimensions.

Because a set of fractal dimensions for each time step can be obtained, we then systematically study how the dimensions change with population (Fig. 5.4b). Previous studies pointed out that the fractal dimension of cities can change with city size. This phenomenon can also be observed by our simulation.

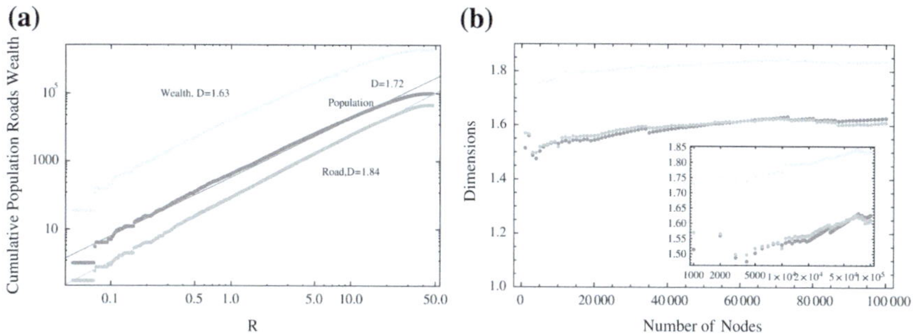

Fig. 5.4 **a** Cumulative population, road volumes and interactions change with distance in simulation by setting $\beta = 0.3$. All these curves can be fitted by power laws with different exponents which are shown in the figure. **b** Fractal dimensions with the number of nodes. The fractal dimensions increase with the number of nodes, indicating a denser city. The inset shows the log-linear plot of the same variables

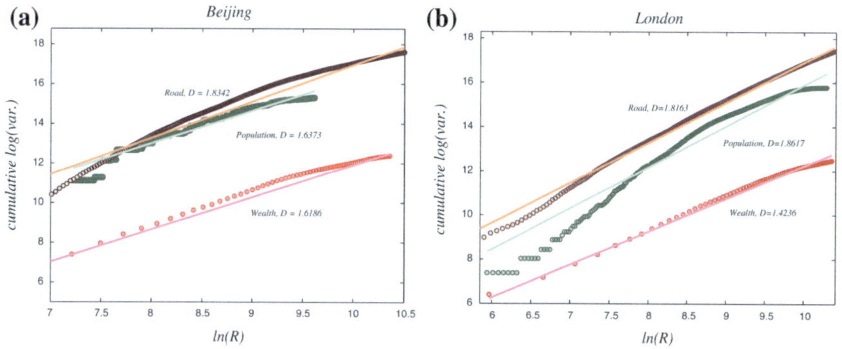

Fig. 5.5 Empirical scaling results for **a** Beijing and **b** London with respect to the distance to the center of city

5.4.1 *Empirical Results*

We study two representative cities, Beijing and London. As Fig. 5.5 shows, all of the scaling laws can be well fitted by an estimated $\beta = 0.36$ for both Beijing and London. We use a residence population distribution with Lower Super Output Area (LSOA) resolution, which may differ from (mainly more decentralized) the assumption in this article.

In this article, all the data sets used for analysis are publicly available. We obtain the employed population data from Beijing Census Bureau and http://data.gov.uk/, respectively. The statistics of GDP are reflected by city lights from NOAA/NGDC (http://ngdc.noaa.gov/eog/viirs/download_monthly.html). The data of road networks are accessible in http://metro.teczno.com/.

5.5 Conclusions

Our model is based on few simple rules, but some complex behaviors that may resemble the real cities can be generated. We can not only obtain all the scaling exponents but also give some insight into the urban dynamics by the basic model. Therefore, this simple model allows to generate complex city. With a slight extension, our model is able to make predictions on both micro and macro parameters and behaviors.

Acknowledgments J. Zhang thanks for the discussions with Prof. Bettencourt in Santa Fe Institute, doctor Wu in Arizona University and Prof. Wang and Chen in Beijing normal university, acknowledges the support from the National Natural Science Foundation of China under Grant No. 61004107 and No. 61174165.

References

1. Montgomery, M.R.: The urban transformation of the developing world. Science **319**, 761–764 (2008)
2. Bettencourt, L.M., Lobo, J., Helbing, D., Kühnert, C., West, G.B.: Growth, innovation, scaling, and the pace of life in cities. Proc. Natl. Acad. Sci. **104**, 7301–7306 (2007)
3. Bettencourt, L.M., Lobo, J., Strumsky, D.: Invention in the city: increasing returns to patenting as a scaling function of metropolitan size. Res. Policy **36**, 107–120 (2007)
4. Bettencourt, L., West, G.: A unified theory of urban living. Nature **467**, 912–913 (2010)
5. Bettencourt, L.M., Lobo, J., Strumsky, D., West, G.B.: Urban scaling and its deviations: revealing the structure of wealth, innovation and crime across cities. PloS ONE **5**, e13541 (2010)
6. Glaeser, E.: *Cities, agglomeration, and spatial equilibrium*. Oxford University Press (2008)
7. Mumford, L.: *The City in history. Its origins, its transformation, and its prospects*. Harcourt, Brace & world (1961)
8. Geddes, P.: *Cities in evolution: an introduction to the town planning movement and to the study of civics*. Williams & Norgate, London (1915)
9. Kostof, S.: The city shaped urban patterns and meaning throughout history. Bulfinch, Boston (1991)
10. Batty, M.: The size, scale, and shape of cities. Science **319**, 769–771 (2008)
11. Zipf, G. K.: *Human behavior and the principle of least effort.* Addison-Wesley Press (1949)
12. Zanette, D. H.: Zipf's law and city sizes: a short tutorial review on multiplicative processes in urban growth. arXiv preprint arXiv:0704.3170 (2007)
13. Eeckhout, J.: Gibrat's law for (all) cities. *American Economic Review*, pp. 1429–1451 (2004)
14. Gabaix, X.: Zipf's law for cities: an explanation. *Quarterly Journal of Economics*, pp. 739–767 (1999)
15. Nordbeck, S.: Urban allometric growth. Geogr. Ann. Ser. B Human Geogr. **53**, 54–67 (1971)
16. Rozenfeld, H.D., Rybski, D., Gabaix, X., Makse, H.A.: The area and population of cities: new insights from a different perspective on cities. Am. Econ. Rev. **101**, 2205–2225 (2011)
17. Batty, M., Ferguson, P.: Defining city size. Environ. Planning B Planning Des. **38**, 753–756 (2011)
18. Batty, M., Longley, P.A.: *Fractal cities: a geometry of form and function*. Academic Press (1994)
19. Makse, H.A., Havlin, S., Stanley, H.: Modelling urban growth. Nature **377**, 19 (1995)
20. Pan, W., Ghoshal, G., Krumme, C., Cebrian, M., Pentland, A.: Urban characteristics attributable to density-driven tie formation. Nat. Commun. **4** (2013)

21. Bettencourt, L.M.: The origins of scaling in cities. Science **340**, 1438–1441 (2013)
22. Doxiadis, C.A.: Ekistics, the science of human settlements. Science **170**, 393–404 (1970)
23. Smeed, R.J.: *The traffic problem in towns*. Manchester Statistical Society (1961)
24. Batty, M., Kim, K.S.: Form follows function: reformulating urban population density functions. Urban Stud. **29**, 1043–1069 (1992)
25. Clark, C. Urban population densities. J. R. Stat. Soc. A (General) **114**, 490–496 (1951)

Chapter 6
Revealing the Relation Between Structure of Chloroplast Genomes and Host Taxonomy

Michael Sadovsky and Anna Chernyshova

Abstract The distribution of chloroplast genomes in 63-dimensional space of triplet frequencies was studied, in connection to the taxonomy correlation to the clusters observed in the distribution. That latter was developed through K-means implementation, for the number of classes varying from 2 to 8. The clade composition of those clusters has been analyzed. Unexpectedly high regularity in clades occupation of different clusters has been found thus proving very high synchrony in evolution of two physically independent genetic entities (chloroplasts vs. nuclear genomes): the proximity in frequency space was determined over the organelle genomes, while the proximity in taxonomy was determined morphologically.

6.1 Introduction

DNA sequences are essentially complex object; a number of various structures are found and described in these latter being defined in different ways. Moreover, the list of structures is not completed yet: new ones could be found. The structures may exhibit a connection to function encoded in DNA moleculae. A relation of a structure and the relevant function is a core issue of the up-to-date system biology. Structures found in DNA sequences are numerous and various; thus, one must carefully fix the type of that former for further consideration. The variety of structures observed in DNA sequences could hardly be outlined here, even in brief; some surveys could be found in [1–6], see also very interesting works [7, 8].

In connection to DNA sequences, three issues make a kind of triad: these are *structure, function* and *taxonomy* of the bearer of sequence. A variety of sequences is tremendous, but frequency dictionary brings very simple structure entity to be identified in DNA (and in symbol sequences of any nature). Thus, it is the most universal and basic one in nucleotide sequences [9–18]. Further, we shall

M. Sadovsky (✉)
Institute of Computational Modelling SB RAS, Krasnoyarsk, Russia
e-mail: msad@icm.krasn.ru

A. Chernyshova
Siberian Federal University, Krasnoyarsk, Russia

S. Battiston et al. (eds.), *Proceedings of ECCS 2014*, Springer Proceedings in Complexity, DOI 10.1007/978-3-319-29228-1_6

concentrate on the study of the frequency dictionaries of the thickness $q = 3$ (i.e. the triplet composition).

Here we concentrate on the study of the relation between *structure* of DNA sequences and the *taxonomy* of their bearers. Both structure and taxonomy could be a matter of discussion; indeed, one may identify a number structures in a sequences, and taxonomy is revised quite often, as new data are incorporated into analysis. Nonetheless, we shall consider the taxonomy rank to be identified as it is shown in official site http://www.itis.gov.

A frequency dictionary W_3 converts a nucleotide sequence into a point in metric space; this allows to define a distance between two (or several) dictionaries. The basic idea of the paper is to figure out whether the clusters of genomes in this metric space could be found, and if yes then check whether the genomes occupying a cluster belong to a relatively close group of species, or not. Let now introduce the problem more exactly and rigidly.

Further, we shall consider continuous symbol sequences from four-letter alphabet $\aleph = \{\mathsf{A}, \mathsf{C}, \mathsf{G}, \mathsf{T}\}$. Number of symbols N in that former is the length of the sequence. No other symbols or gaps in a sequence are stipulated to take place. Any coherent string $\omega = \nu_1\nu_2\nu_3$ of the length 3 makes a triplet. A set of all the triplets occurred within a sequence yields the support of that latter. Counting the numbers of copies n_ω of the triplets, one gets a finite dictionary; changing the numbers n_ω for the frequency

$$f_\omega = \frac{n_\omega}{N}$$

one gets the frequency dictionary W_3 of the thickness 3. This is the main object of our study.

Thus, any genome is mapped into a 64-dimensional metric space, with coordinates corresponding to triplets $\omega_j = \nu_1\nu_2\nu_3$, $j = \mathsf{AAA}, \mathsf{AAC}, \mathsf{AAG}, \ldots, \mathsf{TTG}, \mathsf{TTT}$. There is a linear constraint

$$\sum_j \omega_j = 1\,. \tag{6.1}$$

This constraint actually forces to change the 64-dimensional space for 63-dimensional one. Obviously, two genomes with identical frequency dictionaries $W_3^{(1)}$ and $W_3^{(2)}$ are occupy the same point in the space. Nonetheless, a congruency of two frequency dictionaries $W_3^{(1)}$ and $W_3^{(2)}$ does not guarantee a complete coincidence of the original sequences, while these two sequences are indistinguishable from the point of view of their triplet composition.

Definitely, some genomes may have very proximal frequencies of all the triplets, and others may not. The inequality of triplet frequencies makes a distribution of the genomes in the 63-dimensional space inhomogeneous. The key question of our study is the inhomogeneity in the distribution mentioned above. Previously, this approach has been used to study the family of bacterial genes, with respect to taxonomy [19, 20]. To address the questions, here we made an unsupervised classification of organelle genomes, and matched the taxa composition of the classes observed through the classification.

Here we studied the relation between the structure and the taxonomy for chloroplast genomes. An unsupervised classification could be done for a various number of classes, and we had developed a series of classifications for different number of classes varied from two to eight. The key issue of the study is the transformation of classes when the number of these latter is changed: we investigated the fusion a classes when change the classification for K classes for $K - 1$ classes. The distribution of genera and species over the clusters, as well as the fusion of the clusters observed for various number of classes to be developed due to K-means proves unambiguously the high level of synchrony in evolution of two genetic systems: the former is organelle one, and the latter is nuclear one. The synchrony manifests in very tight and non-random distribution of species over the clusters. The proximity in structure space (i.e. triplet frequency composition) was determined over chloroplast genomes; the proximity in taxonomy was determined morphologically, i.e. through the nuclear genomes.

6.2 Materials and Methods

6.2.1 Genetic Sequences

All genomes have been retrieved from EMBL–bank. There were ~650 chloroplast genomes, while the final database used in our study enlisted 246 chloroplast genomes. The reasons to eliminate some entities from the dataset are provided below. Some entries contain the "junk" symbols (those that fall beyond $\aleph$), then all such symbols have been omitted, and the subsequences appeared due to the elimination of the junk symbols have been concatenated into a bounded entity.

Raw set of chloroplast genomes deposited in EMBL–bank is very inhomogeneous, in terms of the species and clades representation. A number of highly ranked clades are presented with a single species. Such solitariness results in a noticeable bias in classification: the pattern of genomes distribution is hidden in the cloud of signals provided by the isolated genomes. Thus, we eliminated the entries representing highly ranked clades solely, or with a few species. The cut-off level of 3 species and more has been established for database development. Table 6.1 shows the species structure of the database used in the study: it contains 36 genera. Evidently, the species composition is far from a balanced one. Moreover, the database contains both species, and strains belonging to the same genus (e.g., *Ginkgo* spp. and the strains, indeed, as well as the family of genomes of *Olea*; that latter consists of eight strains of *O. europaea* and only one species *O. woodiana*). Same is true for *Ostreococcus tauri* species.

Table 6.1 The species composition of the database of chloroplast genomes; M is the genus abundance in the database

Clade	M	Clade	M	Clade	M
Camellia spp.	9	*Ginkgo* spp.	5	*Orobancheae* spp.	4
Chrysanthemum spp.	3	*Glycine* spp.	9	*Oryza* spp.	12
Chrysobalanaceae spp.	6	*Gossypium* spp.	24	*Ostreococcus* spp.	14
Corymbia spp.	4	*Hordeum* spp.	3	*Phyllostachys* spp.	3
Cucumis spp.	4	*Liliaceae* spp.	3	*Picea* spp.	4
Cupressaceae spp.	4	*Magnolia* spp.	10	*Pinus* spp.	11
Cuscuta spp.	4	*Monodopsidaceae* spp.	3	*Pyropia* spp.	3
Cymbidium spp.	8	*Nannochloropsis* spp.	8	*Silene* spp.	7
Eucalyptus spp.	32	*Nelumbo* spp.	3	*Solanum* spp.	7
Euglena spp.	3	*Nicotiana* spp.	4	*Taxus* spp.	3
Eupatorieae spp.	3	*Oenothera* spp.	5	*Triticum* spp.	4
Fragaria spp.	4	*Olea* spp.	9	*Vitis* spp.	4

6.2.2 Clusterization Methods

Unsupervised K-means classification has been used to develop the classes. As it is said above, we had to reduce the data space dimension to 63: the reduction comes from the constraint (6.1). Formally speaking, any triplet could be excluded from the data set; practically, we excluded the triplet yielding the least standard deviation, determined over the set of genomes under consideration. This choice is evident: such triplet makes the least contribution into the separation of the entities, in the space of frequencies. Thus, the triplet GAC was eliminated; it has the standard deviation $\sigma_{\mathsf{GAC}} = 0.000540$.

One may use various distances to implement K-means; we used Euclidean one. Another essential point in K-mean analysis is the stability of clusterization: since a development of that latter starts *de novo* from a random (and homogeneous) distribution of the points into M classes, then there is no guarantee that the final distribution would be the same. Practically, one meets a situation with few different final distributions. Besides, some genomes may (almost randomly) change their class attribution, in different realizations of the procedure [21].

This point requires to check a stability of the final distribution obtained through K-means implementation. We supposed a final distribution to be stable, if there was not less than 75 identical final distribution patterns, in a series of a hundred of runs. An advanced K-mean implementation also includes a separability of classes checkout, with merging the classes that do not meet a criterion of separability. We checked no class separability, in our study. All the results were obtained with *ViDaExpert* software by A. Zinovyev.[1]

[1]http://bioinfo-out.curie.fr/projects/vidaexpert/.

6.2.3 "Downward" Versus "Upward" Classification

Apart from the distribution stability, K-means also poses another question towards the optimal number of classes to be developed. The most advanced version of K-means allows to get the maximal number of distinguishable classes, starting from the high enough set of these latter. Here we did not check the separability of the developed classes, hence the number of classes becomes an important parameter, in the classification implementation.

Two alternative approaches could be pursued in this situation; we call them "downward" and "upward" one, respectively. They both are based on a standard K-means technique, but differ in the definition of clusters implementation.

6.2.3.1 "Downward" Classification

This kind of classification resembles classical morphology based pattern. It starts from the clusterization of the entire set of genomes (frequency dictionaries) into the minimal number $\mathfrak{M}_c$ of clusters with the given stability of the clusterization. That latter is understood as the given number of volatile genomes, i.e. genomes that may change their cluster attribution with any new clustering realization. Then each of the clusters is to be separated into the similar (i.e. minimal stable subclusters) set of subclusters, etc. The procedure is to be trunked at the given "depth" of the cluster implementation, usually determined by the volatility of a significant part of genomes.

Thus, a "downward" classification yields the tree-like graph structure, so making it close to a standard morphological classification.

6.2.3.2 "Upward" Classification

On the contrary, the upward classification consists in the separation of the entire set of genomes, sequentially, into the series of clusters

$$\mathfrak{C}_2, \mathfrak{C}_3, \mathfrak{C}_4, \ldots, \mathfrak{C}_{K-1}, \mathfrak{C}_K .$$

Here we assume that the clusterization $\mathfrak{C}_2$ is stable. Again, the series is to be trunked at the given number K; we put $K = 8$.

The key question here is the mutual relation between the members of a cluster $\mathfrak{C}(l)_j$ from $\{\mathfrak{C}(i)_j\}$ $(1 \leqslant i \leqslant j)$ clusterization with the clusters from $\{\mathfrak{C}(m)_{j-1}\}$ $(1 \leqslant m \leqslant j-1)$ clusterization (see Fig. 6.1). Here the index l enlists the clusters at the $\{\mathfrak{C}(i)_j\}$ clusterization. There could be (roughly) three options:

- A cluster $\mathfrak{C}(n)_j$ is entirely embedded into the cluster $\mathfrak{C}(l)_{j-1}$, with some l and j;
- The greater part of the members of a cluster $\mathfrak{C}(n)_j$ is embedded into the cluster $\mathfrak{C}(l)_{j-1}$, but the minor part is embedded into the other cluster $\mathfrak{C}(m)_{j-1}$;

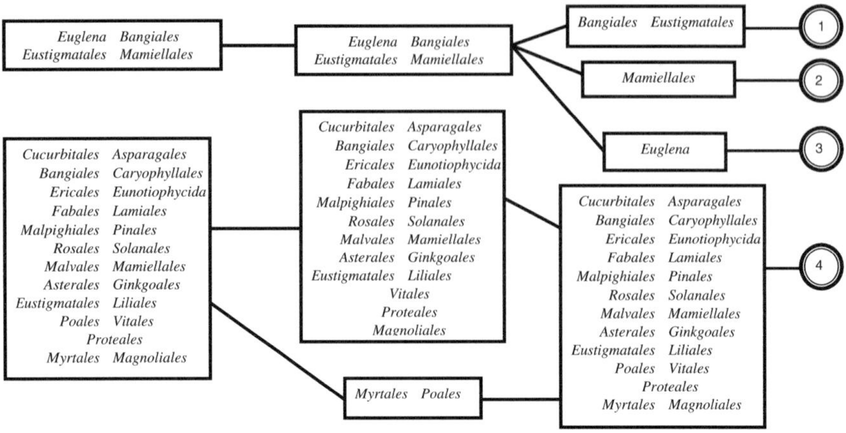

Fig. 6.1 Pattern of the "upward" embedment of various clades of plants in the stable upward classification, developed over chloroplast genomes; the case of two to, four clusters

- A cluster $\mathfrak{C}(n)_j$ is almost randomly spread between the set of clusters $\mathfrak{C}(l)_{j-1}$, $l = 1, 2, \ldots, l^*$.

Thus, an upward classification yields a pattern that is a graph with cycles. The graph may be fully connected, at the worst case, thus no essential structuredness is observed. If the graph has rather small number of cycles, then it reveals the relations between the clusters (determined through the proximity in frequency space), and the taxonomy (determined over the nuclear genome).

6.3 Results

We developed the upward classification through the K-means clusterizations obtained for two, three, four, ..., eight clusters, and studied the composition of each cluster, at the each classification level. The key questions were:

(1) whether the species (and higher clades) tend to keep together, when the number of clusters in a clusterization goes down from 8 to 2, and
(2) whether the "younger" classes tend to merge an "elder" one, or not.

Here "elder" class means a class observed for the clusterization over L clusters, while the "younger" one is that former obtained for $L + 1$ clusters.

Speaking on the upward classification, one has to keep in mind the problem of a stability of that latter. Indeed, volatile genomes must not be too numerous. We checked the stability of the distribution of genomes into the classes; a distribution was stable, if more than a half of the realizations in a series yield the same pattern. Of course, this definition is quite sensitive to the number of realizations. A good

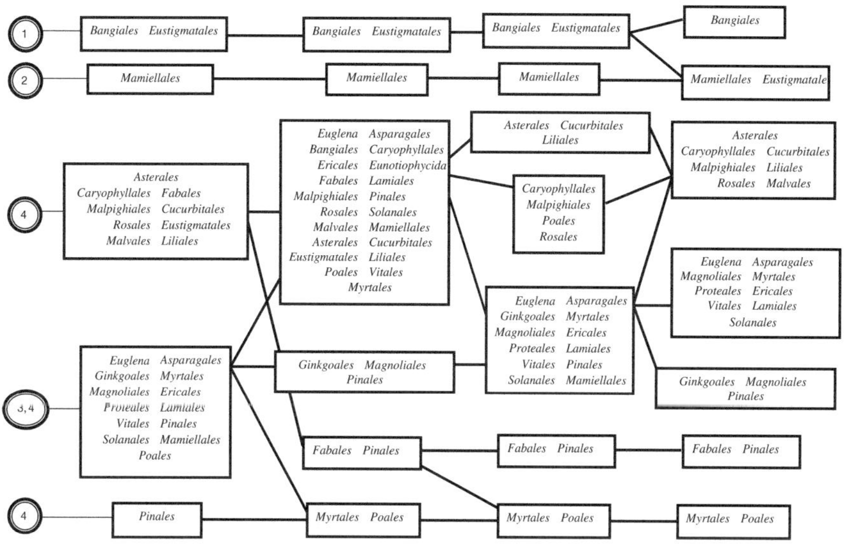

Fig. 6.2 Pattern of the "upward" embedment of various clades of plants in the stable upward classification, developed over chloroplast genomes; the case of five to eight clusters

estimation for the series length is provided by the evaluation $\sqrt{N} \leqslant L \leqslant N$, where L is a series length, but N is the number of genomes under consideration. This, we used a series of a hundred of realizations of K-means, in our work, and present here only stable subset of genomes. That latter enlists 196 entries.

Figures 6.1 and 6.2 answer distinctly and apparently the question. These figures present the clusterization mentioned above. For technical reasons, the entire graph is divided into two parts: Fig. 6.1 shows the clusterization for 2–4 classes, while Fig. 6.2 shows further classification (for five to eight classes). The numbers in circles in these figures show a conjunction edges connecting two figures in an entity. The clades in the boxes correspond to genera, while the species (not shown in the figure) always make a solid group, when changing the number of clusters. Thus, boxes having two upright arrows showing the transfer of entities from $\mathfrak{C}_l$ clusterization to $\mathfrak{C}_{l-1}$ one contain two groups of species belonging the same genus (or family).

First of all, the is not fully connected. This fact makes a strong evidence of a valuable non-randomness in the clades distribution over the classes as the number of these latter grows up. Moreover, the list of clades occupying the classes (see Figs. 6.1 and 6.2) also is very far from a random one.

Careful examination of Fig. 6.2 shows that the "youngest" level of classification (that is for eight classes) actually bears seven classes, only. This is not a mistake: the stable clusterization of the given set of chloroplast genomes yields an empty class, when making the K-means clusterization for eight classes. On the contrary, the subset of genomes comprising an unstably clustering body of the genomes (that is not considered in detail in this paper) shows a reasonable separation of the entries into eight classes.

Both Figures show the existence of rather isolated clades; the point is that this isolation seems be quite unstable, in terms of a number of clusters to be implemented due to K-means. In simple words, some clades may organize a tight cluster observed for some specific number L^* of classes, while they may merge another group for $L = L^* - 1$ or $L = L^* + 1$.

6.4 Discussion

All chloroplasts have the same function; thus, the impact of a function divergence was eliminated, in our study. Evidently, a database structure is crucial in this kind of studies. We have used an unsupervised classification technique to develop a distribution of genomes into few groups. The results of such classification are usually quite sensitive to an original database composition [21]. Luckily, the genetic banks are rapidly enriched with newly deciphered genomes of organelles, so the stable and comprehensive results showing the reliable relation between structure and taxonomy could be obtained pretty soon. Moreover, a growth of genetic database may provide a comprehensive implementation of a "downward" classification.

The data provided in this paper unambiguously prove the strong synchrony of the evolution of two genetic systems: the host one, and the organelle genome. Still further studies are needed to extend and clarify some results. First of all, the clusterization provided by K-means (or K-line) is quite sensitive to the database under consideration. Secondly, a slight bias may take place due to the significant disproportion of the species included into the database; probably, the most numerous genera should be hashed, in addition to the removal of the single-species clades.

A study presented in this paper is done within the scope of population genomics methodology. The most intriguing result of the study is the very high correlation between the statistically identified clusters of genomes, and their taxonomy reference. The key point is that we used organelle genomes to derive the clusterization, while the taxonomy was determined traditionally, through morphology, which is ultimately defined by a nuclear genomes. There is no immediate interaction between the nuclear and organelle genomes. The study has been carried out for both main organelles: chloroplasts and mitochondria.

The approach presented above looks very fruitful and powerful. One can expand the approach for the following problems to be solved:

- To study the clusterizations as described above for the database consisting of the genomes of mitochondria, and chloroplasts, of the same species. This study would unambiguously address the question on the relation between structure and function: since the organelle genomes under consideration would belong the same organisms, one may expect an elimination of the taxonomy impact, on the results.

Meanwhile, this point should be carefully checked, since the results might be sensitive to the list of species involved into the study;

- To study the clusterization of the frequency dictionaries corresponding to the individual genes (or genes combinations) retrieved from the raw genomes of organelles. The clusterization of such genetic entities would address the question on the mutual interaction in a triadic pattern *structure–function–taxonomy*.

6.5 Conclusion

We explored the relation between structure of DNA sequences and the taxonomy of their bearers. Very high correlation between these two issues has been observed, over the study of a set of chloroplast genomes. The correlation means that taxonomically proximal species tend to occupy the same cluster together. The key issue here is that the proximity of clades has been determined morphologically (i.e. over nuclear genomes), while the proximity of structures has been determined over the organelle genomes (the chloroplasts, to be exact). These two genetic systems are physically disconnected, and exhibit very low level of physical interaction. The observed correlation proves the fact of the high synchrony in evolution of these two genetic systems.

Acknowledgments This work was partly supported by a research grant No. 14.Y26.31.0004 from the Government of the Russian Federation.

References

1. Provata, A., Nicolis, C., Nicolis, G.: DNA viewed as an out-of-equilibrium structure. Phys. Rev. E **89**, 052105 (2014)
2. Qin, L., Zhang, Z., Zhao, X., Xiaolong, W., Chen, Y., Tan, Z., Li, S.: Survey and analysis of simple sequence repeats (SSRs) present in the genomes of plant viroids. FEBS Open Bio **4**, 185–189 (2014)
3. Tiwari, A.K., Srivastava, R.: A survey of computational intelligence techniques in protein function prediction. Int. J. Proteomics. 845479 (2014)
4. Huang, Y., Mrázek, J.: Assessing diversity of DNA structure-related sequence features in prokaryotic genomes. DNA Res. **21**(3), 285–297 (2014)
5. Foulongne-Oriol, M., Murat, C., Castanera, R., Ramírez, L., Sonnenberg, A.S.: Genome-wide survey of repetitive DNA elements in the button mushroom Agaricus bisporus. Fungal Genet. Biol. **55**, 6–21 (2013)
6. Sharma, M.K., Sharma, R., Peijian, C., Jenkins, J., Bartley, L.E., Qualls, M., Grimwood, J., Schmutz, J., Rokhsar, D., Ronald, P.C.A.: Genome-wide survey of switchgrass genome structure and organization. PLoS One. **7**(4), e33892 (2012)
7. Fischer, N.O., Tok, J.B., Tarasow, T.M.: Massively parallel interrogation of aptamer sequence, structure and function. PLoS One **3**(7), e2720 (2008)
8. Burge, S., Parkinson, G.N., Hazel, P., Todd, A.K., Neidle, S.: Quadruplex DNA: sequence, topology and structure. Nucleic Acids Res. **34**(19), 5402–5415 (2006)
9. Zinovjev, A.Y., Gorban, A.N., Popova, T.G.: Seven clusters in genomic triplet distribution. Siliko Biol. **3**, 471–482 (2003)

10. Gorban, A.N., Zinovjev, A.Y., Popova, T.G.: Self-organizing approach for automated gene identification. Open Syst. Inf. Dyn. **10**, 321–333 (2003)
11. Gusev, V.D., Nemytikova, L.A., Chuzhanova, N.A.: On the complexity measures of genetic sequences. Bioinformatics **15**, 994–999 (1999)
12. Bugaenko, N.N., Gorban, A.N., Sadovsky, M.G.: Towards the definition of information content of nucleotide sequences. Mol. Biol. Moscow. **30**, 5, 529–541 (1996)
13. Bugaenko, N.N., Gorban, A.N., Sadovsky, M.G.: The information capacity of nucleotide sequences and their fragments. Biophysics **5**, 1063–1069 (1997)
14. Bugaenko, N.N., Gorban, A.N., Sadovsky, M.G.: Maximum entropy method in analysis of genetic text and measurement of its information content. Open Syst. Inf. Dyn. **5**, 2, 265–278 (1998)
15. Popova, T.G., Sadovsky, M.G.: Splicing results in decrease of gene redundancy. Mol. Biol. Moscow **29**(3), 500–506 (1995)
16. Popova, T.G., Sadovsky, M.G.: Introns differ from exons in their redundancy. Rus. J. Genet. **31**(10), 1365–1369 (1995)
17. Gorban, A.N., Popova, T.G., Sadovsky, M.G.: Human viruses genes are less redundant than the human genes. Rus. J. Genet. **32**(2), 281–294 (1996)
18. Sadovsky, M.G.: On the redundancy of viral and prokaryotic genomes. Rus. J. Genet. **38**(5), 695–701 (2002)
19. Gorban, A.N., Popova, T.G., Sadovsky, M.G., Wünsch, D.C.: Information content of the frequency dictionaries, reconstruction, transformation and classification of dictionaries and genetic texts. In: Intelligent Engineering Systems through Artificial Neural Networks, vol. 11. Smart Engineering System Design, pp. 657–663. ASME Press, New York (2001)
20. Gorban, A.N., Popova, T.G., Sadovsky, M.G.: Classification of symbol sequences over their frequency dictionaries: towards the connection between structure and natural taxonomy. Open Syst. Inf. Dyn. **7**, 1–17 (2000)
21. Fukunaga, K.: Introduction to Statistical Pattern Recognition, 2 edn., p. 591. Academic Press, London (1990)

Chapter 7
Complex Synchronization Patterns in the Human Connectome Network

Pablo Villegas, Jorge Hidalgo, Paolo Moretti and Miguel A. Muñoz

Abstract A major challenge in neuroscience is posed by the need for relating the emerging dynamical features of brain activity with the underlying modular structure of neural connections, hierarchically organized throughout several scales. The spontaneous emergence of coherence and synchronization across such scales is crucial to neural function, while its anomalies often relate to pathological conditions. Here we provide a numerical study of synchronization dynamics in the human connectome network. Our purpose is to provide a detailed characterization of the recently uncovered broad dynamic regime, interposed between order and disorder, which stems from the hierarchical modular organization of the human connectome. In this regime—similar in essence to a Griffiths phase—synchronization dynamics are trapped within metastable attractors of local coherence. Here we explore the role of noise, as an effective description of external perturbations, and discuss how its presence accounts for the ability of the system to escape intermittently from such attractors and explore complex dynamic repertoires of locally coherent states, in analogy with experimentally recorded patterns of cerebral activity.

7.1 Introduction

The current mapping of neural connectivity patterns relies on advanced neuroimaging techniques, which have recently allowed for the reconstruction of structural human brain networks, establishing at an individual-based level which brain regions are mutually connected, as well as the strength of pairwise connections. The resulting "human connectome network" [1, 2] has been found to be structured in moduli or compartments—characterized by a much larger intra than inter

P. Villegas · J. Hidalgo · M.A. Muñoz (✉)
Departamento de Electromagnetismo y Física de la Materia e Instituto Carlos I de Física Teórica y Computacional, Universidad de Granada, 18071 Granada, Spain
e-mail: mamunoz@onsager.ugr.es

P. Moretti
Institute of Materials Simulation (WW8), Friedrich-Alexander-University Erlangen-Nünberg, Dr.-Mack-Straße 77, 90762 Fürth, Germany

S. Battiston et al. (eds.), *Proceedings of ECCS 2014*, Springer Proceedings in Complexity, DOI 10.1007/978-3-319-29228-1_7

connectivity—organized in a hierarchical fractal-like fashion across diverse scales [3–8]. On the other hand, functional connections between different brain regions can be inferred e.g. from correlations in neural activity as detected in EEG or fMRI time series. Unveiling the relation between structural and functional networks is a current challenge in modern neuroscience. In this context, a few pioneering works found that the hierarchical-modular organization of structural brain networks has remarkable implications for neural dynamics [7, 9–11]. As opposed to the case of simpler network structures, neural activity propagates in hierarchical networks in a peculiar way. For example, models of neural activity propagation usually exhibit two familiar phases; percolating and non-percolating, respectively; but it has been recently found [12] that when such models operate on top of the "human connectome" structural network a novel intermediate regime, named a "Griffiths phase" [13, 14] emerges. This novel phase originates from the highly-diverse and relatively isolated structural moduli where dynamical activity may remain mostly localized for long time periods [12, 14].

Given that the correct brain functioning requires coherent neural activity at a wide range of scales [15, 16], the study of synchronization among neural populations is one of the central ideas in computational neuroscience [17, 18].

In a recent work [19], some of us scrutinized the special features of synchronization dynamics [20] using the canonical Kuramoto model for phase synchronization [21–23], in the actual human connectome (HC) network [1, 2, 24]. In analogy to what described above for activity propagation, we uncovered the existence of a novel intermediate phase for synchronization dynamics, stemming from the hierarchical modular organization of the HC. Furthermore, we found that the dynamics in such a region presented a plethora of complex and interesting dynamical features [19].

Our goal here is to describe in more detail the complex behavior within such an intermediate regime, both in individual moduli and at a global brain level. We measure the fluctuations of the global order parameter as a function of the overall coupling strength, and we show that there is a broad region (rather than a unique "critical" point) with huge variability and response. Finally, we assess the role of noise and perturbations in the robustness of the metastable stated arising in the intermediate regime, and we show that adding intrinsic fluctuations to the picture of synchronization dynamics in hierarchical modular networks accounts for the ability of the brain to explore different attractors, giving access to the varied functional configurations recorded in experiments [25–27].

7.2 Kuramoto Model in the Human-Connectome Network

The HC network we employ consists of a set of $N = 998$ nodes, each of them representing a population of neurons producing self-sustained oscillations [28], connected pairwise through a precise pattern of symmetric weighted edges, altogether determining a connectivity matrix $\mathbf{W}$ [1, 2].

On top of such a HC network, we implement a noisy Kuramoto dynamics, defined by the set of differential equations [21–23]:

$$\dot{\theta}_i(t) = \omega_i + \alpha\eta_i(t) + k\sum_{j=1}^{N} W_{ij}\sin\left[\theta_j(t) - \theta_i(t)\right] \tag{7.1}$$

where $\theta_i(t)$ is the phase at node i at time t, the intrinsic frequencies ω_i—accounting for region heterogeneity—are extracted from some probability distribution function $g(w)$, W_{ij} are the elements of the $N \times N$ weighted connectivity matrix $\mathbf{W}$, k is an overall coupling parameter and $\eta_i(t)$ is a zero-mean delta-correlated Gaussian noise, tuned by the real-valued amplitude α.

The Kuramoto complex order parameter is defined by $Z(t) = R(t)e^{i\psi(t)} = \langle e^{i\theta_k(t)}\rangle_k$, where $0 \leq R \leq 1$ gauges the overall coherence and $\psi(t)$ is the average phase. It is common wisdom that for an (infinitely) large population of oscillators interacting in a fully connected network, the model exhibits a phase transition at some value of k, separating a coherent steady state ($R > 0$) from an incoherent one ($R = 0$, plus $1/\sqrt{N}$ finite-size corrections) [21–23]. On the other hand, in the absence of frequency heterogeneity the system always reaches a coherent state. Thus, frequency heterogeneity is responsible for frustrating synchronization if the coupling strength is weak. Similarly, in our recent work [19] we argued that the combined effect of frequency heterogeneity *and* network heterogeneity (in particular, a hierarchical modular structure) can lead to much richer and interesting ways of ordering frustration. Here we explore that phenomenology in much deeper detail, introducing external stochastic fluctuations (i.e. noise) as the mechanism accounting for the ability of the system to explore metastable configurations.

7.3 Results

We considered the HC network [1, 2] and employed standard community detection algorithms [8, 29] to identify the underlying modular structure. The optimal partition into communities—i.e. the one maximizing the modularity parameter [30]—turns out to correspond to a division in 12 moduli [19]. At a higher hierarchical level, a separation into just 2 moduli (roughly corresponding to the 2 cerebral hemispheres) also gives a quite high modularity value. As illustrated in Fig. 7.1, 4 (out of the 12) moduli belong to one of the two hemispheres, 5 to the other, while 3 moduli (cyan, blue and red) overlap with both hemispheres. We label these two hierarchical levels as $l = 2$ (2 large moduli) and $l = 1$ (12 smaller moduli), respectively.

We have conducted computational analyses of the noisy Kuramoto model on top of the HC network and performed a number of new computational experiments complementing the analyses in our previous work [19].

As illustrated in Fig. 7.2, beside the aforementioned coherent and incoherent phases (usually encountered in synchronizing systems) there is an intermediate

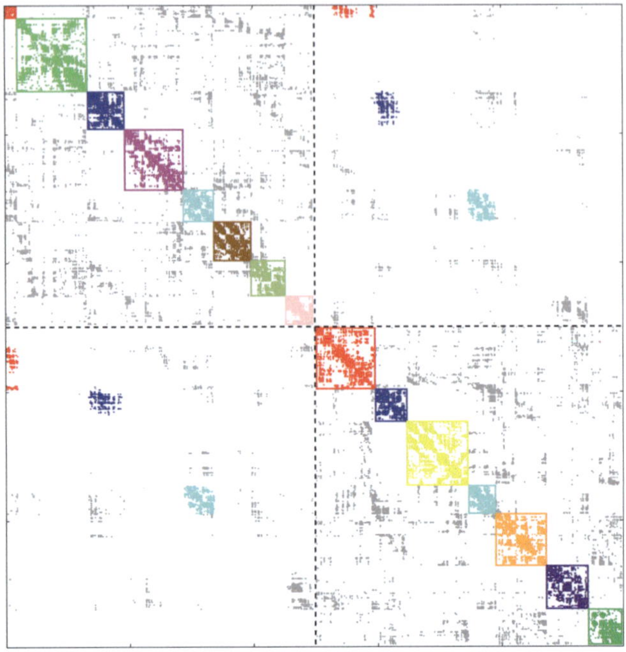

Fig. 7.1 Adjacency matrix of the HC network with nodes ordered to emphasize its modular structure as highlighted by a community detection algorithm (see main text), showing also the partition into the 2 hemispheres (*dashed lines*). 12 moduli can be distinguished (each plotted with a different color); 4 of them correspond to one of the two hemispheres, 5 to the other, and only 3 moduli overlap with both hemispheres (*cyan*, *blue* and *red* moduli). Inter-modular connections (*grey*) are limited to small subsets, acting as interfaces or connectors between moduli

regime between them exhibiting a large variability. Individual trajectories are depicted in the inset, for different values of the coupling strength k; observe in particular the irregular oscillations obtained for intermediate values of k.

The reported values of $\langle R \rangle_t$ in the main plot of Fig. 7.2 correspond to the time-averaged value for a single realization in its steady state, considering up to a fixed maximum time T. The observed variability in the central region means either that (i) larger time windows would be required for the system to self-average or (ii) that ergodicity is broken and for each parameter value the realization ends up in a different type of (stable or metastable) steady state, depending on the initial condition. This last possibility implies that the system may remain trapped in some sort of metastable states, from which it can escape away only after very rare and large fluctuations.

These observations are robust against changes in the frequency distribution, connectivity matrix normalization, and other details, whereas the location and width of the intermediate phase are not universal. For example, Fig. 7.2 has been obtained for a Gaussian frequency distribution but similar curves are obtained for, usually employed, Lorentzian or uniform distributions.

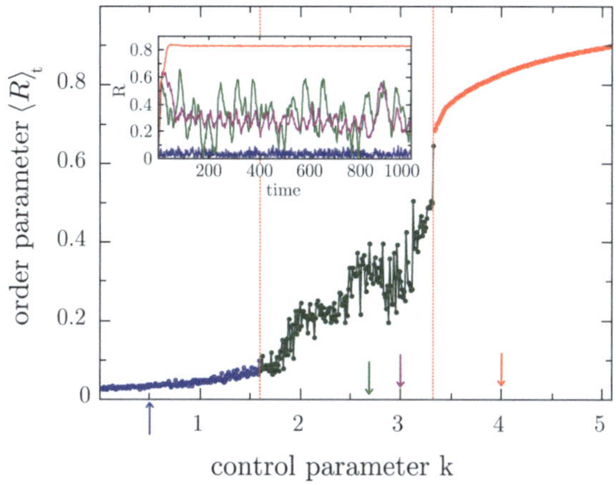

Fig. 7.2 (Main) Time-averaged value of the order parameter for the noisy Kuramoto dynamics running upon the Human Connectome network (998 nodes) with a Gaussian distribution of frequencies. Three different regimes emerge: an incoherent phase ($k < 1.6$), a synchronous one ($k > 3.3$), and an intermediate irregular one. In this last, much larger averaging times would be required to obtain reliable mean values and these would depend upon initial conditions, reflecting metastability. We have chosen not so large measuring times ($t = 100$ for all values of k) to illustrate the variability in the intermediate region. (Inset) Time-series for 4 different k values, indicated by arrows in the x-axis (from left to right: $k = 0.5, 2.7, 3.0$ and 4.0)

As this robust intermediate regime is reminiscent of Griffiths phases in networks—posed in between order and disorder and emerging from rare-region effects [12–14]—it is natural to wonder how the structural network modularity affects synchronization dynamics in general. As a matter of fact, it is straightforward to convince oneself that any network consisting of perfectly isolated moduli, each of them synchronized at different intrinsic frequencies and phases, should exhibit oscillations of the collective order parameter, R, and these oscillations are preserved when the moduli are weakly interconnected [19]. Thus, in large networks without delays or other additional ingredients, time oscillations in the global coherence are the trademark of an underlying modular structure.

To illustrate the role played by internal network modularity on global synchronization, Fig. 7.3 portraits the trajectories of the parameter $Z(t)$ in the complex plane for different values of the control parameter k, measured at different hierarchical levels: two (out of the existing 12) different small moduli (violet and orange curves), the two hemispheres (red and green), and the overall brain (blue). In the incoherent phase (panel a), the real and imaginary parts of Z fluctuate around zero at all scales in the hierarchy. On the other hand, in the coherent phase (panel d), all nodes are synchronized, and trajectories are circles with radii close to unity at all hierarchical levels

A much richer behavior is found in the intermediate region: panel b (left) illustrates a situation in which one modulus (orange) is mostly coherent, while the other (violet) is not; however, hemispheres and global dynamics remain mostly unsynchronized

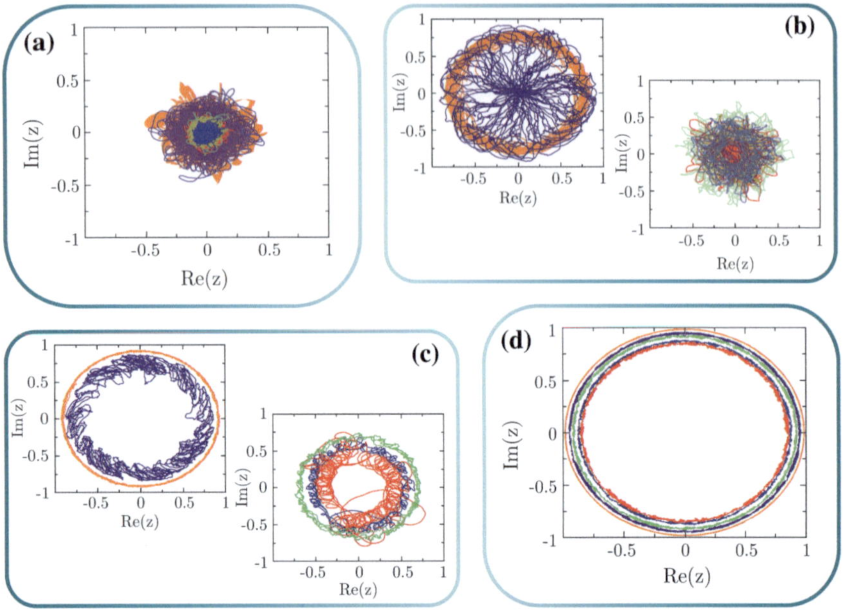

Fig. 7.3 Phase portraits of the complex order parameter $Z(t)$, measured at different scales in the hierarchy for a Gaussian $g(w)$ (different realizations from those in Fig. 7.2): two of the existing moduli are plotted in violet and orange, respectively, the two hemispheres in red and green, and the global scale in blue. Panels (**a**)–(**d**) correspond to values of the control parameter $k = 1, 3, 5$ and 8, respectively (panels (**b**) and (**c**) have been split into two to enhance clarity). **a** In the non-ordered phase, the real and imaginary components of Z fluctuate around zero, not exhibiting synchronization at any scale. **b** In the early region of the intermediate phase, a few moduli are coherent (as the one in *orange*) but most of them remain unsynchronized (*violet*), and the system does not present coherence for upper scales in the hierarchy. **c** Increasing k, more heterogeneity of synchronization among moduli is found, and the system exhibits complex trajectories for the intermediate (hemispheres) and global scale. **d** In the coherent phase, all moduli are synchronized, and trajectories are concentric circles

(panel b (right)). In panel c (left), we have slightly increased the control parameter with respect to panel b, with a subsequent increase of the coherence for all hierarchical levels. Interestingly, as not all moduli exhibit the same state of coherence, chaotic-like oscillations of the order parameter are observed at the global scale.

We are interested in quantifying the observed variability of R in the intermediate phase. To this end, we take a particular realization of frequencies (extracted from a Gaussian $g(w)$) and, starting from an initial—uniformly distributed—random configuration of individual phases, $\{\theta_i(t=0)\}_{i=1}^{N}$, we measure the temporal standard deviation of the global coherence parameter R (after the transient) up to a maximum time $T = 10{,}000$,

$$\sigma = \left(\langle (R - \langle R \rangle_t)^2 \rangle_t\right)^{1/2} \tag{7.2}$$

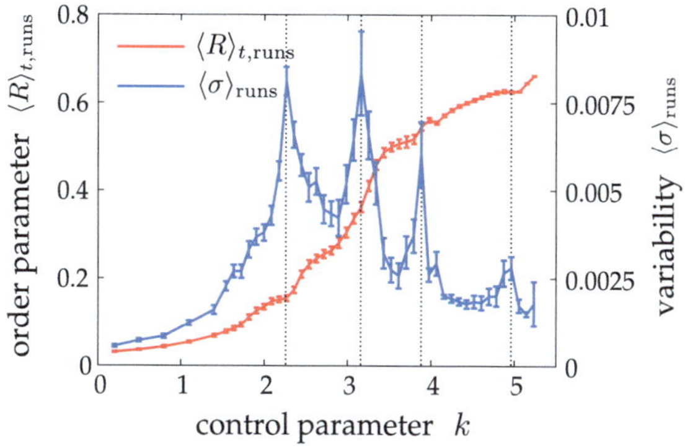

Fig. 7.4 Time-averaged order parameter $R = \langle R(t)\rangle$ and standard deviation of time-series, averaged over realizations with different—uniformly distributed—initial conditions. Maximal variability is found in the intermediate phase, where the system is neither too unsynchronized nor too coherent. Several peaks in the variability can be distinguished (*dashed lines*), which appear at values of the control parameter k for which the system experiments a fast increase in global synchronization. Statistical sampling of different realizations indicate that error bars are larger in the intermediate region, suggesting the existence of several attractors depending on the initial conditions. We have averaged 100 different realizations, each one integrated for 10,000 time steps

as a function of the coupling strength k.[1]

As ergodicity may be broken, different initial conditions may lead to different attractors of the dynamics, therefore we also average σ over 100 different independent realizations of the dynamical process. Results are illustrated in Fig. 7.4, in which we also have plotted the diagram of the order parameter obtained for this particular realization of $g(\omega)$ averaged over the 100 realizations. Let us stress the following salient aspects: (i) averaged time variabilities are small in the non-coherent ($k \lesssim 1$) as well as in the coherent ($k \gtrsim 5$) phases, whereas much larger variabilities are found in the intermediate region ($1 \lesssim k \lesssim 5$); (ii) the curve of time variabilities presents several peaks for the intermediate region, lying in the vicinity of values of the control parameter at which the system experiences a change in its level of coherence (see the corresponding jumps in the derivative of the order parameter); and finally, (iii) error bars are also larger in the intermediate phase; this variability of time variabilities means that different initial conditions can lead to different types of time-series, suggesting a large degree of metastability in the intermediate regime.

[1]Notice that this definition of σ, that we call, "*time variability*" is closely related to the chimera index introduced by Shanahan [31]. While chimera indices are averaged between individual network moduli and measure the onset of local coherence, σ is defined at the global level and records fluctuations of the global order parameter.

7.3.1 Metastability in HMNs

Our previous results vividly illustrate the existence of an intermediate region in which the HC exhibits maximal dynamical variability at the global scale, suggesting metastable behavior. In order to explore more directly whether metastable states exist, we now assess if the dynamics may present different attractors and, for some values of the control parameter k and noise amplitudes, if the system may switch between different global attractors with different levels of coherence.

Figure 7.5a shows a time series of the global parameter, for a fixed realization of internal frequencies and initial phases. It clearly illustrates how the HC spontaneously switches between two different attractors. These type of events, however, are not easy to observe in the HC network. Due to the coarse-grained nature of the HC mapping, different attractors may actually have comparable average values of the coherence R, which makes their discrimination especially difficult at the global scale.

Instead, such events are easier to spot in synthetic hierarchical modular networks (HMN), such as proposed to model brain networks in an efficient way (see [12] and references therein). In such hierarchical networks, the effects of modularity and hierarchy are much enhanced, as they develop across a larger number of hierarchical levels than the one allowed by current imaging techniques for empirically obtained connectomes.

All the previously reported phenomenology is still present in such HMNs (see [19]); in particular, the phase diagram of the synchronization order parameter exhibits a phase transition with an intermediate region, where variability is much enhanced [19]. Figure 7.5b illustrates the bi-stable nature of the global parameter in the intermediate phase for a HMN, in which metastability can be very well appreciated. This switching behavior closely resembles "up and down" states, which are well known to appear in certain phases of sleep or under anaesthesia (see [32] and refs. therein).

We hypothesize that hierarchical modular networks in general (and the HC in particular) enable the possibility of a large repertoire of attractors, with different degrees of coherence and stability. Such metastability can be made evident and quantified by performing the following type of numerical test. Starting from a fixed random initial condition and considering a vanishing noise amplitude (i.e. $\alpha = 0$), the system might deterministically fall into a number of different attractors, each of them with an associated value of the global coherence depending on the initial conditions, the network structure, and the choice of natural frequencies. Once this attractor A is reached, the system is perturbed by switching on a non-vanishing noise amplitude ($\alpha > 0$) during a finite time window. The system may remain stable in the same attractor A if the noise is weak enough ($\alpha \ll 1$). However, if larger values of the noise amplitude are chosen, the system may jump into another close, more stable, attractor. If the noise amplitude is very large ($\alpha \gg 1$), the system can in principle jump to any attractor, but, very likely, will also escape from it, wandering around a large fraction of the configuration space. After the perturbation time-window is over, we let the system relax once again, and check if the new resulting steady state B has changed with respect to A. In that case, we can conclude that the systems was in

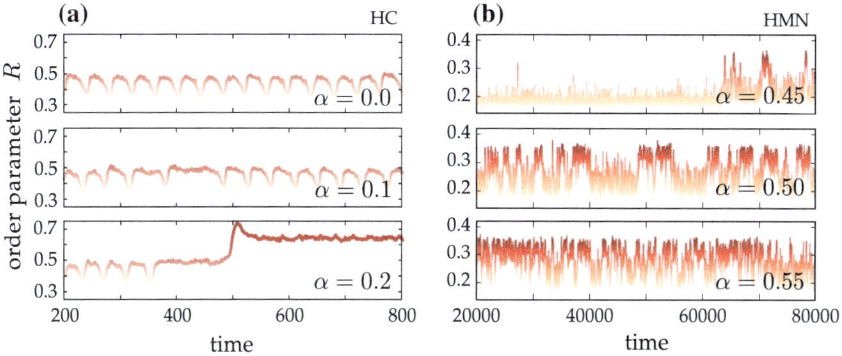

Fig. 7.5 Time series exhibit metastability of the global synchronization in the Human Connectome and in HMNs, in the intermediate region. **a** Times series of the noisy Kuramoto dynamics in the HC with Gaussian $g(w)$ in the intermediate region: for low noise amplitudes ($\alpha = 0$ and 0.1), the system stays stable in the same attractor. But, for sufficiently large values, such as $\alpha = 0.2$, the system is able to "jump" to another more coherent attractor, where it settles. **b** In HMNs (size $N = 1024$, 9 hierarchical levels), we observe the same phenomenology, but much enhanced: when noise is very low ($\alpha \leq 0.45$), the system tends to remain stable in a certain attractor (with a few exceptions after very large waiting times). Choosing a higher α ($\alpha \geq 0.5$), the system exhibits bi-stable behavior, switching intermittently between two different attractors. For large enough α ($\alpha \geq 0.55$), the dynamics becomes too erratic to appreciate metastability. Here, frequencies were extracted from a Lorentz distribution

a metastable state A before the perturbation, and has reached another state B after it—potentially a metastable state itself.

We have carried out this type of test using an artificial HMN (see Fig. 7.6) for a specific value of the control parameter k, belonging in the intermediate region. Natural frequencies are sampled from the a Lorentzian distribution $g(\omega)$ (as above, our main results are not sensible to this choice). Starting from a random initial configuration of phases, we integrate Eq. (7.1) up to time 500 with $\alpha = 0$. After this, we introduce the external perturbation by switching the noise coefficient α to a certain non-zero value during a time window of duration 100. Finally we revert to $\alpha = 0$ and continue the integration up to time $t = 1000$. The last steady state value is averaged over 10^4 realizations of initial conditions, networks, and intrinsic frequencies.

As illustrated in Fig. 7.6, for low as well as for high values of the noise amplitude, the system has the same average order parameter close to $\langle R \rangle_{t,\text{runs}} \simeq 0.2$, as could have been anticipated. However, a resonant peak emerges for intermediate values of the noise, where the system switches to states with different levels of coherence. This plot explicitly illustrates the existence of metastability and noise-induced jumps between attractors. As noise is enhanced, progressively more stable states are found, but above some noise threshold, the system does not remain trapped in a single attractor but jumps among many, resulting in a progressive decrease of the overall coherence.

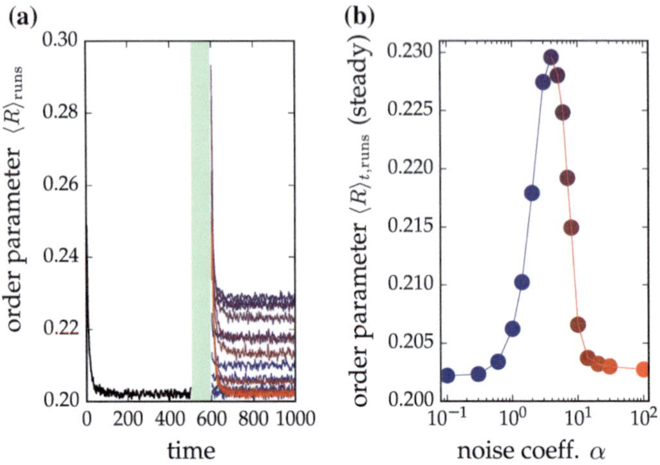

Fig. 7.6 Perturbations can lead the system to more coherent attractors in the intermediate non-coherent phase. **a** Order parameter R averaged in time over 10^4 realizations. A noise pulse of amplitude α is applied during the green interval. This same protocol is repeated for different values of α. **b** Average order parameter in the final steady state (after the noise pulse) as a function of α. For intermediate values of α, a resonant peak emerges for $1 < \alpha < 10$, illustrating that the system can jump to a close, more coherent on-average attractor. Simulations are run on HMN networks of size $N = 1024$, with 9 hierarchical levels

7.4 Discussion

It is well established that in the absence of frequency dispersion, the Kuramoto dynamics leads to a perfectly coherent state, which is progressively achieved in time by following a bottom-up ordering dynamics in which increasingly larger communities become synchronized [33].

If a hierarchical modular networks is loosely connected, this type of "matryovska-doll" synchronization process is constrained at all levels by structural bottlenecks, bringing about anomalously-slow synchronization dynamics as recently reported [19].

In the presence of intrinsic frequency dispersion the above slow ordering process is further frustrated [19]. For small values of k the system may remain trapped into metastable states in which the loose connectivity between some moduli does not allow them to overcome intrinsic-frequency differences and achieve coherence. While persistence in metastable states may extend indefinitely, experimental evidence suggests that the brain is able to switch between a rich repertoire of attractors [25–27]. We have shown that a simple description of neural coherence dynamics based on the noisy Kuramoto model may suffice to reproduce a very rich phenomenology, in hierarchical modular networks and in particular in the human connectome. The introduction of small fluctuations (exemplifying external perturbations, stimuli, or intrinsic stochasticity) allow the system to escape from metastable states and sample the configuration space, proving a paradigmatic modeling tool for the attractor surfing behavior suggested by experiments. Additional ingredients, such as explicit phase frustration [31] or time delays [28, 34], should only add complexity to the structural frustration effect reported here, providing a finer description of brain activity.

Acknowledgments We acknowledge financial support from J. de Andalucía P09-FQM-4682 and the Spanish MINECO FIS2012-37655-C02-01 and FIS2013-43201-P.

References

1. Hagmann, P., Cammoun, L., Gigandet, X., Meuli, R., Honey, C.J., Wedeen, V.J., Sporns, O.: Mapping the structural core of human cerebral cortex. PLoS Biol. **6**(7), e159 (2008)
2. Honey, C.J., Sporns, O., Cammoun, L., Gigandet, X., Thiran, J.P., Meuli, R., Hagmann, P.: Predicting human resting-state functional connectivity from structural connectivity. Proc. Natl. Acad. Sci. **106**(6), 2035–2040 (2009)
3. Bullmore, E., Sporns, O.: Complex brain networks: graph theoretical analysis of structural and functional systems. Nat. Rev. Neurosci. **10**, 186–98 (2009)
4. Sporns, O.: Networks of the Brain. MIT Press, Cambridge (2010)
5. Kaiser, M.: A tutorial in connectome analysis: topological and spatial features of brain networks. NeuroImage **57**(3), 892–907 (2011)
6. Meunier, D., Lambiotte, R., Bullmore, E.: Modular and hierarchically modular organization of brain networks. Front. Neurosci. **4**, 200 (2010)
7. Zhou, C., Zemanova, L., Zamora-López, G., Hilgetag, C.C., Kurths, J.: Hierarchical organization unveiled by functional connectivity in complex brain networks. Phys. Rev. Lett. **97**(23) (2006)
8. Ivković, M., Amy, K., Ashish, R.: Statistics of weighted brain networks reveal hierarchical organization and Gaussian degree distribution. PLoS ONE **7**(6), e35029 (2012)
9. Zhou, C., Zemanova, L., Zamora-López, G., Hilgetag, C.C., Kurths, J.: Structure-function relationship in complex brain networks expressed by hierarchical synchronization. New J. Phys. **9**(6), 178–178 (2007)
10. Kaiser, M., Goerner, M., Hilgetag, C.: Criticality of spreading dynamics in hierarchical cluster networks without inhibition. New J. Phys. **9**, 110 (2007)
11. Kaiser, M., Hilgetag, C.C.: Optimal hierarchical modular topologies for producing limited sustained activation of neural networks. Front. Neuroinform. **4**(8) (2010)
12. Moretti, P., Muñoz, M.A.: Griffiths phases and the stretching of criticality in brain networks. Nat. Commun. **4**(2521) (2013)
13. Vojta, T.: Rare region effects at classical, quantum and nonequilibrium phase transitions. J. Phys. A **39**(22), R143–R205 (2006)
14. Muñoz, M.A., Juhász, R., Castellano, C., Ódor, G.: Griffiths phases on complex networks. Phys. Rev. Lett. **105**, 128701 (2010)
15. Steinmetz, P.N., Roy, A., Fitzgerald, P.J., Hsiao, S.S., Johnson, K.O., Niebur, E.: Attention modulates synchronized neuronal firing in primate somatosensory cortex. Nature **404**(6774), 187–190 (2000)
16. Kandel, E.R., Schwartz, J.H., Jessell, T.M.: Principles of Neural Science. McGraw-Hill, New York (2000)
17. Buzsáki, G.: Rhythms of the Brain. Oxford University Press, NY (2006)
18. Breakspear, M., Stam, C.J.: Dynamics of a neural system with a multiscale architecture. Phil. Trans. R. Soc. Lond. B **360**(1457), 1051–1074 (2005)
19. Villegas, P., Moretti, P., Muñoz, M.A.: Frustrated hierarchical synchronization and emergent complexity in the human connectome network. Sci. Rep. **4**(5990) (2014)

20. Rosenblum, M.G., Pikovsky, A., Kurths, J.: Synchronization—A universal concept in nonlinear sciences. Cambridge University Press, Cambridge (2001)
21. Kuramoto, Y.: Self-entrainment of a population of coupled nonlinear oscillators. Lect. Not. Phys. **39**, 420–422 (1975)
22. Strogatz, S.H.: From kuramoto to crawford: exploring the onset of synchronization in populations of coupled oscillators. Physica D **143**(1), 1–20 (2000)
23. Acebrón, J.A., Bonilla, L.L., Pérez-Vicente, C.J., Ritort, F., Spigler, R.: The Kuramoto model: a simple paradigm for synchronization phenomena. Rev. Mod. Phys. **77**, 137–185 (2005)
24. Arenas, A., Díaz-Guilera, A., Kurths, J., Moreno, Y., Zhou, C.: Synchronization in complex networks. Phys. Rep. **469**(3), 93–153 (2008)
25. Chialvo, D.R.: Emergent Complex Neural Dyn. Nat. Phys. **6**, 744–750 (2010)
26. Deco, G., Jirsa, V.K.: Ongoing cortical activity at rest: criticality, multistability, and ghost attractors. J. Neurosci. **32**(10), 3366–3375 (2012)
27. Haimovici, A., Tagliazucchi, E., Balenzuela, P., Chialvo, D.R.: Brain organization into resting state networks emerges at criticality on a model of the human connectome. Phys. Rev. Lett. **110**, 178101 (2013)
28. Cabral, J., Hugues, E., Sporns, O., Deco, G.: Role of local network oscillations in resting-state functional connectivity. NeuroImage **57**(1), 130–139 (2011)
29. Duch, J., Arenas, A.: Community detection in complex networks using extremal optimization. Phys. Rev. E **72**, 027104 (2005)
30. Newman, M.: The structure and function of complex networks. SIAM Rev. **45**(2), 167–256 (2003)
31. Shanahan, M.: Metastable chimera states in community-structured oscillator networks. Chaos **20**(1), 013108 (2010)
32. Eckmann, J.P., Feinerman, O., Gruendlinger, L., Moses, E., Soriano, J., Tlusty, T.: The physics of living neural networks. Phys. Rep. **449**(1–3), 54–76 (2007)
33. Arenas, A., Díaz-Guilera, A., Pérez-Vicente, C.: Synchronization reveals topological scales in complex networks. Phys. Rev. Lett. **96**, 114102 (2006)
34. Wildie, M., Shanahan, M.: Metastability and chimera states in modular delay and pulse-coupled oscillator networks. Chaos **22**(4), 043131 (2012)

Chapter 8
Structure of a Media Co-occurrence Network

V.A. Traag, R. Reinanda and G. van Klinken

Abstract Social networks have been of much interest in recent years. We here focus on a network structure derived from co-occurrences of people in traditional newspaper media. We find three clear deviations from what can be expected in a random graph. First, the average degree in the empirical network is much lower than expected, and the average weight of a link much higher than expected. Secondly, high degree nodes attract disproportionately much weight. Thirdly, relatively much of the weight seems to concentrate between high degree nodes. We believe this can be explained by the fact that most people tend to co-occur repeatedly with the same people. We create a model that replicates these observations qualitatively based on two self-reinforcing processes: (1) more frequently occurring persons are more likely to occur again; and (2) if two people co-occur frequently, they are more likely to co-occur again. This suggest that the media tends to focus on people that are already in the news, and that they reinforce existing co-occurrences.

8.1 Introduction

Complex networks have been a prominent research topic for the past decade. One of the reasons is that complex networks appear in a multitude of scientific disciplines, varying from neurology [6, 18], ecology [13, 16] to international relations [9, 14, 22] and human mobility [15, 30] providing a unified theoretical framework for analysis. Although many properties seem to be (nearly) universal (e.g. degree distribution, clustering) [2], there are also some noteworthy differences between different types of networks (e.g. assortativity, weak links) [1, 17, 28].

V.A. Traag (✉)
CWTS, Leiden University, Leiden, The Netherlands
e-mail: v.a.traag@cwts.leidenuniv.nl

R. Reinanda
Faculty of Science, University of Amsterdam, Amsterdam, The Netherlands

G. van Klinken
KITLV, Leiden, The Netherlands

S. Battiston et al. (eds.), *Proceedings of ECCS 2014*, Springer Proceedings in Complexity, DOI 10.1007/978-3-319-29228-1_8

Social networks can nowadays be relatively easily scraped from online services such as Facebook or Twitter [8, 11, 33]. In addition, traditional media (i.e. newspapers) are also increasingly being digitised. Whereas online social media are open to the general public, and the large masses use them intensively, traditional media are biased towards the more influential members of society. Therefore, we might learn something about the elite of a society by studying how they appear in the newspapers. In this study, we focus on who co-occurs with whom. We aim to understand the structure of the co-occurrences. Do frequently occurring people co-occur mostly with each other, or not? How strongly do frequently occurring people connect to each other?

Given the possibilities of using media co-occurrence reports for constructing social networks, there have been surprisingly few earlier studies of media co-occurrence networks [19, 27, 29, 31]. The analyses were relatively succinct, and showed that such networks were both scale-free and a small world, properties we also find here. However, the scale-free and small world nature of these co-occurrence networks can be expected, and also show up in a randomised graph. In fact, various common properties are quite close to what can be expected in a randomised graph. Nonetheless, three deviations from the random graph stand out. First of all, the average degree is much lower than expected, while the average weight is much higher than expected. Secondly, high degree nodes attract disproportionately much weight. Thirdly, much of the weight is concentrated in between high degree nodes.

These observations suggest that people repeatedly occur with the same people, or at least more so than expected at random. To explain this, we create a model that concentrates more of the co-occurrences in fewer people, thus explaining this deviation. The model consists of only two simple ingredients: (1) more frequently occurring people tend to occur more frequently in the future; and (2) people that co-occur more frequently tend to co-occur more frequently in the future. In addition, high degree nodes are more likely to occur with already existing neighbours.

8.2 Data and Network

We use newspaper articles to construct a social network. The idea is that people are linked if they co-occur in the same sentence. We use two corpora in our current study: (1) a corpus from an Indonesian news service called *Joyo*[1]; and (2) a corpus from the New York Times[2] (NYT). The Joyo dataset covers roughly 2004–2012 and contains 140,263 articles, while the NYT dataset covers 1987–1988 and contains 210,645 articles. The Joyo dataset is a selection of political news (in English) from both domestic and foreign sources that is relevant to the politics of Indonesia, while

[1] http://www.joyonews.org.

[2] See https://catalog.ldc.upenn.edu/LDC2008T19 for the corpus. We only used the first two years of the dataset.

the NYT is a complete corpus of all the articles of that newspaper. We scan the whole text and automatically identify entities by using a technique known as named entity recognition (NER) [12]. The technique automatically identifies different entities, and classifies them into three distinct categories: persons, organisations and locations. Although it is not perfect, the error rate tends to be relatively low [12].

We have only included persons in our media co-occurrence network, and only people that occurred in more articles than on average. We thereby exclude people that only appear quite infrequently, which are presumably less influential, and thus less of interest. Although this skews the results more towards people that appear more prominently in the media, it still includes many less prominent people. Once all persons have been identified, we have to disambiguate them. There are generally two types of errors that can be made with names [23]: (1) a single name corresponds to two different persons (e.g. "Bush" can refer to the 43rd or 41st US president); and (2) two different names refer to the same person ("President Clinton" or "Bill Clinton" both refer to the 42nd US president). The second problem appears much more prominent than the first problem in our corpus, as people are generally referred to in many different ways in journalistic prose (including or not positions, titles, initials, maiden names, etc…).

We disambiguated these names by using a combination of similarity measures based on Wikipedia matching, string similarity and network similarity (using Jaccard similarity). The more prominent people often have a Wikipedia page, and various spelling variants are redirected to the same entity (e.g. "President Clinton" and "Bill Clinton" both redirect to the same Wikipedia page). Each similarity is normalised to fall between 0 and 1 (with 1 being identical), which we threshold at 0.75, such that we only take it into account if the similarity is at least 0.75. We then take the average of the similarities that are higher than 0.75 (which is then also at least 0.75). We then find clusters of names such that each cluster has an average internal similarity of 0.85 (all similarity measures are between 0 and 1), using a technique called the Constant Potts Model [32].

Once the names have been disambiguated, we create a link for all the unique names in a sentence (i.e. repeated names have no effect). We do this for every sentence, and simply count in how many sentences such a co-occurrence was observed. We take only the largest connected component of the network, and this constitutes the media co-occurrence network we analyse in this paper.

Of course, what co-occurrence exactly implies is not always clear: two people might be mentioned together for example because they collaborate, or because they are contestants in an election. A co-occurrence might not coincide with any one single definition of a "relationship" in the sociological sense [20]. Hence, we cannot say if two people that co-occur have any more significant relationship: do they know each other? Have they ever communicated? Have they met face to face? Are they close friends? Sworn enemies? We simply cannot tell. This is essential to bear in mind when drawing any conclusions: the network is based on co-occurrence, not on "actual relationships".

8.3 Results

We denote the undirected network by $G = (V, E)$ where $V = \{1, \ldots, n\}$ is the node set, representing the people, and $E \subseteq V \times V$ constitutes the edge set with $m = |E|$ edges, representing the co-occurrences between people. Hence, if node $i \in V$ and node $j \in V$ co-occurred in some sentence, then there is an edge $(ij) \in E$. Each edge has a weight associated to it, which represents the number of times the two nodes i and j have co-occurred, which we denote by w_{ij}. Finally, the adjacency matrix is denoted by A, such that $A_{ij} = 1$ if there is an edge between node i and j and zero otherwise. Since the network is undirected, we have that $A_{ij} = A_{ji}$ and $w_{ij} = w_{ji}$. Notice that this network is a projection of a bipartite network of people and sentences. Let us denote by B the bipartite adjacency matrix, so that $B_{is} = 1$ if person i occurs in sentence s. Then $w_{ij} = \sum_s B_{is} B_{js}$ and $A_{ij} = 1$ if $w_{ij} > 0$ and $A_{ij} = 0$ otherwise. In other words, $w = BB^T$. We denote the degree of node i by $k_i = \sum_j A_{ij}$ and the strength by $s_i = \sum_j A_{ij} w_{ij}$.

We compare our results to a bipartite randomisation of the network. Formally, let P be a list of persons and S a list of sentences, such that $(P_e S_e)$ is a bipartite edge, so that the length of the list equals the number of bipartite edges. If σ is a random permutation, then $(P_{\sigma_e} S_e)$ for all e are the randomized edges. Simply put, this randomisation takes the list of occurrences of people in sentences and shuffles the complete edge list, so that people occur in a random sentence. This preserves the number of times somebody occurs in a sentence, and preserves the number of people that occur in a sentence. We then take a projection of this network as we did for the empirical observations. In other words, $\hat{w} = \hat{B}\hat{B}^T$ and $\hat{A}_{ij} = 1$ if $\hat{w}_{ij} > 0$, where $\hat{B}_{P_{\sigma_e} S_e} = 1$ for all e and zero otherwise. We create a 100 different randomisations and use it to compare to our empirical observations.

The Joyo network has $n = 9{,}467$ nodes, an average degree of $\langle k_i \rangle \approx 12.22$ and with an average weight per edge of $\langle w_{ij} \rangle \approx 2.95$ the average strength is $\langle s_i \rangle = \langle k_i \rangle \langle w_{ij} \rangle \approx 36.07$. So, in Joyo, a person co-occurs on average about 3 times with 12 other people, so in total about 36 times. Based on the randomisation, we would expect roughly 22 people to occur about 1.2 times, giving a total strength of around 28. Hence, people co-occur roughly 2.5 times more often with the same people as expected, and co-occur with almost 2 times less people. The NYT corpus contains $n = 31{,}093$ nodes and has an average degree of about $\langle k_i \rangle \approx 22.35$. The average weight $\langle w_{ij} \rangle \approx 2.01$, which gives an average strength of about $\langle s_i \rangle \approx 44.91$. In summary, a person in the NYT co-occurs about 2 times with about 22 different people. Similar as for Joyo, based on the randomisation, we would expect around 45 people to occur about 1.1 times, giving a total strength of about 50. So, in NYT people co-occur almost 2 times more frequently with the same people, with roughly 2 times fewer people. We provide an overview of some of the key statistics in Table 8.1.

The degree, strength and weight are all heterogeneously distributed, as frequently observed in complex networks. They follow approximately powerlaws [7] in both networks, where we use MLE techniques for estimating the powerlaws. The degree and strength in the Joyo corpus is more broadly distributed compared to the NYT

Table 8.1 Summary overview

	Empirical	Random	Model
Joyo			
Nodes	9567	9114 ± 418.0	9481 ± 93.5
Avg. Degree	12.4	22.1 ± 8.0	$12.3 \pm 8.8 \times 10^{-2}$
Avg. Weight	2.9	1.2 ± 0.050	$3.0 \pm 8.9 \times 10^{-3}$
Avg. Strength	36.5	27.7 ± 11.0	36.4 ± 0.36
Assortativity	−0.067	$-0.13 \pm 8.3 \times 10^{-3}$	$-0.094 \pm 1.3 \times 10^{-3}$
Clustering	0.29	$0.32 \pm 2.2 \times 10^{-4}$	$0.25 \pm 8.2 \times 10^{-4}$
W. Clustering	0.33	$0.34 \pm 2.3 \times 10^{-4}$	$0.26 \pm 6.9 \times 10^{-4}$
Path Length	3.45	$2.51 \pm 4.3 \times 10^{-4}$	$3.34 \pm 1.5 \times 10^{-3}$
Diameter	10	$5.7 \pm 4.9 \times 10^{-2}$	$8 \pm 5.8 \times 10^{-2}$
Radius	6	3	4.5
Exponents			
Degree	$2.46 \pm 1.5 \times 10^{-2}$	$2.29 \pm 1.4 \times 10^{-2}$	$2.03 \pm 1.1 \times 10^{-2}$
Weight	$2.21 \pm 1.2 \times 10^{-2}$	$2.56 \pm 1.6 \times 10^{-2}$	$2.45 \pm 6.0 \times 10^{-3}$
Strength	$2.06 \pm 1.1 \times 10^{-2}$	$2.17 \pm 1.2 \times 10^{-2}$	$1.89 \pm 9.1 \times 10^{-3}$
Degree-Strength	$1.30 \pm 2.6 \times 10^{-3}$	$1.42 \pm 1.4 \times 10^{-3}$	$1.39 \pm 1.8 \times 10^{-3}$
NYT			
Nodes	31093	30860 ± 69	31015 ± 11
Avg. Degree	22.3	45.2 ± 0.19	22.1 ± 0.14
Avg. Weight	2.01	$1.11 \pm 6.6 \times 10^{-4}$	1.94 ± 0.025
Avg. Strength	44.9	50.1 ± 0.24	43.1 ± 0.27
Assortativity	0.062	$-0.090 \pm 6.4 \times 10^{-4}$	$-0.17 \pm 2.4 \times 10^{-3}$
Clustering	0.32	$0.26 \pm 4.2 \times 10^{-4}$	$0.36 \pm 7.7 \times 10^{-3}$
W. Clustering	0.35	$0.26 \pm 4.6 \times 10^{-4}$	$0.36 \pm 7.7 \times 10^{-3}$
Path Length	3.7	$2.7 \pm 5.9 \times 10^{-3}$	$3.2 \pm 4.6 \times 10^{-3}$
Diameter	10	6	9
Radius	5	3	4.5
Exponents			
Degree	$3.99 \pm 1.7 \times 10^{-2}$	$2.63 \pm 9.3 \times 10^{-3}$	$1.77 \pm 4.4 \times 10^{-3}$
Weight	$2.42 \pm 8.1 \times 10^{-3}$	$2.90 \pm 1.1 \times 10^{-2}$	$2.71 \pm 2.9 \times 10^{-3}$
Strength	$2.95 \pm 1.1 \times 10^{-2}$	$2.51 \pm 8.6 \times 10^{-3}$	$1.72 \pm 4.1 \times 10^{-3}$
Degree-Strength	$1.48 \pm 3.5 \times 10^{-3}$	$1.22 \pm 4.8 \times 10^{-4}$	$1.48 \pm 1.1 \times 10^{-3}$

We report various properties for both datasets and the randomisations and model. Refer to the main text for details about the randomisation procedures and the model

corpus, but the weight is distributed similarly. However, these distributions are also expected to be heterogeneous based on the randomisation. Although there are some deviations, they seem mainly due to the lower average empirical degree and the higher average empirical weight.

The network shows signs of a small world network. The average path length is relatively low, while the clustering is relatively high. Again, this does not deviate much from what is expected at random. The (weighted) clustering is almost the same, and the path length is even slightly longer than expected at random. The longer path length is probably due to the lower average degree. With fewer neighbours on average, there are fewer possibilities for paths, thus leading to somewhat longer paths.

We find that the strength scales superlinearly with the degree as $s_i \sim k_i^\beta$, with an exponent of $\beta \approx 1.30$ for Joyo and $\beta \approx 1.48$ for NYT. This means that high degree nodes attract more weight and low degree nodes less weight. In this context, people that co-occur with many other people, also tend to co-occur more often. A similar superlinear scaling was found in transportation and technological networks [5, 26, 34]. This contrasts with for example mobile phones where there is a sublinear growth [25], suggesting that people who call many people, do so less frequently than people who call few people. This is quite different from what is expected at random, especially when looking at this from the perspective of the average weight s_i/k_i, which increases for large degree k_i empirically, but which increases only very slightly in the randomisation. We have plotted the behaviour of the average weight and the strength in Fig. 8.1.

The average weighted neighbour degree [4] increases with larger degree. This implies that high degree nodes connect relatively stronger to other high degree nodes. For the NYT the weighted neighbour degree starts to increases more clearly for a degree larger than about 200. This suggest that relatively much weight is in between high degree nodes. See Fig. 8.1 for plots of the weighted neighbour degree. The ordinary neighbour degree decreases for Joyo, as evidenced by the negative assortativity [24], whereas this increases for NYT (Table 8.1), pointing out a difference between the two datasets.

8.3.1 Model

Three phenomena deviate from what can be expected from a random graph. First, the degree is higher than expected and the weight is lower than expected. Secondly, high degree nodes attract disproportionately much weight. Thirdly, much of the weight is between the hubs. These observations suggest that people tend to co-occur repeatedly with the same people. We therefore introduce a very simple stylistic model that is able to reproduce most of the observations in the empirical network qualitatively. The model consists of two key ingredients: (1) more frequently occurring people have a higher probability of occurring; and (2) two more frequently co-occurring people have a higher probability of co-occurring.

More specifically, we employ the following procedure. We start out with an empty graph. Each time step, we draw a random sentence s, with degree k_s (i.e. the number of people occurring in a sentence) drawn from the empirical sentence degree distribution. We then choose k_s nodes in the following way. With probability q we introduce a new person into the graph, which is chosen so that the expected number of nodes

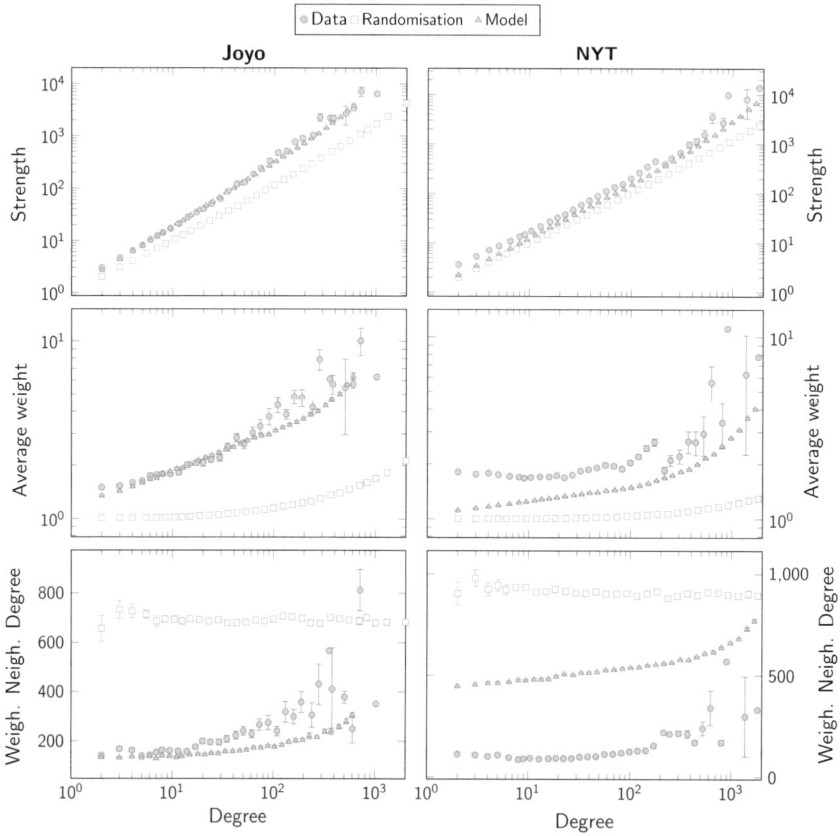

Fig. 8.1 Properties. The first column shows the properties for Joyo, the second for NYT. This clearly shows that the strength increases faster than expected (*first row*), which is even more clear when looking at the average strength (*second row*). The model however, captures quite well this increase, especially for Joyo. The weighted neighbour degree increases, whereas this remains nearly constant for the randomisation. The model again, shows a relatively similar increase, especially for Joyo

equals the number of nodes n in the empirical graph. That is, if p_k is the probability a sentence has degree k, then the total expected number of nodes occurring in sentences will be $n_s \sum_k kp_k$, with n_s the number of sentences. So, if $q = \frac{n}{n_s \sum_k kp_k}$ we generate on average about n nodes.

If we don't introduce a new node, we pick a random node i in the sentence. Then with probability $(k_i + 1)^{-\beta}$ we choose an already existing node, where k_i is the degree of node i and β a tunable parameter. The probability a node is selected is proportional to its degree, so that $\Pr(\text{choose } j) = k_j / \sum_l k_l$. If we don't pick an existing node, we choose a random neighbour of i with probability proportional to the weight, so that $\Pr(\text{choose } j|i) = w_{ij} / \sum_k w_{ik}$. After we picked all k_s nodes, we create an edge for all combinations of persons in the sentence. If an edge already exists, we increase its weight.

The node sampling is very similar to the preferential attachment model from Barabási and Albert [3] and similar models [10], as nodes that have more links are more likely to receive additional links. However, our model differs in several important ways from the model by Barabási and Albert [3]. First of all, it tends to generate a superlinear scaling of the strength with the degree. Secondly, it generates a much higher clustering coefficient. This latter effect is mainly a result of sampling neighbours, which relates to triadic closure, which has also been used in other models [21].

Besides preferential attachment, our model also has a counter tendency. Higher degree nodes are increasingly more likely to co-occur with already existing neighbours. The idea behind the scaling $(k_i + 1)^{-\beta}$ is based on the idea that higher degree nodes have a higher than linear strength. In other words, hubs are more likely to repeatedly co-occur with their neighbours, more so than on average.

A crude argument shows that indeed this model should result in superlinear scaling of the strength. Consider $k_i(s_i)$, the degree of node i as a function of the strength s_i, and suppose that all sentences have only degree 2. Every time we add a co-occurrence for i, s_i increases, although k_i does not necessarily increase. Now if i was the first node to be chosen, it will get a new neighbour with probability $(k_i + 1)^{-\beta}$. If i was the second node to be chosen, it implies it is chosen by another node. The probability that this node selected a new neighbour (which by definition then is node i, since we already know it was chosen) is then $(k_j + 1)^{-\beta}$ given that node j was chosen first. But the probability that node j was chosen first is proportional to k_j. Then the probability i gets a new neighbour if s_i increases is

$$(k_i + 1)^{-\beta} + \sum_j \frac{k_j}{\sum_l k_l}(k_j + 1)^{-\beta}. \tag{8.1}$$

Taking a mean-field approach, we approximate $k_i \approx \langle k \rangle$, and simply write $k = \langle k \rangle$ for ease of writing, we obtain that

$$\Delta k = (k + 1)^{-\beta} + \sum_j \frac{k}{\sum_l k}(k + 1)^{-\beta} = 2(k + 1)^{-\beta}. \tag{8.2}$$

Taking then the approximation that $\partial k/\partial s \approx \Delta k$, we obtain the solution that $s \sim k^{1+\beta}$ so that the strength increases superlinearly with k. This is of course a rather crude argument, but it nonetheless shows that using this approach we should indeed expect a superlinear scaling of the strength with the degree.

This contrasts with using a constant probability for choosing a new neighbour. In that case, essentially each time that the strength increases, the probability that the degree increases is a fixed probability ρ. This then results in $k \sim \rho s$, showing only linear scaling.

Additionally, this model has the tendency to create a relatively high clustering coefficient. Every time that a new sentence is introduced with k_s persons, all these

k_s persons will be connected amongst each other, creating small cliques. In addition, these cliques are also reinforced by the mechanism of adding neighbours to sentences. Interestingly, despite such reinforcement, the model still generates a dissortative structure [24]. Although this is congruent with Joyo, it contrasts with the assortative structure for NYT.

In order to estimate the parameter β, we compare the average degree and average weight to the empirically observed values. That is, for each parameter value β, we compare the average degree $\langle \hat{k}_i \rangle$ and average weight $\langle \hat{w}_{ij} \rangle$ of the model, and compare that to the average degree $\langle k_i \rangle$ and average weight $\langle w_{ij} \rangle$ of the empirical network. We use a simple squared error for the fitting,

$$E(\beta) = (\langle \hat{k}_i \rangle - \langle k_i \rangle)^2 + (\langle \hat{w}_{ij} \rangle - \langle w_{ij} \rangle)^2. \tag{8.3}$$

We used 100 replications for each parameter value. The best parameter fit differs quite a bit between Joyo and NYT. For Joyo we find an optimal parameter of $\beta \approx 0.46$, while for NYT we find $\beta \approx 0.22$. See Fig. 8.2 for the fitting of the parameter.

As expected, we can fit the empirically observed average degree and weight very well. Additionally, we indeed observe a very similar increase in the average weight for high degree nodes, as was already argued above. Finally, the model also shows an increase in the average weighted neighbour degree, similar as empirically. Nonetheless, although the average degree is quite well fit for NYT, the distribution of the degree is more broader, leading to a higher average weighted neighour degree. Nonetheless, both the model and the empirical network show an increase. The fit of the model for Joyo is especially striking in Fig. 8.1. Perhaps this is because Joyo has a more particular focus on politics, while the NYT includes also other subjects such as culture, arts and sports.

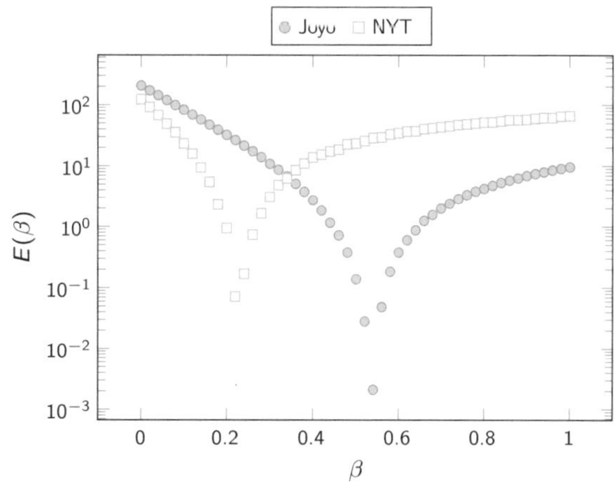

Fig. 8.2 Model fit. The fit of the parameter β in the model. The error $E(\beta)$ it taken with respect to the average degree and the average weight

8.4 Conclusion

In this paper we analysed two networks based on the co-occurrence of people in newspapers. We have analysed various properties of this network, and whereas many properties are in line with what could be expected from such a co-occurrence network, a few deviations stand out. First, people occur with fewer people than expected and more often with those people than expected. Secondly, high degree nodes attract disproportionately much weight, so that the hubs co-occur much more often than their degree justifies. Third, much of the weight concentrates between these hubs.

This suggests that people repeatedly co-occur with the same people. We constructed a model that tries to reproduce these observations. It is based on two simple processes: (1) people that occur in the media are more likely to occur again; (2) two people that co-occur are more likely to co-occur again. Moreover, people with a higher degree co-occur more often with people with whom they already co-occur. This seems to explain the observations quite well, although some deviations remain.

There are some clear differences between the Joyo and the NYT corpus. Whether this is reflective of differences between Indonesia and the US, or the more politically oriented corpus of Joyo, or a difference in time periods, is difficult to ascertain. Further analysis and comparison of these networks should provide more insight.

Acknowledgments VT would like to thank Fabien Tarissan for interesting comments and remarks on an earlier version of this manuscript. This research is funded by the Royal Netherlands Academy of Arts and Sciences (KNAW) through its eHumanities project (http://www.ehumanities.nl/computational-humanities/elite-network-shifts/).

References

1. Amaral, L.A.N., Scala, A., Barthélémy, M., Stanley, H.E.: Classes of small-world networks. Proc. Natl. Acad. Sci. USA **97**(21), 11149–11152 (2000)
2. Barabási, A.L.: Scale-free networks: a decade and beyond. Science **325**(5939), 412–413 (2009)
3. Barabási, A.L., Albert, R.: Emergence of scaling in random networks. Science **286**(5439), 509–512 (1999)
4. Barrat, A., Barthélemy, M., Pastor-Satorras, R., Vespignani, A.: The architecture of complex weighted networks. Proc. Natl. Acad. Sci. USA **101**(11), 3747–3752 (2004)
5. Barthélemy, M., Barrat, A., Pastor-Satorras, R., Vespignani, A.: Characterization and modeling of weighted networks. Physica A **346**(1–2), 34–43 (2005)
6. Bullmore, E., Sporns, O.: Complex brain networks: graph theoretical analysis of structural and functional systems. Nat. Rev. Neurosci. **10**(3), 186–98 (2009)
7. Clauset, A., Shalizi, C.R., Newman, M.E.J.: Power-law distributions in empirical data. SIAM Rev. Soc. Ind. Appl. Math. **51**(4), 661–703 (2009)
8. Corten, R.: Composition and structure of a large online social network in the netherlands. PLoS ONE **7**(4), e34760 (2012)
9. Cranmer, S.J., Menninga, E.J., Mucha, P.J.: Kantian fractionalization predicts the conflict propensity of the international system. arXiv:1402.0126 [physics] (2014)
10. Dorogovtsev, S.N., Mendes, J.F.F., Samukhin, A.N.: Structure of growing networks with preferential linking. Phys. Rev. Lett. **85**(21), 4633–4636 (2000)

11. Ferrara, E.: A large-scale community structure analysis in facebook. EPJ Data Sci. **1**(1), 9 (2012)
12. Finkel, J.R., Grenager, T., Manning, C.: Incorporating non-local information into information extraction systems by gibbs sampling. In: Proceedings of the 43rd Annual Meeting on Association for Computational Linguistics. p. 363–370. Association for Computational Linguistics, Stroudsburg, PA, USA (2005)
13. Garlaschelli, D., Caldarelli, G., Pietronero, L.: Universal scaling relations in food webs. Nature **423**(6936), 165–8 (2003)
14. Garlaschelli, D., Loffredo, M.I.: Structure and evolution of the world trade network. Physica A **355**(1), 138–144 (2005)
15. González, M.C., Hidalgo, C.A., Barabási, A.L.: Understanding individual human mobility patterns. Nature **453**(7196), 779–82 (2008)
16. Guimerà, R., Stouffer, D.B., Sales-Pardo, M., Leicht, E.A., Newman, M.E.J., Amaral, L.A.N.: Origin of compartmentalization in food webs. Ecology **91**(10), 2941–2951 (2010)
17. Guimerà, R., Sales-Pardo, M., Amaral, L.A.N.: Classes of complex networks defined by role-to-role connectivity profiles. Nat. Phys. **3**(1), 63–69 (2007)
18. Hagmann, P., Cammoun, L., Gigandet, X., Meuli, R., Honey, C.J., Wedeen, V.J., Sporns, O.: Mapping the structural core of human cerebral cortex. PLoS Biol. **6**(7), e159 (2008)
19. Joshi, D., Gatica-Perez, D.: Discovering groups of people in google news. In: Proceedings of the 1st ACM International Workshop on Human-centered Multimedia, pp. 55–64. HCM '06, ACM, New York, NY, USA (2006)
20. Knoke, D., Yang, S.: Social Network Analysis. In: Quantitative Applications in the Social Sciences, vol. 154, 2nd edn. SAGE Publications, Inc, Cambridge, Mass (2007)
21. Kumpula, J.M., Onnela, J.P., Saramäki, J., Kaski, K., Kertész, J.: Emergence of communities in weighted networks. Phys. Rev. Lett. **99**(22), 228701 (2007)
22. Maoz, Z., Terris, L.G., Kuperman, R.D., Talmud, I.: What is the enemy of my enemy? causes and consequences of imbalanced international relations, 1816–2001. J. Politic. **69**(01), 100–115 (2008)
23. Milne, D., Witten, I.H.: Learning to link with wikipedia. In: Proceedings of the 17th ACM Conference on Information and Knowledge Management, pp. 509–518. CIKM '08, ACM, New York, NY, USA (2008)
24. Newman, M.E.J.: Assortative mixing in networks. Phys. Rev. Lett. **89**(20), 208701 (2002)
25. Onnela, J.P., Saramäki, J., Hyvönen, J., Szabó, G., de Menezes, M.A., Kaski, K., Barabási, A.L., Kertész, J.: Analysis of a large-scale weighted network of one-to-one human communication. New. J. Phys. **9**(6), 179–179 (2007)
26. Ou, Q., Jin, Y.D., Zhou, T., Wang, B.H., Yin, B.Q.: Power-law strength-degree correlation from resource-allocation dynamics on weighted networks. Phys. Rev. E **75**(2), 021102 (2007)
27. Özgür, A., Bingol, H.: Social network of co-occurrence in news articles. In: Aykanat, C., Dayar, T., Korpeoglu, I. (eds.) Computer and Information Sciences—ISCIS 2004, pp. 688–695. No. 3280 in Lecture Notes in Computer Science. Springer Verlag, Heidelberg (2004)
28. Petri, G., Scolamiero, M., Donato, I., Vaccarino, F.: Topological strata of weighted complex networks. PLoS ONE **8**(6), e66506 (2013)
29. Pouliquen, B., Tanev, H., Atkinson, M.: Extracting and learning social networks out of multilingual news. In: Social Networks and application tools (2008)
30. Simini, F., González, M.C., Maritan, A., Barabási, A.L.: A universal model for mobility and migration patterns. Nature **484**(7392), 96–100 (2012)
31. Steinberger, R., Pouliquen, B.: Cross-lingual named entity recognition. Ling. Inv. **30**(1), 135–162 (2007)
32. Traag, V.A., Van Dooren, P., Nesterov, Y.: Narrow scope for resolution-limit-free community detection. Phys. Rev. E **84**(1), 016114 (2011)
33. Traud, A.L., Mucha, P.J., Porter, M.A.: Social structure of facebook networks. Physica A **391**(16), 4165–4180 (2012)
34. Wang, W.X., Wang, B.H., Hu, B., Yan, G., Ou, Q.: General dynamics of topology and traffic on weighted technological networks. Phys. Rev. Lett. **94**(18), 188702 (2005)

Chapter 9
Spatial Effects of Delay-Induced Stochastic Oscillations in a Multi-scale Cellular System

Dmitry Bratsun and Andrey Zakharov

Abstract The combined spatial effect of time delay and intrinsic noise on gene regulation is studied numerically. It is based on the multi-scale chemo-mechanical model of the epithelium. The protein fluctuations in each cell are described by a single-gene auto-repressor model with constant delay. It is found that time delay, noise and spatial signaling can result in the protein pattern formation even when deterministic description exhibits no patterns.

9.1 Introduction

Variance is increased, if the number of elements in the set is reduced. A large variance indicates that numbers in the set are far from the mean and each other, while a small variance indicates the opposite. This is why the small number of reactant molecules involved in gene regulation can lead to significant fluctuations in mRNA and protein concentrations, and there have been numerous studies devoted to the consequences of such noise at the regulatory level [1–4].

In fact, the transcriptional and translational processes are compound multistage reactions involving the sequential assembly of long molecules. It can provoke a time lag in gene regulation processes. The combined effect of time delay and intrinsic noise on the temporal dynamics have been explored in [5, 6]. It was found that quasi-regular oscillations can arise in such a stochastic system even when its deterministic counterpart exhibits no oscillations. Several years ago, it was declared that "*space is the final frontier in stochastic simulations of biological systems*" [7, 8]. If the spatial stochastic simulations of Markovian processes then has made considerable progress [4, 8–10], examples of stochastic simulation of non-Markovian processes in space are almost absent in the literature.

Generally, many biological processes, ranging from gene expression, cell proliferation to higher-order processes such as vision, memory, and learning, necessitate that a cell be aware of its environment. These processes involve transmission of signals

D. Bratsun (✉) · A. Zakharov
Theoretical Physics Department, Perm State Pedagogical University, Perm, Russia
e-mail: dmitribratsun@rambler.ru

S. Battiston et al. (eds.), *Proceedings of ECCS 2014*, Springer Proceedings in Complexity, DOI 10.1007/978-3-319-29228-1_9

across the plasma membrane. It is well-established that the exchange of such signals is a necessary condition for a self-organization at the cellular level. Early epithelial models (for example, the model of epidermal wound healing [11]) were based on reaction-diffusion equations describing cell motion and proliferation in response to a diffusive chemical signal. This approach turned out to be inadequate for the description of spreading epithelia, as it could not explain proliferation in the absence of external injury of tissue and many other features observed in later experiments [12]. The same situation was observed in multiscale modelling of tumour growth [13]. It became clear that a more realistic approach should involve the modeling of the forces transmitted by adjoining cells. One of such new kind models was proposed in [14]. Authors have constructed a minimal phenomenological model taking into account the effect of both chemical and mechanical factors on collective cell motion and compatible with available observations. A very recent two-dimensional chemo-mechanical model [15] was suggested to describe a growth of cancer tumour induced by circadian rhythm disruption in epithelial tissue. In all these works however it was supposed that cells exchange signals within the deterministic description.

In this paper, we explore the spatial effects of time delay and intrinsic noise on gene regulation in a milticellular system. This approach helps to avoid the problem of the lack of a reliable algorithm for spatial stochastic processes of the non-Markovian nature. The cellular system is constructed within the multiscale chemo-mechanical model of the epithelium tissue suggested primarily in [14] and then applied to reproduce the carcinoma growth in [15]. At a single cell level, we use a single-gene auto-repressor model with constant delay which is relatively simple yet but still maintains a high degree of biological relevance. The numerical simulation of these stochastic fluctuations in each cell is performed using the modified Gillespie algorithm proposed in [5].

9.2 Mechanics of Epithelial Tissue

Epithelial tissue is a layer of cells covering the surface of an organ or body. The cells always remain attached to each other forming a continuous two-dimensional epithelial surface (Fig. 9.1). The curvature of the layer (presumed small compared to the cell size) and thickness inhomogeneities are neglected. The model includes the calculation of separate cells dynamics, which are presented in the form of polygons.

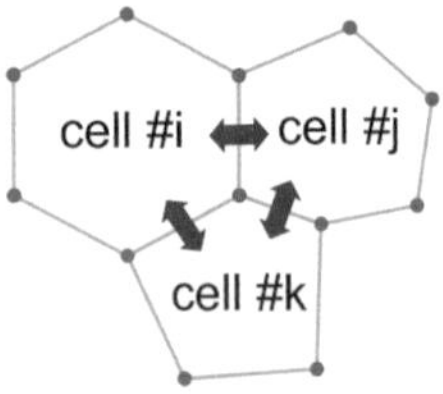

Fig. 9.1 Elements of the chemo-mechanical model of an epithelial tissue

The initial configuration is a regular hexagonal lattice. But in the course of evolution, the structure is distorted, and the polygons with different number of vertices appear.

The mechanical model is based on the elastic potential energy U of the tissue, defined by summing up the contributions of the perimeter L and the area A of each cell [14–16]:

$$U = \frac{1}{2}\sum_{cells}\left(\kappa L^2 + \eta(A - A_0)^2\right), \tag{9.1}$$

where κ is attributed to the action of active contractile forces, η is the elastic constant and A_0 is the reference cell area.

The tissue evolves by moving the cell nodes (indicated by blue points in Fig. 9.1). The mechanical force acting on any jth node is defined as

$$\mathbf{F}_j = -\frac{\partial U}{\partial \mathbf{R}_j}, \tag{9.2}$$

where $\mathbf{R}_j$ stands for radius vector of jth node.

Since the motion is strongly overdamped [14], the appropriate equation governing the displacement velocities $\mathbf{V}_i$ should have the form similar to the Darcy law with the mobility coefficient K:

$$\mathbf{V}_i = \frac{d\mathbf{R}_i}{dt} = K\mathbf{F}_i H\left(|\mathbf{F}_i| - F_0\right), \tag{9.3}$$

where H is the Heaviside function and F_0 is the threshold below which the node remains immobile. Altogether, (9.1–9.3) define the mechanics of the tissue.

9.3 Single-Gene Auto-Repressor Model with Dimerization

Let us consider a single gene protein synthesis with negative auto-regulation. This is a popular motif in genetic regulatory circuits, and its temporal dynamics has been analyzed both deterministic and stochastic framework [1]. The generalized version of this system taking into account that transcription of auto-repressor protein takes a finite amount of time τ has been studied in [5, 6].

Suppose that protein can exist both in the form of isolated monomers X and dimers X_D. Both forms actively interact with each other via the reactions dimerization and undimerization:

$$X + X \xrightarrow{k_2} X_D, \quad X_D \xrightarrow{k_{-2}} X + X, \tag{9.4}$$

where k_2, k_{-2} stand for reaction rates.

We denote the unoccupied and occupied state of the promoter site of the gene as D_0 and D_1 respectively. Let us postulate that the chemical state of the operator sites $D \in \{D_0, D_1\}$ determines the production of corresponding protein at time $t + \tau$. If the operator at time t is unoccupied (D_0) then the protein may be produced at time $t + \tau$. Otherwise, if the operator is occupied (D_1), the production at time $t + \tau$ is blocked. The transitions between operator states for each protein occur with rates $k_{\pm 1}$, when some dimer binds to the promoter or unbinds from it respectively:

$$D_0 + X_\mathrm{D} \xrightarrow{\mathrm{k}_1} D_1, \quad D_1 \xrightarrow{\mathrm{k}_{-1}} D_0 + X_\mathrm{D}. \tag{9.5}$$

An important role in this model comes from the delay in the synthesis reaction:

$$D_0^\mathrm{t} \xrightarrow{\mathrm{A}} D_0^{\mathrm{t}+\tau} + X^{\mathrm{t}+\tau}, \tag{9.6}$$

where A is the rate of a time-delayed production of protein monomer.

Finally, the system should be supplemented by the effect of protein degradation with rate B:

$$X \xrightarrow{\mathrm{B}} \emptyset. \tag{9.7}$$

Thus, (9.4–9.7) define the kinetics of a gene regulation in each cell.

9.4 Deterministic Description for a Single Cell

Let us assume that $D_0(t)$ and $D_1(t)$ are continuous variables of time standing for the average number of unoccupied and occupied operator sites at time t with obvious relation between them:

$$D_0 + D_1 = 1. \tag{9.8}$$

The main approximation we make here is an assumption that the reactions of dimerization (9.4) and binding/unbinding (9.5) are fast in comparison with production/degradation of protein (9.6, 9.7), i.e. $k_i \gg A, B$ [5, 6]. Thus, we can suppose that dynamics of operator-site and dimers quickly enters into a local equilibrium, where concentrations of reagents become

$$X_D = \varepsilon X^2, \quad D_1 = \varepsilon \delta D_0 X^2, \tag{9.9}$$

where $\varepsilon \equiv k_1/k_{-1}$, $\delta \equiv k_2/k_{-2}$.

By taking into account (9.8) and (9.9), delay differential equations derived from (9.4–9.7) can be reduced to a single equation for the slow variable:

$$(1+4\varepsilon X(t))\,\frac{dX(t)}{dt} = \frac{A}{1+\varepsilon\delta X^2(t-\tau)} - BX(t). \tag{9.10}$$

The equation (9.10) has a unique positive stationary solution:

$$X^* = -\frac{B}{2A\varepsilon\delta} + \sqrt{\frac{B^2}{4A^2\varepsilon^2\delta^2} + \frac{1}{\varepsilon\delta}}. \tag{9.11}$$

By linearizing (9.10) near (9.11), and looking for a solution of the form $X \sim e^{\lambda t}$, where $\lambda = \chi + i\omega$, we obtain the explicit formulas for the eigenvalue:

$$\chi = \frac{1}{\tau} Re(W(-2\tau A\varepsilon\delta X^* e^{\tau B})) - B, \tag{9.12}$$

$$\omega = \frac{1}{\tau} Im(W(-2\tau A\varepsilon\delta X^* e^{\tau B})), \tag{9.13}$$

where $W(z)$ stands for the Lambert function defined as $W(z)e^{W(z)} = z$.

The condition for Hopf bifurcation then can be found by solving (9.12), (9.13) for $\chi = 0$ and $\omega \neq 0$. The neutral curve derived in this way is plotted in Fig. 9.2 in the plane of dimensionless parameters of the production τA and degradation τB. The instability domain in the figure is above the curve. Numerical solution of (9.10) reveals quite a common picture for dynamics near the Hopf bifurcation: all trajectories are attracted to the steady state and limit cycle below and above the bifurcation point respectively. The black squares in Fig. 9.2 indicate the parameter values for which the results of nonlinear numerical simulations will be presented below.

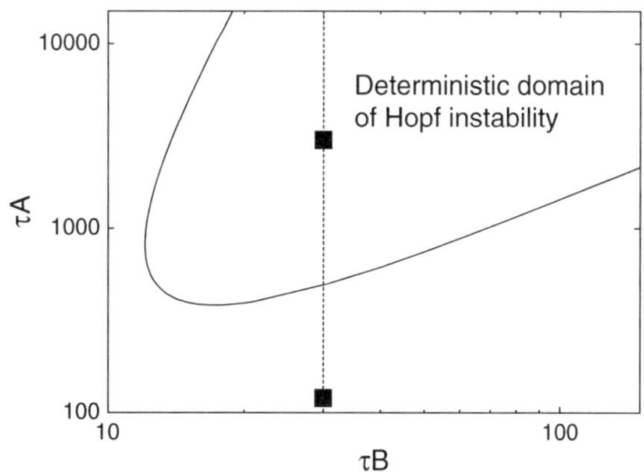

Fig. 9.2 Deterministic Hopf bifurcation curve for $\varepsilon = 0.1$, $\delta = 0.2$

9.5 Deterministic Description of Intercellular Signaling

In order to describe the effect of intercellular signaling, we introduce the signaling species T positively regulated by the of dimers X_D. Let the chemical state of the operator site $D^T \in \{D_0^T, D_1^T\}$ determines the production of T. If the operator is occupied (D_1^T) then the transport protein may be produced immediately with a certain probability A_T in a unit time. Otherwise if the operator is unoccupied (D_0^T), the production of signaling protein T is blocked. The transitions between operator states with rates k_3, k_{-3} are

$$D_0^{\mathrm{T}} + X_{\mathrm{D}} \xrightarrow{\mathrm{k_3}} D_1^{\mathrm{T}}, \quad D_1^{\mathrm{T}} \xrightarrow{\mathrm{k_{-3}}} D_0 + X_{\mathrm{D}}. \tag{9.14}$$

The production reaction with rate A_T then can be written as

$$\emptyset \xrightarrow{A_T D_T(t)} T. \tag{9.15}$$

We assume also that once a signal has come in a certain cell, it is converted into X protein:

$$T \xrightarrow{B_T} X. \tag{9.16}$$

The chemical interactions here include the activation linear in the signal concentration with the constant B_T and the linear decay with the constant B (9.7). The activation is assumed to be linear because this is the simplest activation form commonly accepted within phenomenological models (see, for example, [14]).

Thus, the reactions (9.15–9.16) have positive feedback with (9.14) through the reaction rate in (9.15). We assume also that

$$D_0^T + D_1^T = N_T, \tag{9.17}$$

where the integer constant N_T stands for the copy number of the signaling species. The protein copy number indicates how many copies of the protein monomer is synthesized in a single act of transcription/translation.

Taking into account that binding/unbinding reactions (9.14) are fast in comparison with production/degradation (9.15, 9.16), we arrive to

$$\frac{dX_j}{dt} = \frac{1}{(1+4\varepsilon X_j)}\left(B_T T_j + \frac{A}{1+\varepsilon\delta X_j^2(t-\tau)} - BX_j(t)\right), \tag{9.18}$$

$$\frac{dT_j}{dt} = \frac{N_T A_T \sigma X_D^j}{1+\sigma X_D^j} - B_T T_j + \sum_{i\in adj(j)} \alpha L_{ij}(T_i - T_j), \tag{9.19}$$

where the subscripts refer to ith and jth cells, α is the transfer coefficient, $\sigma = k_3/k_{-3}$ and $adj(j)$ stands for "adjacent to j-cell".

It is assumed that the signaling species T is transported diffusively from one cell to the other, whereas its flux does not depend on the distance between the two cells i and j but is proportional to the boundary length L_{ij} (see Fig. 9.1). This implies that the transport is limited by the transfer though cell membranes.

The initial configuration of the system is a regular hexagonal lattice comprising 1560 cells. The shape and location of each cell is defined by its nodes. The tissue as a whole has the form of a stripe with two free borders with periodic boundary conditions applied there. The typical values of the parameters governing the mechanics of the tissue are as follows: $\kappa = 1.0$, $\eta = 1.0$, $A_0 = 3\sqrt{3}/2$, $K = 1.0$, $F_0 = 0.02$ [15]. In all calculations presented below, the cell division was turned off since it generates the extrinsic noise.

The set of delay differential equations (9.18–9.19) have been solved using the explicit Euler method, whose stability was warranted by a sufficiently small time step $\Delta t_{chem} = 0.005$. The time step for the calculation of the molecular processes in cells was synchronized with the step $\Delta t_{mech} = 0.01$ of calculating the mechanical evolution of the tissue governed by (9.1–9.3).

Figure 9.3 presents the results of numerical simulation of (9.18–9.19) for parameter values from the balloon of instability in Fig. 9.2: $A = 500$, $B = 5$, $\tau = 6$. The evolution of the system starts from random phase distribution. Then the nonlinear dynamics demonstrates a spiral traveling wave pattern which arises against the background a synchronized oscillation field. The oscillation period is approximately equal to the double delay time, i.e. $\omega^* = 0.524$, which corresponds to the result of the linear stability analysis (9.13).

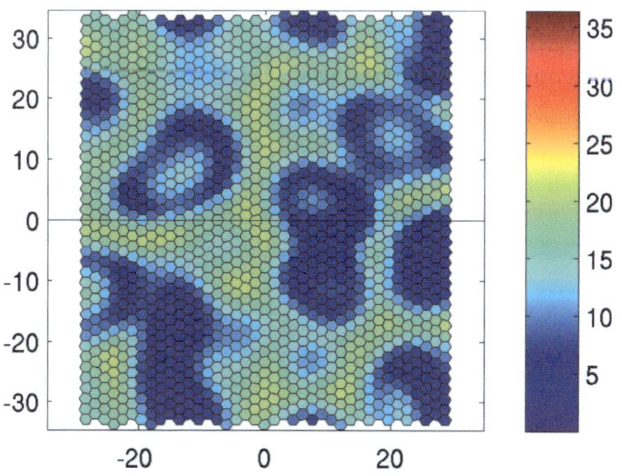

Fig. 9.3 Typical pattern formed by the protein X in the epithelial tissue consisting of more than 1500 cells based on the deterministic description at $t = 310$. $A = 500$, $B = 5$, $\tau = 6$, $\varepsilon = 0.1$, $\delta = 0.2$, $N_T = 1$, $\sigma = 0.1$, $A_T = 500$, $\alpha = 0.05$

9.6 Stochastic Description

In order to describe the spatial stochastic effects, we use a hybrid model, which is constructed as follows. The dynamics of the protein X in a single cell is obtained by performing direct Gillespie simulations of single-gene auto-repressor model given by the reactions (9.4–9.7). The modified version of the Gillespie algorithm which accounts for the non-Markovian properties of random biochemical events with delay was developed in [5].

The signaling between cells is still organized as diffusive transport from one cell to the other. For simplicity, we assume that monomers of the basic protein X seep through cell membranes according to finite-difference formula:

$$X_j^{t+\Delta t} = X_j^t + \left\lceil \Delta t_{diff} N_T \sum_{i \in adj(j)} \alpha L_{ij}(X_i^t - X_j^t) \right\rceil, \tag{9.20}$$

where $\lceil ... \rceil$ stands for the ceiling function which maps the smallest integer not less than the function argument.

The approximation (9.20) is justified if the protein X regulates the relatively rapid synthesis of a transport protein with the copy number N_T. The latter parameter evidently plays an important role in increasing the possible fluctuations in the system.

Generally speaking, systems containing a time delay can be highly sensitive to fluctuations, and the effect of noise should be explored in detail. It has been shown previously that time delay in gene expression coupling with noise can cause a system to be oscillatory even when its deterministic counterpart exhibits no oscillations [5, 6].

Let us consider two examples of spatial stochastic simulations according to the aforementioned algorithm. Figure 9.4 presents the typical pattern formed by the monomers of the X protein at the parameter values chosen so that the same system

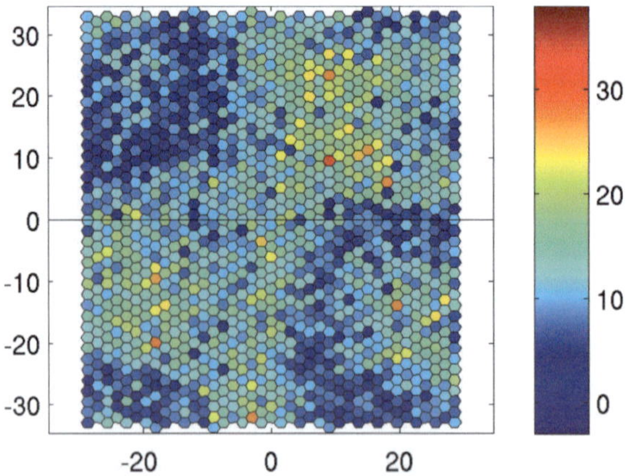

Fig. 9.4 Typical pattern formed by the protein X in the epithelial tissue based on the stochastic description at $t = 310$. $A = 500$, $B = 5$, $\tau = 6$, $k_1 = 100$, $k_{-1} = 1000$, $k_2 = 200$, $k_{-2} = 1000$, $N_T = 5$, $\alpha = 0.05$

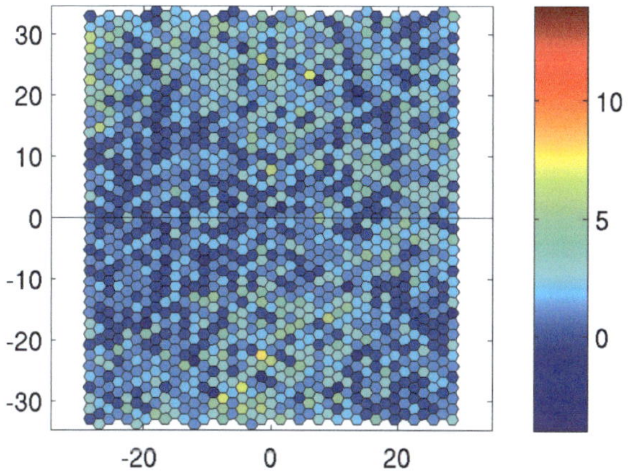

Fig. 9.5 Typical pattern formed by the protein X in the epithelial tissue consisting of more than 1500 cells based on the stochastic description $t = 800$. $A = 20$, $B = 5$, $\tau = 6$, $k_1 = 100$, $k_{-1} = 1000$, $k_2 = 200$, $k_{-2} = 1000$, $N_T = 8$, $\alpha = 0.05$

demonstrates the oscillatory behavior under deterministic description (it is indicated by the upper black square in Figs. 9.2 and 9.3). We found that nonlinear dynamics of spatially extended system consists of two distinct oscillatory modes. One is a quasi-standing wave pattern oscillating with ω^*. The second oscillatory mode is a traveling wave which arise from some selected cells (Fig. 9.4). In fact, stochastic pattern looks very similar to its deterministic counterpart obtained for the same parameter values (compare with Fig. 9.3).

Consider now the case when the deterministic description of the system predicts the stationary behavior (it is indicated by the lower black square in Fig. 9.2). Starting with random initial conditions, the system fairly quickly falls into a fully synchronized mode oscillations with a common frequency $\omega \approx \omega^*$.

We found also that depending on the copy number N_T one can observe the effect of clustering when the cells form two approximately equal communities, which collectively oscillate in anti-phase (Fig. 9.5). For example, the numerical simulation of the system for copy number $N_T = 4$ has showed that the clustering is not observed within reasonably long integration times. In contrast to that, at $N_T = 8$ this effect gradually manifests itself after a sufficiently long integration (about 150–200 periods of basic oscillations).

In fact, the clustering in the system with a large amount of elements exchanging chemical signals has become at the center of attention of many scientists recently. For example, the group of synthetic genetic oscillators has been studied in [17]. Tissue clustering has been found to be divided into two types of oscillating cells in time. It is believed that the clustering is likely to be the most important characteristic of most communities and could be the reason of further cells differentiation in organs.

9.7 Conclusions

It is known that the noise during gene expression comes about in two ways. The inherent stochasticity of biochemical processes such as transcription and translation generates *intrinsic* noise. On the other hand, *extrinsic* noise refers to variation in identically-regulated quantities between different cells. Since in this paper all cells are considered to be identical, we have focused on intrinsic noise.

An important result of the present work is the demonstration of how the excitation of quasi-regular subcritical fluctuations found in [5], manifests itself in space. We show that there may be observed both a spatial synchronization of oscillations and clustering of cell community. In the supercritical range of parameters it is observed the formation of the stochastic pattern which is quite similar to the pattern obtained within a deterministic description of the system.

Acknowledgments The research has been supported by the Ministry of Education and Science of Perm Region (grant C-26/004.4) and grant of Russian Fund for Basic Research (14-01-96022r_ural_a).

References

1. Kepler, T.B., Elston, C.: Stochasticity in transcriptional regulation: origins, consequences, and mathematical representations. Biophys. J. **81**, 3116–3136 (2001)
2. Hasty, J., Collins, J.J.: Translating the noise. Nat. Gen. **31**, 13–14 (2002)
3. Rosenfeld, N., Young, J.W., Alon, U., Swain, P.S., Elowitz, M.B.: Gene regulation at the single-cell level. Science **307**, 1962–1965 (2005)
4. Tsimring, L.S.: Noise in biology. Rep. Prog. Phys. **77**, 026601 (2014)
5. Bratsun, D., Volfson, D., Hasty, J., Tsimring, L.S.: Delay-induced stochastic oscillations in gene regulation. Proc. Natl. Acad. Sci. U.S.A. **102**, 14593–14598 (2005)
6. Bratsun, D., Volfson, D., Hasty, J., Tsimring, L.: Non-Marcovian processes in Gene Regulation. In: Kish, L.B., Lindenberg, K., Gingl, Z. (eds.) Noise in Complex Systems and Stochastic Dynamics III. Proceedings of the SPIE, vol. 5845, pp. 210–219 (2005)
7. Gillespie, D.T.: Exact stochastic simulation of coupled chemical reactions. J. Phys. Chem. **81**, 2340–2361 (1977)
8. Lemerle, C., Di Ventura, B., Serrano, L.: Space as the final frontier in stochastic simulations of biological systems. FEBS Lett. **579**, 1789–1794 (2005)
9. Li, C.-W., Chen, B.-S.: Stochastic spatio-temporal dynamic model for gene/protein interaction network in early drosophila development. Gene Regul. Syst. Biol. **3**, 191–210 (2009)
10. Burrage, K., Burrage, P.M., Leier, A., Marquez-Lago, T., Nicolau, D.V.: Stochastic simulation for spatial modelling of dynamic processes in a living cell. In: Koeppl, H., Densmore, D., Setti, G., Di Bernardo, M. (eds.) Design and Analysis of Biomolecular Circuits: Engineering Approaches to Systems and Synthetic Biology, pp. 43–62. Springer, Heidelberg (2011)
11. Tranquillo, R.T., Murray, J.D.: Continuum model of fibroblast driven wound contraction: inflammation—mediation. J. Theor. Biol. **158**, 135–172 (1992)
12. Poujade, M., Grasland-Mongrain, E., Hertzog, A., Jouanneau, J., Chavrier, P., Ladoux, B., Buguin, A., Silberzan, P.: Collective migration of an epithelial monolayer in response to a model wound. Proc. Natl. Acad. Sci. U.S.A. **104**, 15988–15993 (2007)
13. Deisboeck, T.S., Stamatakos, G.S.: Multiscale Cancer Modeling. Chapman and Hall/CRC, Boca Raton (2011)

14. Salm, M., Pismen, L.M.: Chemical and mechanical signaling in epithelial spreading. Phys. Biol. **9**, 026009–026023 (2012)
15. Bratsun, D.A., Merkuriev, D.V., Zakharov, A.P., Pismen, L.M.: Multiscale modeling of tumour growth induced by circadian rhythm disruption in epithelial tissue. J. Biol. Phys. 42, 107–132 (2016)
16. Farhadifar, R., Röper, J.C., Aigouy, B., Eaton, S., Jülicher, F.: The influence of cell mechanics, cell-cell interactions, and proliferation on epithelial packing. Curr. Biol. **17**, 2095–2104 (2007)
17. Koseska, A., Ullner, E., Volkov, E., Kurths, J., Garcia-Ojalvo, J.: Cooperative differentiation through clustering in multicellular populations. J. Theor. Biol. **263**, 189–202 (2010)

Chapter 10
An Agent-Based Modelling Approach to Biological Invasion by Macroalgae in European Coastal Environments

James T. Murphy, Mark P. Johnson and Frédérique Viard

Abstract Introductions of species to new continents and oceans by human activities cause fundamental and irreversible changes to natural communities and ecosystems worldwide, resulting in systematic homogenization of biota at regional and global scales and substantial changes in ecosystem functioning. Seaweeds are major primary producers in coastal areas, and large-scale substitution of dominant native seaweeds with non-native species can consequently alter coastal productivity and food web structure, and therefore impact ecosystem services. In this study, an agent-based modelling approach is taken, in association with data already gathered by the host institution from field studies, ecological experiments and molecular work, to study the impact of the Asian kelp seaweed *Undaria pinnatifida*, introduced to Europe in the 1970s, on native biodiversity under variable climatic conditions. Our model framework can be used to explicitly represent complex spatial and temporal patterns of invasion in order to be able to predict quantitatively the impact of these factors on the invasion dynamics of *U. pinnatifida*. This would be a useful tool for making accurate risk assessments of invasion potential under different environmental conditions and for choosing optimal management strategies in order to minimise future control costs.

J.T. Murphy (✉) · F. Viard
Sorbonne Universités, UPMC Univ Paris 6, UMR 7144,
Station Biologique de Roscoff, Place Georges Teissier, 29680 Roscoff, France
e-mail: james.murphy@nuigalway.ie

J.T. Murphy · F. Viard
CNRS, UMR 7144, Equipe Div&Co, Station Biologique de Roscoff,
Place Georges Teissier, 29680 Roscoff, France

J.T. Murphy · M.P. Johnson
Ryan Institute, National University of Ireland Galway, Galway, Ireland

S. Battiston et al. (eds.), *Proceedings of ECCS 2014*, Springer Proceedings in Complexity, DOI 10.1007/978-3-319-29228-1_10

10.1 Introduction

Biological invasion refers to the introduction by human activities of non-native species of plants or animals which adversely affect local ecosystems and transform their structure and species composition. It has been identified in the Millennium Ecosystem Assessment as one of the principal environmental problems influencing future economic and social development in the world [1]. Invasive seaweeds represent one of the largest groups of introduced marine species in Europe, and constitute between 20 and 29 % of all non-native marine species [2]. Seaweeds are major primary producers in coastal areas, and large-scale substitution of dominant native seaweeds with introduced species can consequently alter coastal productivity and food web structure, and impact on biodiversity [3].

In this study, an agent-based modelling approach is taken, in association with data already gathered by the host institution from field studies, ecological experiments and molecular work, to study the impact of the Asian kelp seaweed *Undaria pinnatifida* (an emblematic invader in European waters) on native biodiversity under variable climatic conditions [4]. One of the advantages of the agent-based modelling approach is that it allows us to trace back the system behaviour to that of its individual components [5]. For example, to understand the underlying factors that lead to the observed population growth rate. This approach was taken in order to understand how the basic physiological interactions between the *U. pinnatifida* individuals and their environment contribute to their ability to adapt to and invade new habitats.

U. pinnatifida has a characteristic life cycle consisting of two generations: a microscopic haploid gametophyte stage and a macroscopic diploid sporophyte stage (Fig. 10.1). Each life cycle stage has specific environmental requirements for growth and reproduction, with particular sensitivity to changes in water temperature, light intensity and photo-period (light:dark ratio) noted in the literature [6]. However, there is only limited understanding about how the combined effects of these environmental parameters affect its potential expansion into non-native habitats such as northern European coastal waters.

In order to integrate the basic knowledge on the physiological responses of *U. pinnatifida* to changing environmental conditions (collected from many years of observations of both natural and cultivated populations in the literature) a computational modelling approach was taken. The theory of autonomous agents is a useful approach for the modelling of algal populations as it allows large-scale population models to be derived from simple rules dictating the growth and interactions of the individual life stages (gametophytes and sporophytes) of the population. We integrated quantitative data from the literature on the responses of these individual stages to environmental factors such as light and temperate in order to build up a model of the overall population growth. This can then be used to explore the effects of changing environmental conditions on growth dynamics and make predictions about potential expansion into new habitat ranges.

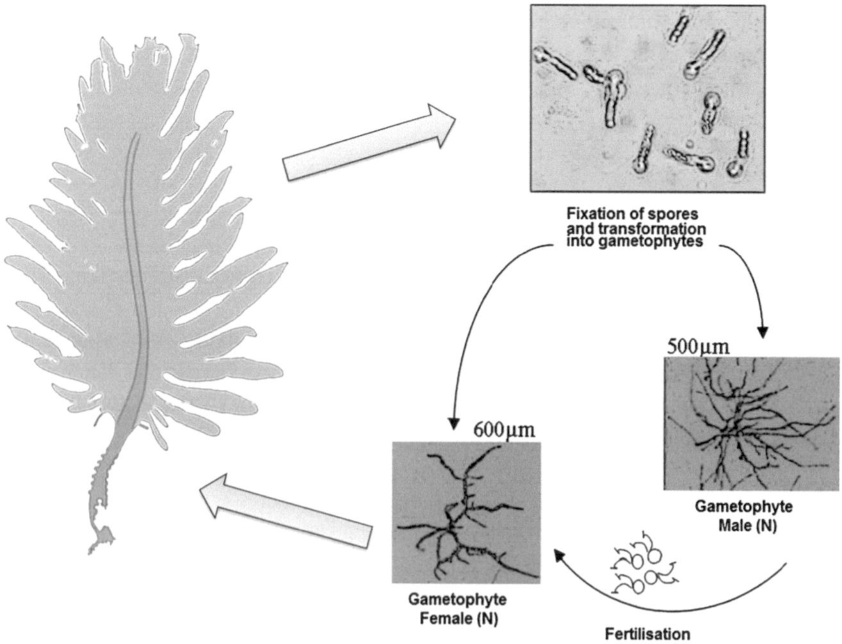

Fig. 10.1 Diagrammatic representation of the *U. pinnatifida* annual life cycle consisting of independent macroscopic sporophyte and microscopic gametophyte stages. Haploid gametophytes reproduce sexually to form a new spore-producing diploid sporophyte generation. *Photos* Daphné Grulois-Station Biologique Roscoff

10.2 Model and Simulation Overview

We have developed a novel agent-based model of *Undaria pinnatifida* to simulate population spread in coastal habitats. This is based upon an underlying generic agent-based modelling framework developed in C++ to represent biological agents in a discrete, two-dimensional environment [7, 8]. The advantages of this framework are that it is fully parallelisable to take advantage of distributed computing architectures and it represents a robust and adaptable tool to simulate spatially and temporally heterogeneous phenomena [7]. A detailed individual-based model of the life history of *U. pinnatifida* (including distinct microscopic gametophyte and macroscopic sporophyte stages) was then built upon this basic framework.

There are two principal types of agents in the model, corresponding to the gametophyte and sporophyte stages in the life cycle of *U. pinnatifida* respectively (Fig. 10.1). These differ in their response to environmental cues and have distinct growth parameters. The growth rate and maturation of the gametophyte and sporophyte agents are functions of local environmental parameters such as irradiance, day length and temperature. In order to quantify this relationship, a review of the literature was carried out to gather quantitative data and build mathematical descriptions of these interactions at the individual level. The overall population dynamics is therefore an emergent property of the interactions between these components and the environmental parameters.

10.2.1 Gametophyte/Sporophyte Agents

The growth and maturation of the gametophyte agents is a function of irradiance, day length (hours of sunlight) and temperature [9, 10]. In a study by Choi et al. [9] the growth rates of gametophytes were measured under various levels of irradiance and day length. We fitted a hyperbolic photosynthesis-irradiance curve to this data by least squares regression ($R^2 > 0.99$) in order to determine the growth rate of gametophyte agents in the model under different irradiance levels and day lengths [11].

The effect of temperature on the relative growth rate of gametophyte agents was calculated using experimental results from Morita et al. [10] and a thermal performance curve, based on the non-linear equation of Stevenson et al. [12], was fitted to this data ($R^2 > 0.99$) [12]. The fitted parameter values for the Stevenson equation are listed in Table 10.1. This curve determines the temperature range in which the gametophyte agents can survive and its effect on their relative growth rate.

The sporophyte stage of the *U. pinnatifida* life cycle is represented as a distinct agent with its own independent growth parameters. Pang and Wu [13] carried out detailed measurements of the growth of juvenile sporophytes in culture [13]. A power law function was fitted to this data ($R^2 = 0.99$) to determine the base growth rate of sporophyte agents as a function of their length.

Similar to the gametophyte agents, thermal performance and photosynthesis-irradiance curves were calculated for the sporophyte stage by fitting to empirical data from the literature [10, 14, 15]. The parameter values estimated using least squares regression are listed in Table 10.1.

The principal means of spatial expansion of the population is by the release of spores in the water column from mature sporophytes. In the model, spores are represented as particles subject to a discretized implementation of Fick's first law of diffusion [16]. When a spore comes into contact with a suitable substrate (represented by a flag in the model) then it may form a new gametophyte agent at that location according to a pre-determined probability of attachment/germination (Table 10.1).

Experimental studies have shown that the maturation of gametophyte agents and the release of gametes (in order to sexually reproduce and form a new sporophyte generation) are a function of day length and water temperature. This empirical data was used to calculate the probability of maturation of the gametophyte agents. In order to achieve this a logistic sigmoidal function was fitted to the temperature data ($R^2 = 0.97$) and a Weibull curve to the day length data ($R^2 > 0.99$) from the literature [9, 10].

Table 10.1 Input parameters for CoastGEN simulations of *Undaria pinnatifida* in simulated coastal environment

Input parameter	Value	
Length of simulation loop (h)	1	
Grid size (No. of cells)	514 × 482	
Cell size (m^2)	0.25	
Depth in water (m)	1.0	
[1]k_{dPAR}	0.6	
Sporophytes		
Initial size (μm)	20	
Base growth rate ($loop^{-1}$)	$3.615 \times 1^{-0.407}$	
Day length response (hyperbolic curve):		
P_{max}	1.56	
α	0.13	
I_c	0.0	
Thermal performance curve [12]:	*Gameto*	*Sporo*
K_1	35.67	21.09
K_2	0.158	0.213
K_3	0.015	0.006
CT_{min}	4.45	1.62
CT_{max}	28.24	28.28
Scale	10.63	3031
Photosynthesis-irradiance curve [11]:	*Gameto*	*Sporo*
P_{max}	$0.29e^{0.11d}$	$0.4\ln(l) - 0.596$
α	$0.029d - 0.2$	$0.51^{-0.33}$
I_c	0.0	$2.5\ln(l) - 19.9$
Maturation of gametophytes		
Prob. fertilisation ($loop^{-1}$)	0.0002	
Temperature response (log curve):		
x_0	17.6	
k	0.82	
Day length response (Weibull):		
α	4.5	
β	10.96	
Spores		
Half-life (hours)	24	
Rate of release ($agent^{-1}$ $loop^{-1}$)	2.0×10^7	
Total spore stock ($agent^{-1}$)	10^{10}	
Diffusion coefficient (fick)	0.15	
Prob. of germination	10^{-9}	

Input values were estimated by fitting to data from the literature. l = plant length (μm), d = day light hours, Gameto = Gametophyte, Sporo = Sporophyte.
[1]k_{dPAR} = diffuse attenuation coefficient for photosynthetically available solar radiation [17]

10.2.2 Program Structure

The coastal environment is represented by a discrete, two-dimensional grid with each grid element corresponding to 0.25 m^2 of surface area. This allows for heterogeneity in the environmental conditions and spatial distribution of organisms, as opposed to assuming a completely homogeneous, mixed environment. The model structure is constructed using the object-oriented programing paradigm of C++, and more information on the basic framework can be found in previous publications [7, 8].

The initial phase of the program involves the creation and initialisation of an array of *U. pinnatifida* agents which are stored in an array data structure. The input parameters for the simulation are entered via a text input file (Table 10.1). These specify physical parameters such as the size and scale of the environment as well as parameters for the *U. pinnatifida* agents such as the initial number of seedlings and their responses to environmental cues.

The main program loop consists of a series of steps representing the main biotic and abiotic functions of the system (Fig. 10.2). Each loop represents a discrete hour of real-time during which all alive agents are updated and interact with the environment concurrently according to their defined behavioural rules.

The first step of the loop is to implement a diffusion algorithm for spore dispersal according to Fick's first law of diffusion. The second step in the loop is to update the energy reserves of each *U. pinnatifida* agent by subtracting a survival cost that represents energy expended during normal metabolic processes such as respiration. Agent reproduction involves the germination of spores to form new gametophyte agents. Steps 4 and 5 are optional steps: movement refers to the possible movement of agents, e.g. via passive dispersal in the water. Control strategies refer to potential human intervention/removal of *U. pinnatifida* agents. Finally, the production phase refers to new growth of the agents, which is a function of the local environmental parameters.

10.3 Results and Discussion

Simulations have been carried out using environmental parameters (light, temperature and day length) representative of Brittany, France in order to validate the model against real-world data collected by researchers at the Station Biologique de Roscoff, France. Initial results are promising and indicate that the model can accurately predict the growth dynamics of *U. pinnatifida* populations under different environmental conditions.

Figure 10.3 shows model predictions for the overall population recruitment of an *U. pinnatifida* invasion in a harbour setting. The model shows an annual pattern of growth and decay characteristic of *U. pinnatifida* populations in nature, in response to seasonal variations in light and temperature levels. For validation purposes, this was compared to real-world field results from the port of Brest in France during the years

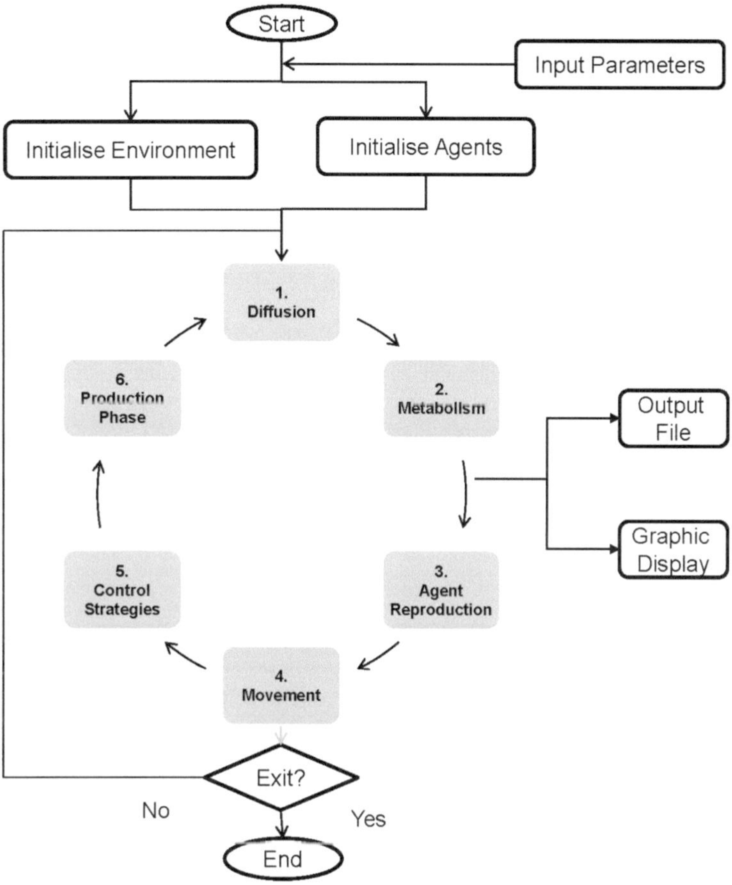

Fig. 10.2 Program flow structure for agent-based model of *Undaria pinnatifida* population dynamics coded in C++ programming language. One program loop = 1 hour real-time

2003–2006 [18]. The model exhibits the same annual pattern of recruitment observed in the real-world population found in Brest: with annual peaks in March/April (along with a secondary peak in November), and minimum recruitment in July/August.

As a test case, a comparison was made to see how closely the model fits the data from a population growing in Brest harbour in the 12 months between August 2005 and July 2006 (Fig. 10.4). The R^2 value was 0.84 when comparing the model predictions and the real-world measurements of population growth over the course of the 12 months. This indicates that the model closely matches the real-world patterns of growth at the population level using only data on the basic physiological processes of the individual algae. Some variation from the real-world results is to be expected since factors such as competition and the physical configuration of Brest harbour were not taken into account. Future work will involve extending the base model to

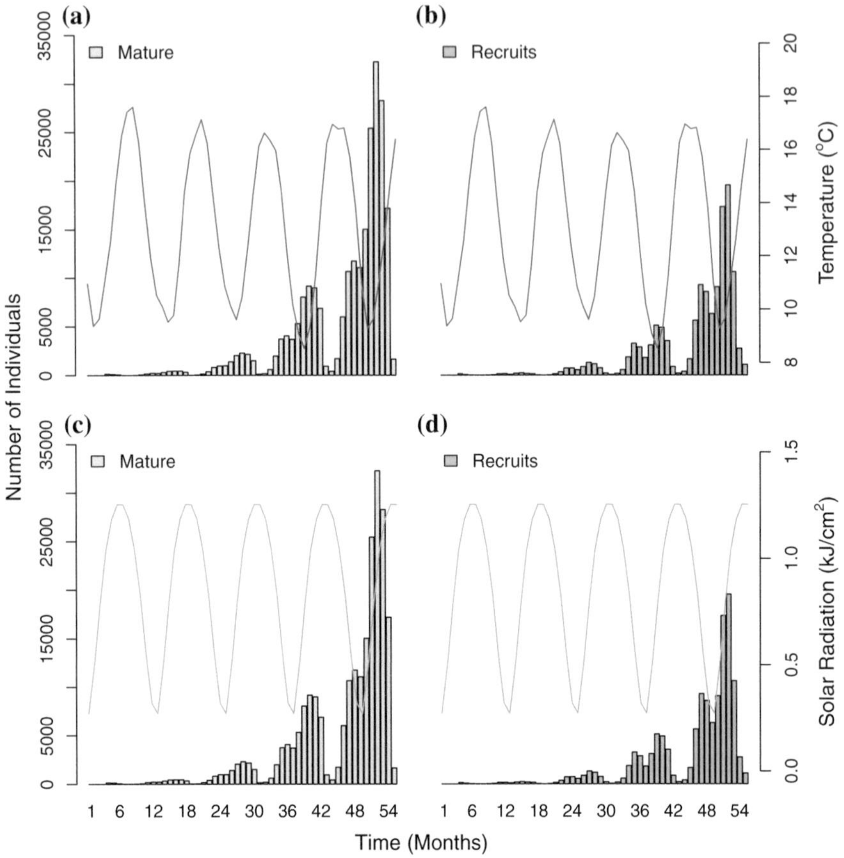

Fig. 10.3 Predicted development of population of *U. pinnatifida* after initial "seeding" of 4000 gametophytes at random locations in the simulated harbour. Environmental conditions (i.e. seasonal changes in light, temperature, and day length) representative of Brest harbour, Brittany were used as input. **a** and **c** show the total population size of sporophytes plotted on a monthly basis versus temperature and solar radiation respectively. **b** and **d** show monthly recruitment of new juvenile sporophytes (>5 cm in length) plotted against temperature and solar radiation respectively

incorporate competition for light/space with other species to explore how this impacts the results.

Figure 10.5 shows the relationship between the maturation of the microscopic gametophyte stage and the formation of new sporophytes. There is a seasonal peak in the numbers of fertile gametophyte agents in Nov/Dec each year. This marks the beginning of a recruitment phase for new sporophyte agents which lasts several months into the Spring. However, due to differences in the rate of growth of the sporophytes, depending on the environmental conditions at the time of recruitment, there will be different survival rates for each individual agent, which results in the characteristic growth dynamics observed in Fig. 10.4

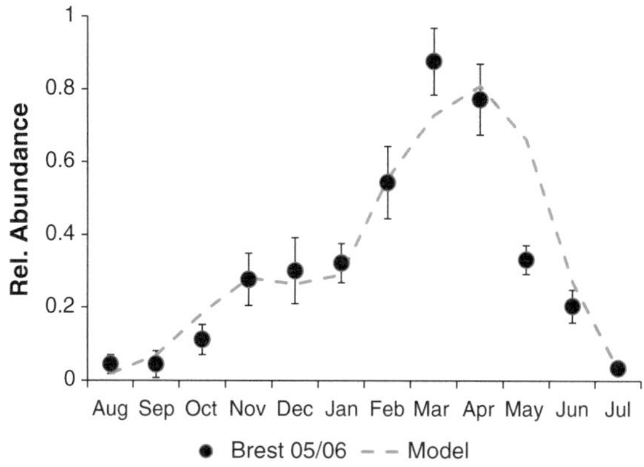

Fig. 10.4 Comparison between model predicted annual growth curve and relative abundance data from Brest harbour, France (August 2005–September 2006). Abundance data normalised relative to peak annual abundance

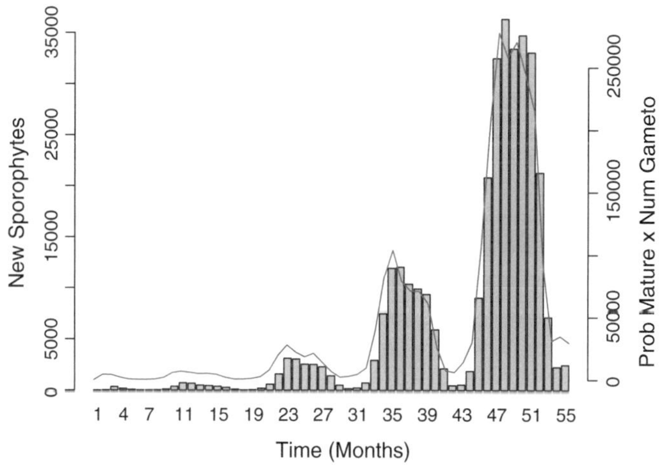

Fig. 10.5 The relationship between the predicted density of fertile gametophytes and the appearance of new juvenile sporophyte agents. *Bar plots* represent the numbers of new sporophytes (<5 cm length) per month. *Line plot* represents the relative probability of fertilisation of the gametophytes (calculated as the total number of gametophytes × probability of maturation)

Voisin [18] investigated the relationship between water temperature and the recruitment of new sporophytes [18]. To test this relationship we compared model predictions with the field-work results of Voisin (Fig. 10.6). Sporophyte recruitment was expressed in terms of water temperature 2 months prior to their appearance, which corresponds with the expected time delay between gamete fertilisation and the appearance of the new sporophyte recruit. The model predicts peak

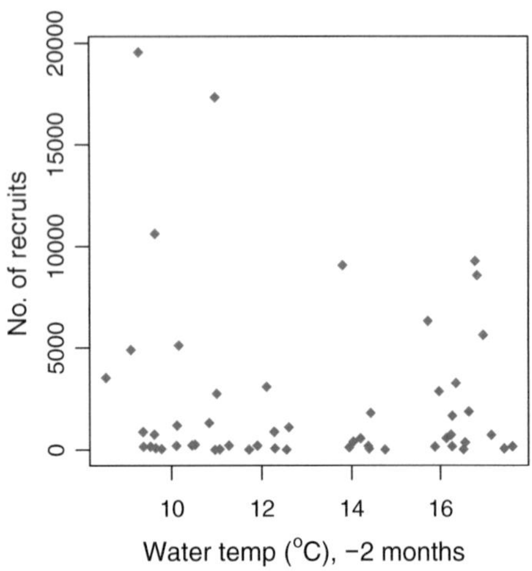

Fig. 10.6 Predicted relationship between temperature and growth dynamics of simulated *U. pinnatifida* population using seasonal temperature values representative of Brest harbour, France (8–18 °C). Number of new recruits (length >5 cm) versus the temperature of the water 2 months before their appearance (i.e. when fertilisation of the gametes occurred)

recruitment occurs at lower temperatures, which agrees with empirical data from the literature indicating significantly higher recruitment of sporophytes at temperatures <15 °C [19].

These initial results show the potential for this type of modelling approach to understand how the population dynamics of an invasion play a key role in the rate and pattern of spread of invasive species. The problem of biological invasion and resulting biodiversity loss in coastal habitats is a complex question that cannot be answered by either purely theoretical or empirical means. It requires a tightly integrated combined approach whereby theoretical studies inform experimental design and vice-versa.

Future work will include using this model framework to explicitly represent complex spatial and temporal patterns of invasion in order to be able to understand the impact of these factors on invasion dynamics. This would be a useful tool for making accurate risk assessments and designing optimal control solutions on a case by case basis in order to be able to minimise the negative impacts on ecosystem services.

Acknowledgments This research is supported by an Irish Research Council ELEVATE international career development fellowship co-funded by Marie Curie Actions (European Unions Seventh Framework Programme).

References

1. Gewin, V.: Industry lured by the gains of going green. Nature 436(7048), 173 (2005); Gewin, Virginia England. Nature. **436**(7048), 173 (2005)
2. Schaffelke, B., Smith, J., Hewitt, C.: Introduced macroalgae a growing concern. J. Appl. Phycol. **18**(3–5), 529–541 (2006)
3. Molnar, J.L., Gamboa, R.L., Revenga, C., Spalding, M.D.: Assessing the global threat of invasive species to marine biodiversity. Front. Ecol. Environ. **6**(9), 485–492 (2008)

4. Voisin, M., Engel, C.R., Viard, F.: Differential shuffling of native genetic diversity across introduced regions in a brown alga: aquaculture vs. maritime traffic effects. Proc. Natl. Acad. Sci. USA **102**(15), 5432–5437 (2005)
5. Jennings, N.R., Sycara, K., Wooldridge, M.: A roadmap of agent research and development. Auton. Agent. Multi-Agent Syst. **1**(1), 7–38 (1998)
6. Floc'h, J.Y., Pajot, R., Wallentinus, I.: The japanese brown alga Undaria pinnatifida on the coast of France and its possible establishment in european waters. J. du Conseil: ICES J. Marine Sci. **47**(3), 379–390 (1991)
7. Murphy, J.T., Johnson, M.P., Walshe, R.: Modeling the impact of spatial structure on growth dynamics of invasive plant species. Int. J. Mod. Phys. C **24**(07), 1350042 (2013)
8. Murphy, J.T., Walshe, R., Devocelle, M.: A computational model of antibiotic-resistance mechanisms in methicillin-resistant *Staphylococcus aureus* (MRSA). J. Theoret. Biol. **254**(2) (2008) 284–293 349AK Times Cited:6 Cited References Count:38
9. Choi, H., Kim, Y., Lee, S., Park, E., Nam, K.: Effects of daylength, irradiance and settlement density on the growth and reproduction of Undaria pinnatifida gametophytes. J. Appl. Phycol. **17**(5), 423–430 (2005)
10. Morita, T., Kurashima, A., Maegawa, M.: Temperature requirements for the growth and maturation of the gametophytes of Undaria pinnatifida and U. undarioides (Laminariales, Phaeophyceae). Phycol. Res. **51**(3), 154–160 (2003)
11. Jassby, A.D., Platt, T.: Mathematical formulation of the relationship between photosynthesis and light for phytoplankton. Limnol. Oceanogr. **21**(4), 540–547 (1976)
12. Stevenson, R.D., Peterson, C.R., Tsuji, J.S.: The thermal dependence of locomotion, tongue flicking, digestion, and oxygen consumption in the wandering garter snake. Physiol. Zool. 46–57 (1985)
13. Shao-jun, P., Chao-yuan, W.: Study on gametophyte vegetative growth of Undaria pinnatifida and its applications. Chin. J. Oceanol. Limnol. **14**(3), 205–210 (1996)
14. Pang, S., Luning, K.: Photoperiodic long-day control of sporophyll and hair formation in the brown alga Undaria pinnatifida. J. Appl. Phycol. **16**(2), 83–92 (2004)
15. Campbell, S.J., Bit, J.S., Burridge, T.R.: Seasonal patterns in the photosynthetic capacity, tissue pigment and nutrient content of different developmental stages of *Undaria pinnatifida* (Phaeophyta: Laminariales) in Port Phillip Bay, south-eastern Australia (1999)
16. Fick, A.: V. on liquid diffusion. Philos. Mag. Series 4 **10**(63), 30–39 (1855)
17. Saulquin, B., Hamdi, A., Gohin, F., Populus, J., Mangin, A., d'Andon, O.F.: Estimation of the diffuse attenuation coefficient k_{dPAR} using MERIS and application to seabed habitat mapping. Remote Sens. Environ. **128**, 224–233 (2013)
18. Voisin, M.: Le processus d'invasions biologiques en milieu côtier marin: le cas de l'algue brune *Undaria pinnatifida*, cultivée et introduite à l'échelle mondiale. Ph.D. thesis (2007)
19. Thornber, C.S., Kinlan, B.P., Graham, M.H., Stachowicz, J.J.: Population ecology of the invasive kelp *Undaria pinnatifida* in California: environmental and biological controls on demography (2004)

Chapter 11
Characterisation of the Idiotypic Immune Network Through Persistent Entropy

Matteo Rucco, Filippo Castiglione, Emanuela Merelli and Marco Pettini

Abstract In the present work we intend to investigate how to detect the behaviour of the immune system reaction to an external stimulus in terms of phase transitions. The immune model considered follows Jerne's idiotypic network theory. We considered two graph complexity measures—the *connectivity entropy* and the *approximate von Neumann entropy*—and one entropy for topological spaces, the so-called *persistent entropy*. The simplicial complex is obtained enriching the graph structure of the weighted idiotypic network, and it is formally analyzed by persistent homology and persistent entropy. We obtained numerical evidences that *approximate von Neumann entropy* and *persistent entropy* detect the activation of the immune system. In addition, persistent entropy allows also to identify the antibodies involved in the immune memory.

11.1 Introduction

Complex systems are system typically characterised by a number of not identical agents whose aggregate activity is nonlinear and often exhibits hierarchical self-organisation under selective pressures. Although complex systems science is a relatively young research area, it attracts lots of interest from researchers mainly due to the emerging of new techniques in several fields, e.g., physics, mathematics, data analysis and computer sciences [12, 22]. Classical data analysis (both descriptive

M. Rucco (✉) · E. Merelli
School of Science and Technology, University of Camerino, Camerino, Italy
e-mail: matteo.rucco@unicam.it

E. Merelli
e-mail: emanuela.merelli@unicam.it

F. Castiglione
Institute for Applied Mathematics (IAC) CNR, Rome, Italy
e-mail: f.castiglione@iac.cnr.it

M. Pettini
Centre de Physique Théorique, Aix-Marseille University, Marseille, France
e-mail: pettini@cpt.univ-mrs.fr

S. Battiston et al. (eds.), *Proceedings of ECCS 2014*, Springer Proceedings in Complexity, DOI 10.1007/978-3-319-29228-1_11

and exploratory) can not be sufficient for analyzing the huge amount of data that usually characterize a complex system. Persistent homology, a branch of computational topology, is nowadays largely applied for the study of complex systems [5]. Ibekwe et al., used Topological Data Analysis (TDA) for reconstructing the relationship structure of *E. coli* O157, they also prove that the non-O157 is in 32 soils (16 organic and 16 conventionally managed soils) [11]. De Silva used TDA for the analysis of sensor network [4]. TDA has been successfully applied for the study of viral evolution in biological complex systems [2]. Rucco et al., used a set of topological data analysis techniques for improving the pulmonary embolism detection in [21]. Petri et al., used an homological approach for studying the characteristics of functional brain networks at the mesoscopic level [18].

In this paper we intend to study the behaviour of a biological complex system: the idiotypic network of the mammal immune system from an information-theoretic viewpoint. In order to accomplish our aim we used both a classical approach graph-based and an innovative approach topology-based. The rest of this paper is organised as follows. Section 11.2 reports the explanation the theory of our case study and the theoretical background related to our work. In Sect. 11.3 we summarised the analysis of our *in silico* experiment. Section 11.4 provides concluding remarks.

11.2 Background

11.2.1 Case Study: The Immune Network Theory

In 1974 Niels Jerne suggested the *idiotypic network theory* to explain the phenomena of antigenic recognition by the immune cells in terms of a network of interacting cells and antibodies. Jerne's model introduced several features of immune system (I.S.), briefly when the antigen is presented to the organism, the I.S. reacts following two possible pathway: suppression or immunity. The class Ab1 of antibodies, elicited directly by the antigen, elicits the production of new antibodies Ab2 which in turn induces Ab3 and so on. This phenomenon is known as the *idiotypic cascade*. During the onto-genesis the I.S. learns which antibodies should be produced and the system remembers this decisions for a long time. This phenomena, called *immunological memory* is a property of the network of cells as a whole, rather than of the individual cells [9].

11.2.2 Graph Entropy Measures

Given a dynamical system and its graph representation there are several measures for its characterization as suggested by Reidys [15]. Even if one needs a global measure for culling the dramatic leap in the dynamics of the system, some classical measures are not sufficient for detecting whenever the system reacts to a stimulus: e.g., the *density* of a graph and defined as the ratio between the present links and the number of all possible links in the graph can not significantly change during the stimulus. For this reason we argue that *entropies* are more meaningful. In this study we report on the application of *connectivity entropy* and *approximate von Neumann entropy* [16, 17]. Both measures can be interpreted as a complexity measure of a graph, in fact both are depending by the number and the degree of the vertices.

Connectivity entropy has been used by Ortiz et al., for analyzing the structural properties of a social network and then for identifying the set of key players in the network. We repeated the experiment using the idiotypic network instead of the social network [16].

Approximate von Neumann entropy has been used by Han and collaborators on a graph classification and characterization tasks. In our approach we used this entropy measures for distinguish graphs corresponding to the same system but in different conditions, namely before, during and after a stimulus [8].

Consider now the following definitions.

11.2.2.1 Connectivity Entropy

Let $G = (V, E)$ be a graph, with $V = \{v_1, v_2, \ldots, v_n\}$ the set of vertices and E the edges. We can define [16]:
Connectivity of a node $v_i \in V$ *in a graph such as:*

$$\chi(v_i) = \frac{deg(v_i)}{2N}, N > 0 \tag{11.1}$$

where $deg(v_i)$ isthe number of incident edges to vertex v_i and N the total number of edges in the graph. Because, $0 \leq \chi(v_i) \leq 1$, and $\Sigma_{i=1}^{N}\chi(v_i) = 1$, $\chi(v_i)$ is known as *connectivity probability distribution* of the graph.
The Connectivity entropy H_{co} of a graph G is:

$$H_{co}(G) = -\Sigma\chi(v_i)log_2\chi(v_i) \tag{11.2}$$

11.2.2.2 Approximated von Neumann Entropy

Let $G = (V, E)$ be a graph, with $V = \{v_1, v_2, \ldots, v_n\}$ the set of vertices and E the edges. Let $deg(u)$ the degree for the vertex u and defined as the sum of the weight of its incident edges. The *approximate von Neumann entropy* is defined as [8]:

$$H_T^U = ln|V| - \frac{1}{2|V|} \sum_{(u,v)\in E} \frac{1}{deg(u)deg(v)} \tag{11.3}$$

11.2.3 Topological Data Analysis

Topological data analysis (TDA) is based on the concept of computational homology, a tool that transforms *local data* into *global algebraic structure*. Roughly speaking, the idea is to infer a "shape of the data", building the so-called *simplicial complexes*. A simplicial complexes is a nested collection of simplices (vertices, line segment, triangle, etc. …). Then, homology associates to the simplicial complexes a sequence of abelian groups $H_k(\mathbb{X})$, $k \in \mathbb{X}$. The vector containing the ranks of each groups is a topological invariant, the so-called **Betti numbers**. Betti numbers are the numbers of *holes* in a space with different dimensions. *Persistent homology* is one of the most used techniques for computing the topological invariants of a topological space. It returns a parametrized version of the Betti numbers: the Betti barcodes (see for example Fig. 11.1) [5]. The barcodes are equipped with the *generators* of the topological feature (connected components, holes, voids, etc.). Generators are the set of nested simplices forming the topological features.

11.2.4 Simplicial Complexes from Graph

Given a graph directed or undirected, it is possible to construct a simplicial complex from it following several approaches [10, 13]. In this paper we apply the *clique weight rank persistent homology* (CWRPH) [19]. This innovative techniques is based on the concept of a flag complex: given a graph G, the simplices of a clique complex C(G) are the complete subgraphs of G and the 0-simplices of C(G) are the vertices of G, (i.e., the complete subgraph complex). The maximal simplices are given by the collection of vertices that make up the cliques of G. In the literature, a clique complex is also referred to as a flag complex. CWRPH describes a formal procedure for building a C(G) but taking into account the weights of the links of G.

11.2.4.1 Improved jHoles

jHoles is the first Java high-performance implementation of the CWRPH algorithm [1]. The first release of *jHoles* implemented the following standard clique weight rank descending persistent homology:

1. Extract the descending (ascending) list W of all weights w_t indexed by the discrete parameter t;
2. List all maximal cliques of each connected component in G with the Bron-Kerbosch algorithm [23];

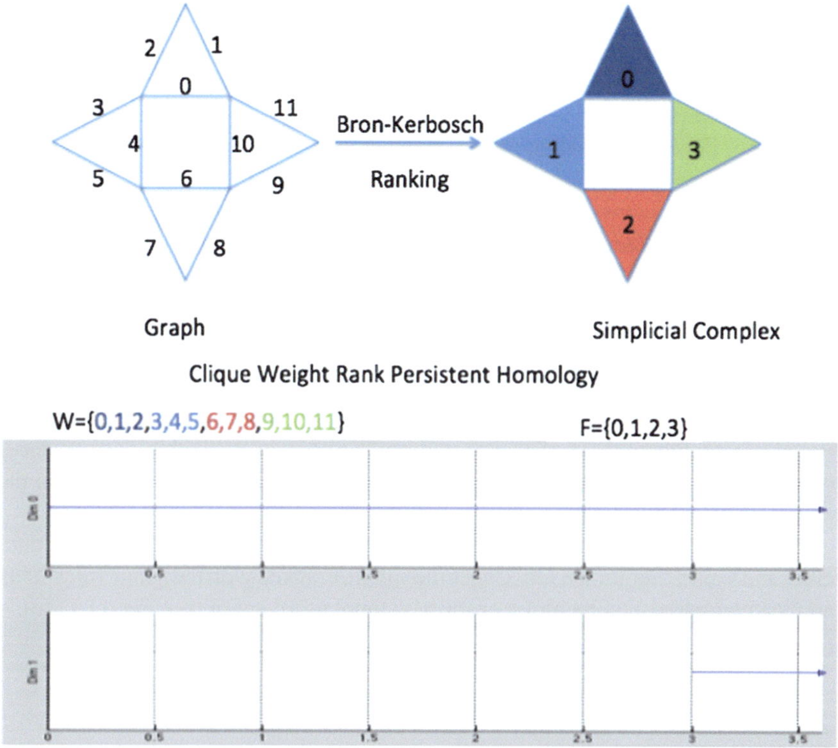

Fig. 11.1 jHoles application example. Construction of a topological space from an undirected weighted graph G (*up*). The Betti barcode representing the evolution of topological invariants (*down*). Where W is the set of weights for the *graph* G, and F denotes the set of filter values used during the computation of persistent homology for the simplicial complex. The resulting simplicial complex is characterized by the tuple $\beta_0 = 1$, and $\beta_1 = 1$. $\beta_0 = 1$ indicates that there is only 1 connected components formed by the four 2-simplices (*filled triangles*), while $\beta_1 - 1$ indicates that the simplices are arranged in a circular motif that corresponds to a persistent homological loop of dimension 1

3. Find all the combinations of each clique: a $n - clique$, with $n \geq 3$ must be tessellated with a set of $3 - cliques$, because a simplex is just the generalization of a n-dimensional triangle;
4. For each combination and clique, rank it according to the index t of the minimum (maximum) weight of the face;
5. The resulting structure is a clique simplicial complex over which persistent homology can be computed; output barcodes, intervals and generators.

Here we propose an *improved* version of *jHoles* that implements the following algorithm:

1. Extract the descending (ascending) list W of all weights w_t indexed by the discrete parameter t;

2. (Parallel) List all maximal cliques of each connected component in G with the Bron-Kerbosch algorithm [23];
3. (Parallel) Find recursively all permutations of each clique (clique tessellation with a set of 3-cliques);
4. (Parallel) For each permutation and clique, rank it according to the index t of the minimum (maximum) weight of the face;
5. The resulting structure is a clique simplicial complex over which persistent homology can be computed; output barcodes, intervals and generators.

Each permutation is a simplex belonging to the complex, while each maximal clique is a largest simplex. Steps from 2 to 4 are tasks that can be executed in parallel respectively on each connected component or face and are the core of the filtration. Step 4 ranks each face according to the index of the minimum (maximum) weight of its edges for the standard descending (ascending) filtration. The use of permutations instead of combinations in step 3 significantly improves algorithm performances and memory usage (the number of permutations of a set is strictly smaller than the number of its combinations).

Roughly speaking, the persistent homology algorithm, spans over the set F of filter values and at each iteration it introduces the simplex ranked with the actual filter value and then computes the homology of the new topological space (see Fig. 11.1) [6].

11.2.5 Persistent Entropy

Diaz et al., defined an entropy based on the persistent barcode (Definition 3 of [3]). The aim of their paper is an algorithm entropy-driven for finding the best filtration of a set of simplices. We argue that when the filtration is given their entropy can be easly extended without loosing the interpretation *à* la Shannon. Here we propose to use the maximum of the filtration value of a persistent barcode plus one as upper bound, let call this quantity m.

Definition 1 (*Persistent entropy*) Given a filtered topological space equipped with an ascending filtration algorithm, the set of filtration value F and the corresponding persistence barcode $B = [a_j\ ;\ b_j] : j \in J$. A persistence line in a barcode is conventionally represented as $[a_j\ ;\ \infty)$ here it is substitute with $[a_j\ ; m)$ where m $= (\max\{F\} + 1)$.

$$E(F) = -\sum_{j \in J} p_j log(p_j) \tag{11.4}$$

where $p_j = l_j/L, l_j = b_j - a_j$, and $L = \sum_{j \in J} l_j$

11.3 Analysis of the Idiotypic Network

11.3.1 Simulation Results

In this section, we provide a detailed study of the simulation of the *idiotypic network* obtained with *C-ImmSim* that is the agent-based simulator simulation of the immune system [20]. The computational ABM employed is discrete in mathematical terms because it represents biological entities as individuals in a heterogeneous population of cells and molecules. In particular, the major classes of cells of the lymphoid lineage and some of the myeloid lineage. Moreover the model accounts for various interleukins and messengers. This "discreteness" confers the model the character of being "easily scalable" in terms of introducing new biological complexity, at variance with corresponding equation based models. The model is stochastic, meaning it can naturally display biological "controlled" variability: for example, it is possible to separate the sorting of repertoire specificities from the random occurrences (encounters, binding, cell death, cell replacement, diffusion, cell division) during the running of the response. In other words, each repertoire expresses a private specificity, and by repeating runs with random events, the impact of different repertoires can be compared and their variations statistically determined, at the same time increasing the significance of results. The ABM model, is a polyclonal model, since all lymphocytes are equipped with a receptor represented as a binary string. This allows for a number of immunological features such as expressed and potential repertoire definition, specific recognition/binding, antigens peptides presentation, specific clonal memory, hypermutation, etc. In our configuration a simulation has a lifespan of 2190 ticks, where a tick $= 8\,\text{h}$, and a repertoire of at most 10^{12} antibodies and antigen volume equal to $V = 10\,\mu\text{L}$. The results are the average over 100 runs. In the simulator each idiotype (both antigens and antibodies) is represented with a bit-string, in our case of 12 bit length. An idiotype interacts with each other if and only if their Hamming distance is $11 \le d(A_j, A_k) \le 12$. The pair-wise distances are stored in a matrix, the so-called *Affinity matrix*: $J_{i,k}$. We defined a weight function for the idiotypic network, the *coexistence function* between antibodies (see 11.5):

$$C_{Ab_{j,k}}(t) = \frac{d(Ab_j(t), Ab_k(t)) * [Ab_j(t)] * [Ab_k(t)]}{\sum_{l=1}^{n}[Ab_l(t)]} \tag{11.5}$$

where $[Ab_l(t)]$ is the concentration of the lth antibody Ab_l at tick t. The meaning of (11.5) is that *for lower values of affinity the concentration must be more significant because the match between antibodies is less probable* (Fig. 11.2).

For each simulation we computed the coexistence function (see 11.5) and we used the weighted idiotypic network to calculate the graph entropies (see 11.2, and 11.3) as input for the persistent homology computation. In this work we used the *clique weight rank persistent homology* (CWRPH) algorithm implemented in the new version of *jHoles*, and described in par.: Sect. 11.2.4. The output of the persistent homology has been used for computing both the *persistent entropy* (see 11.4) and for identifying

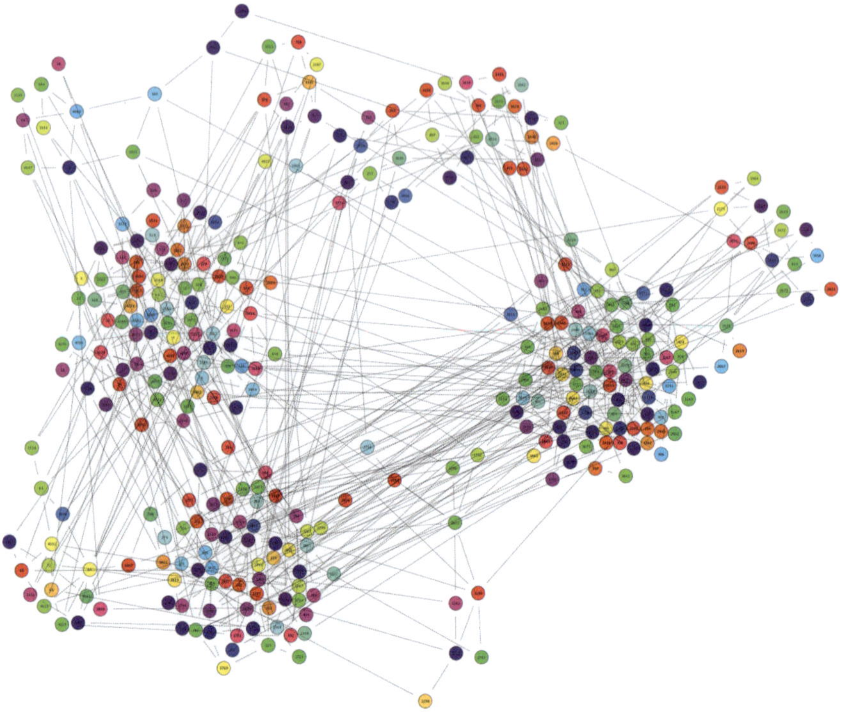

Fig. 11.2 Example of immune network at the end of the simulation (tick = 2190). The nodes represent the antibodies, a link exists if and only if two antibodies are affine. The node color represents the antibodies classes

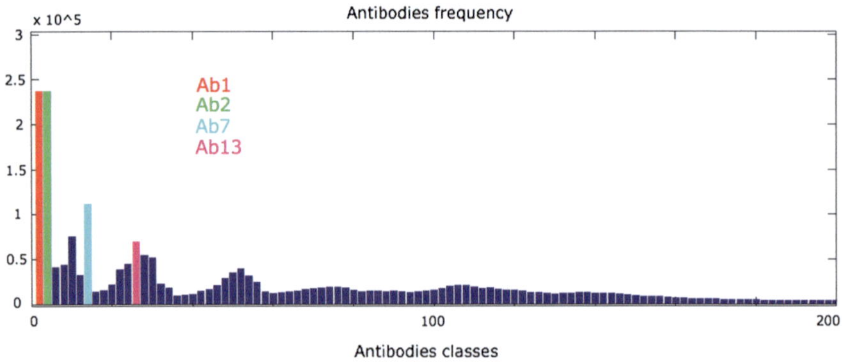

Fig. 11.3 Antibodies frequency

the persistent holes and their generators, namely the persistent antibodies that govern the evolution of the idiotypic network during the virgin state, the activation and the *immune memory* (see Fig. 11.3).

11.4 Concluding Remarks

The charts for each simulations highlight interesting emerging features. In all cases the peak corresponding to the *activation* of the *immune response* has been identified. However the *connectivity entropy* (see Fig. 11.4) does not distinguish between the activation and the *immune memory* states. The *connectivity entropy* highlighted that after time step 199 some new higher-degree antibodies are involved in the dynamics of the system. While, both the *approximated von Neumann entropy* (see Fig. 11.5) and the *persistent entropy* (see Fig. 11.6) are able to recognize the activation of the immune system: the peaks in the charts point out the *immune activation* that is following by a transient that represents the *immune response*. During the immune response the antibodies play a dual role: they can simultaneously elicit and suppress each other. After this transient there is a plateau that represents the persistent *immune network activation* corresponding to the immune memory. Persistent entropy is directly computed from the result of *persistent homology*: the Betti numbers. The analysis of the generators of the homological classes allows to identify the real number of antibodies that have been used: 203 instead of 4096 (see Fig. 11.3). The analysis of the persistent Betti numbers reveals that there is a subset of antibodies arranged in a 1-dimensional hole that is present both in the *activation state* and in the *memory state*. This 1-dimensional hole is formed by the antibodies: Ab_1, Ab_2, Ab_7, Ab_{13}. This hole is formed by the most active antibodies, see the histogram in Fig. 11.3. The removal of this 1-dimensional hole from the barcodes will flatten the entropy, that means this cycle is formed by the most specialized antibodies for the antigen that has been injected. Both the *approximated von Neumann entropy* and the *persistent entropy* can be thought as *complexity measures* for graphs or for simplicial complexes. The reason is evident in their mathematical definitions: von Neumann entropy depends on the total number of vertices and the degree of linked vertices, while persistent entropy depends by the *topological noise* and by the *persistent topological features*.

To conclude, we suggest that *persistent entropy* and in general *topological data analysis* are useful tools for the analysis of dynamical complex systems. Topological data analysis and persistent entropy can be used for discovering hidden patterns among antibodies. The transformation of a graph into a simplicial complex of dimension greater than 1 allows to discover new patterns and than it allows to extract new knowledge that otherwise can not be captured. The dimension of these patterns is equivalent to the dimension of the relation among antibodies. A relation of dimension 2, typically expressed by a graph, represents a classical 2-bodies problem while a n-ary relation represents a n-(anti)bodies problem that makes sense if and only if all the (anti)bodies are communicating simultaneously. Generally, a n-ary relation can not be decomposed in a set of 2-bodies problems: e.g. a filled triangle, that is a simplicial complex of dimension 2, represents a 3-body problems, it exists if and only if simultaneously are present the three vertices and three edges, otherwise it is represented as an empty triangle (Fig. 11.1).

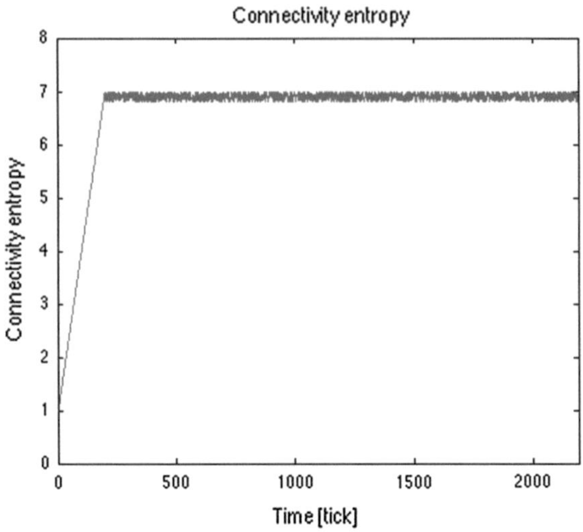

Fig. 11.4 Connectivity entropy. The maximum is reached at tick = 199

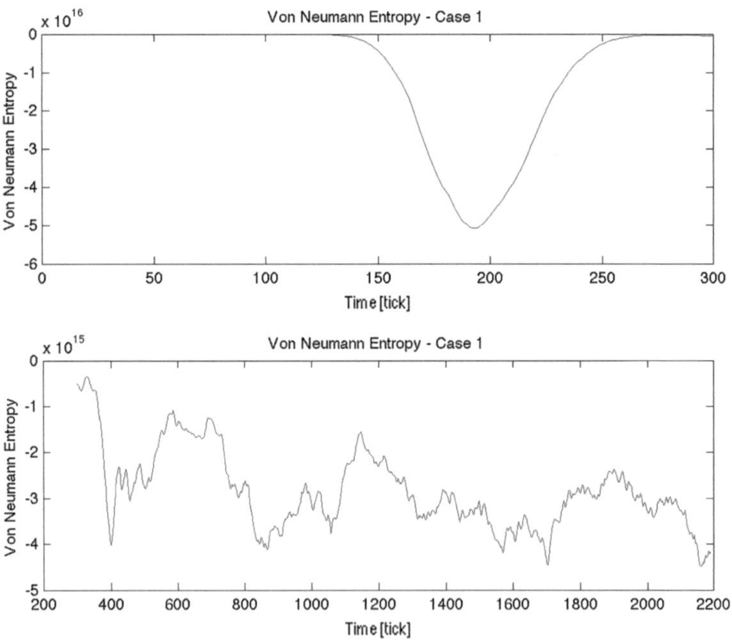

Fig. 11.5 Approximate von Neumann entropy. The minimum is reached at tick = 199

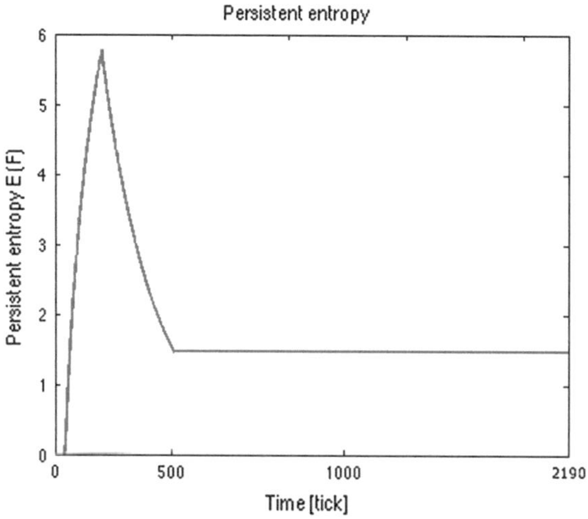

Fig. 11.6 Persistent entropy. The maximum is reached at tick = 199

Future investigation will be focusing on the use of the *persistent entropy* for characterising the S[B] systems [14], and on the use of other entropy measures, like the ones proposed by Felice [7].

Acknowledgments We acknowledge the financial support of the Future and Emerging Technologies (FET) programme within the Seventh Framework Programme (FP7) for Research of the European Commission, under the FP7 FET-Proactive Call 8—DyMCS, Grant Agreement TOPDRIM, number FP7-ICT-318121.

References

1. Binchi, J., Merelli, E., Rucco, M., Petri, G., Vaccarino, F.: jHoles: A tool for understanding biological complex networks via clique weight rank persistent homology. Elect. Notes Theoret. Comput. Sci. **306**, 5–18 (2014)
2. Chan, J.M., Carlsson, G., Rabadan, R.: Topology of viral evolution. Proc. Natl. Acad. Sci. **110**(46), 18566–18571 (2013)
3. Chintakunta, H., Gentimis, T., Gonzalez-Diaz, R., Jimenez, M.-J., Krim, H.: An entropy-based persistence barcode. Pattern Recogn. **48**(2), 391–401 (2015)
4. de Silva, V., Ghrist, R.: Coverage in sensor networks via persistent homology. Algebraic Geom. Topol. **7**(339–358), 24 (2007)
5. Edelsbrunner, H., Harer, J.: Computational Topology: An Introduction. American Mathematical Soc. (2010)
6. Edelsbrunner, H., Letscher, D., Zomorodian, A.: Topological persistence and simplification. Discrete Comput. Geom. **28**(4), 511–533 (2002)
7. Felice, D., Mancini, S., Pettini, M.: Quantifying networks complexity from information geometry viewpoint. arXiv preprint arXiv:1310.7825 (2013)
8. Han, L., Escolano, F., Hancock, E.R., Wilson, R.C.: Graph characterizations from von neumann entropy. Pattern Recogn. Lett. **33**(15), 1958–1967 (2012)

9. Hoffmann, G.W.: A theory of regulation and self-nonself discrimination in an immune network. Eur. J. Immunol. **5**(9), 638–647 (1975)
10. Horak, D., Maletić, S., Rajković, M.: Persistent homology of complex networks. J. Stat. Mech.: Theory Exp. **2009**(03), P03034 (2009)
11. Ibekwe, A.M., Ma, J., Crowley D.E., Yang, C.-H., Johnson, A.M., Petrossian, T.C., Lum, P.Y.: Topological data analysis of Escherichia coli O157: H7 and non-O157 survival in soils. Frontiers Cell. Infect. Microbiol. **4** (2014)
12. Jankowski, A., Skowron, A.: Practical Issues of Complex Systems Engineering: Wisdom Technology Approach (2014)
13. Jonsson, J.: Simplicial Complexes of Graphs, vol. 1928. Springer (2008)
14. Merelli, E., Pettini, M., Rasetti, M.: Topology driven modeling: the IS metaphor. Nat. Comput. 1–10 (2014)
15. Mortveit, H., Reidys, C.: An Introduction to Sequential Dynamical Systems. Springer Science & Business Media (2007)
16. Ortiz-Arroyo, D., Akbar Hussain, D.M.: An information theory approach to identify sets of key players. In: Intelligence and Security Informatics, pp. 15–26. Springer (2008)
17. Passerini, F., Severini, S.: The von Neumann entropy of networks. arXiv preprint arXiv:0812.2597 (2008)
18. Petri, G., Expert, P., Turkheimer, F., Carhart-Harris, R., Nutt, D., Hellyer, P.J., Vaccarino, F.: Homological scaffolds of brain functional networks. J. R. Soc. Interface **11**(101), 20140873 (2014)
19. Petri, G., Scolamiero, M., Donato, I., Vaccarino, F.: Topological strata of weighted complex networks. PloS ONE **8**(6), e66506 (2013)
20. Rapin, N., Lund, O., Castiglione, F.: Immune system simulation online. Bioinformatics **27**(14), 2013–2014 (2011)
21. Rucco, M., Falsetti, L., Herman, D., Petrossian, T., Merelli, E., Nitti, C., Salvi, A.: Using Topological Data Analysis for diagnosis pulmonary embolism. arXiv preprint arXiv:1409.5020 (2014)
22. Stein, D.L., Newman, C.M.: Nature versus nurture in complex and not-so-complex systems. In: ISCS 2013: Interdisciplinary Symposium on Complex Systems, pp. 57–63. Springer (2014)
23. Tomita, E., Tanaka, A., Takahashi, H.: The worst-case time complexity for generating all maximal cliques and computational experiments. Theoret. Comput. Sci. **363**(1), 28–42 (2006)

Chapter 12
Interests Propagation in Computer Science Research Community

Gregorio D'Agostino and Antonio De Nicola

Abstract This work proposes a framework to study the propagation of individual interests in scientific social networks. We analyze the domain of computer science and we profile members of the social network by means of semantic techniques. We model the evolution of interests as a diffusion process and we measure individual features, such as members' susceptibilities and authorities. The DBLP (Digital Bibliography and Library Project) dataset has been selected as main source since it provides an extensive list of scientific publications in this field.

12.1 Introduction

This paper proposes to model the fundamental mechanisms underlying the evolution of interests in scientific social networks. We assumed the process to be driven by diffusion. We propose a method to estimate some individual features characterizing social network (SN) members such as susceptibility and authority. This means we can measure the tendency of members to be influenced by their connections and their capability to influence others.

A *social network* consists of a community of "members" linked together with some kind of relationships (e.g., friendship, coauthorship, co-working). A SN is a human organization reflecting people common activities. We study the temporal evolution of human interests as a dynamic phenomenon arising in an anthropic system.

G. D'Agostino · A. De Nicola (✉)
ENEA-CR Casaccia, Rome, Italy
e-mail: antonio.denicola@enea.it

G. D'Agostino
e-mail: gregorio.dagostino@enea.it

A. De Nicola
University of Rome Tor Vergata, Rome, Italy

G. D'Agostino
Center for Polymer Studies, Boston University, Boston, MA, USA

S. Battiston et al. (eds.), *Proceedings of ECCS 2014*, Springer Proceedings in Complexity, DOI 10.1007/978-3-319-29228-1_12

Our hypothesis is that this phenomenon results from the combined action of several factors: people connections, general trends, pre-existing interests and both the attitudes of people to be influenced by or to influence others. Furthermore, we assume that the temporal evolution of interests depends on the topics, since people can be susceptible to some specific information more than to others.

We treat the SN as a physical system and we model interests dynamics as a diffusion process. Like a thermic system, a thermodynamic equilibrium is reached after a certain time period when no heat source is applied. Similarly in SNs, arising of new topics can be considered as a heat source that prevents the equilibrium, thus allowing diversity of views among people.

We introduced a general Markov process and we tested it against a co-authorships network in computer science.

This work proposes a framework consisting of the following building blocks: a modelling approach for social networks, to give an explicit specification of SN knowledge concerning people, their relationships and their interests; a diffusion theory, to describe the interest propagation phenomena and to make predictions about them; a method and a software application to assess the theory and to measure individual features (i.e., people susceptibility and authority).

In the following the above-mentioned building blocks are briefly presented along with the main outcomes of the analysis of the DBLP dataset. Section 12.2 presents the related work in the field. The modelling approach to represent the implicit knowledge of social networks is described in Sect. 12.3. Section 12.4 briefly presents the interest propagation theory. Section 12.5 describes the case study and Sect. 12.6 provides conclusions.

12.2 Related Work

The interest propagation phenomenon in social networks has been already studied by different disciplines [24] by different means: data mining, complexity science, semantic, and social science.

In [21], the authors propose a data mining approach to study the chain propagation of events (e.g. threads) and to identify leading influential members. Most of the efforts in the data mining community have been devoted to define progressive models. In such models, once a member becomes active (i.e. interested in a topic), it remains active. The most important propagation models are the Independent Cascade Model (ICM) and the Linear Threshold Model (LTM). Both of the previous models were first introduced in [12]. The key characteristic of ICM is that diffusion events along every arc in the social network are mutually independent; while the key characteristic of LTM is that members adapt their behaviour upon exposition to multiple independent sources. Another data mining approach is presented in [11] where the authors propose models and algorithms to extract influence probabilities parameters from a "social graph" and a log of actions by the users.

Complexity science includes the study of complex networks [3]. Among the phenomena treated by this discipline, epidemics [17, 24] studies the spread of viral processes in networks. The complexity science is mainly focusing on human infectious diseases and software malware spread. However there is a growing interest in studying topics diffusion in social networks [26], social dynamics [5, 9] or even non consensus dynamics [22].

Merging the topological and semantic analysis of social networks represents a new and potentially fruitful research field which is providing promising results [4, 6, 14]. Our work shares the use of a semantic conceptual representation of a Domain of Interest in the social network context with the formers.

A social science approach is presented in [2] where an experiment on Facebook allows to estimate influential and susceptible members of social networks with respect to some social features, such as age and sex. Another interesting issue considered by the social science community is homophily (i.e., the tendency for individuals to choose friends with similar tastes and preferences) [1, 13]. Present work does not deal with such issues.

12.3 Knoweldge Representation

Social networking platforms (SNPs) are one of the most important substrates to support the activities of a real social network (SN) in modern society. Here we introduce a "semantic social network" (SSN) consisting of a social network (SN), a semantic network (SeN) [23], and a weighted interest graph (WIG) connecting them.

A SN can be represented by a directed graph $SoN = (H, F)$, where the set H of nodes $\{h_i\}$ represents the members of the social community $H = \{h_1, h_2, \ldots, h_{|H|}\}$ and the set F of links $f_{i,k}$ represents relationships between members as ordered pairs $F = \{f_{1,1}, f_{1,2}, \ldots, f_{|F|}\}$.

Expressions of interest are events (e.g., publishing a paper) demonstrating a positive attention by a member to a product. All possible products form the Domain of Interest (DoI). It is worth mentioning that the term product here is employed in its broad sense, referring not only to goods, but also to cultural events and scientific products such as articles, books, etc.

Conceptual images of products can be expressed in terms of a finite number of concepts belonging to a semantic network representing the DoI. A semantic network can be seen as a graph $SeN = (\Lambda, R)$ where the set $\Lambda = \{\lambda_1, \lambda_2, \ldots, \lambda_{|\Lambda|}\}$ of nodes are concepts (logos) and $R = \{r_1, r_2, \ldots, r_{|R|}\}$ are the links that represent semantic relationships of different types as subsumption, meronymy and similarity [20] between the different concepts.

Given the semantics structure, we further assume that there exists a set of elementary concepts, that we name "topics" $C = \{c_1, c_2, \ldots, c_{|C|}\}$, such that one can associate a subset of topics to each product (or its abstraction). The identification of this set of basic topics plays a fundamental role and is a critical issue dealt with by the ontology engineering discipline [8]; the latter involves both automatic procedures (such as natural language processing) and human validation.

A *Interest Graph* (IG) represents an abstraction of a community of people together with their interests. It can be represented as a bipartite graph consisting of two sets of nodes, one representing people and the other representing the topics, together with a set of relationships I representing the interests of people in topics. Consequently, $IG = (H, C, I)$, where $I = \{i_1, i_2, \ldots, i_{|I|}\}$, and $i_i = (h_j, c_k)$ with $h_j \in H$ and $c_k \in C$.

A *Weighted Interest Graph* WIG is an IG with weights assigned to the links between people and topics. Such links may be viewed as either the probability to be interested or the degree of interest in a topic. Consequently, $WIG = (H, C, I, W(I))$, where $w(I)$ is a mapping from the set of relationships I to the [0, 1] range ($w(I) : I \rightarrow [0, 1]$).

We define *semantic profiling* the process of associating interests to the members of the SN, that is, inferring links and their relative weights of the WIG. The set of interests characterizing a member h_i is defined as her/his *semantic profile*. Since the *semantic profile* of a member is not static, one needs to account also for its temporal evolution:

$$S_{h_i}(t) = \{c_k : (h_i, c_k) \in I(t)\} \tag{12.1}$$

where $c_k \in C(t)$, $k \in (0, |C|)$, $h_i \in H(t)$ and $w_k(t) = w((h_i, c_k))$.

Given the basic set of interests c_k, one possible choice to provide a member h_i with a semantic profile is to attribute a likelihood $L_{h_i}(c_k)$.

A *semantic social network SSN* represents the relationships between members, the semantics of the Domain of Interest and the actual interests of the community of members with their weights. From the mathematical point of view it can be formally written as a set of six entities: $SSN = (H, F, \Lambda, R, I, W_I)$ where H represents the set of members (humans); F represents the relationships between members; Λ represents the set of concepts; R represents the set of semantic relationships; I represents the interest of people on topics; and W_I represents the degree of interests in topics of people.

Semantic social networks are dynamic entities: they are born, grow, shrink and, finally, die (close). Appearance of new nodes may describe both inclusion of new members or emergence of novel topics. Similarly, disappearances of nodes may represent the cease of participation of people to the community or the obsolescence of topics. Moreover interests of members on topics may change their intensity during the time. To model the dynamics of the latter entities, we define a *dynamic semantic social network* $SSN_t = (H(t), F(t), \Lambda(t), R(t), I(t), W_I(t))$.

12.4 Interest Propagation Dynamics

In this Section we assess a model of *interest propagation* to predict the evolution of the interests in a *semantic social network*. It accounts for the structure of the social network and its evolution [16] without predicting it. Consequently, it has the aim to estimate the probability for a person h_i to be interested in a topic c_k at a given time.

Our model is based on the following four assumptions:

- As a person, each member tends to keep her/his own beliefs.
- Each member is partly influenced by others interacting with her/him (one to one interaction).
- Each member is partly influenced by trends (one to all interaction).
- The evolution mechanism is markovian.

The evolution equations resulting from the above assumptions can be approximated for short time increments:

$$L_{h_i}(c_k, t + \Delta t) = [1 - x_i(c_k) - x_{is}(c_k)] \cdot L_{h_i}(c_k, t) + \frac{1}{|N_{h_i}|} \cdot \sum_{h_j \in N_{h_i}} x_{ij}(c_k) \cdot L_{h_j}(c_k, t) + x_{is}(c_k) \cdot L_s(c_k, t) \qquad (12.2)$$

The three terms at the right hand side model three different features: the personal tendency of a person to keep interest in a topic c_k, the influence of the neighbours, and that of the environment. In particular, $L_{h_i}(c_k, t + \Delta t)$ represents the probability of person h_i to be interested in the topic c_k at time $t + \Delta t$. $L_{h_i}(c_k, t)$ represents the probability of person h_i to be interested in the topic c_k at time t. $L_s(c_k, t)$ is the probability for the environment to provide some information on topic c_k at time t. We refer to this quantity as the "source term". $x_i(c_k)$ and $x_{ij}(c_k)$ are parameters (to be experimentally determined) characterizing the different individuals. We assume that when all neighbours share the same interests (i.e. their profiles) the interest profile should not experience any variation, therefore:

$$x_i(c_k) = \frac{1}{|N_{h_i}|} \sum_{h_j \in N_{h_i}} x_{ij}(c_k) \qquad (12.3)$$

and similarly, when the single member profile equals the trends source, no influence is expected.

Key concepts in the interest propagation theory are the individual features, i.e., *susceptibility* and *authority*, characterizing a person within a specific Domain of Interest. According to Merriam-Webster,[1] *susceptibility* is defined as the "state of being easily affected, influenced, or harmed by something". Here, in particular, there are three different parameters related to it: $x_{ij}(c_k)$, $x_i(c_k)$, and $x_{is}(c_k)$. $x_{ij}(c_k)$ is a positive number representing the attitude of a member h_i to be influenced by each of her or his neighbours h_j with respect to the topic c_k. The $x_i(c_k)$ parameter measures the susceptibility of a member h_i to her/his neighbours' total solicitation with respect to the topic c_k. It is given by the average of x_{ij} over all j's (as in 12.3). Finally, $x_{is}(c_k)$ represents the attitude of a member to be influenced by the general trends (i.e., environment or *trends susceptibility*). According to Merriam-Webster,

[1] http://www.merriam-webster.com/.

the authority is the "power to influence or command thought, opinion, or behavior". We may introduce a_i that measures individual authority as following:

$$a_i \stackrel{def}{=} \sum_{h_j \in N_{h_i}} x_{ji}(c_k) \tag{12.4}$$

12.5 The Computer Science Case Study

The analysis of DBLP publications is a representative application of our general framework. In this context, according to our knowledge representation approach, we identify the members of the social network with the authors connected by the co-author relationships (representing the edges). Other types of relationships, such as working for the same institution and participating to shared projects, can also be considered [10, 19].

The expressions of interest can be of different types. We account for the most significant that is publication of new scientific products (e.g., paper, book); however there are others such as the citation of a work, the invitations to conferences, attendance to talks, seminars, conferences and other presentations, etc. that we do not take into account.

The DBLP dataset is our main source of information. It provides a list of scientific papers in journals, conferences and workshops in the field of computer science. In order to perform the analysis, we need to acquire the information about the topics defining the scope of the domain and the evolution dynamics of both the social relationships and the interests of the authors. The above information, in principle, should be derived from different sources, however we can extract from the DBLP dataset both of them.

Results presented in this work refer to the DBLP dataset as published by november 2013. The observation period has been limited to years from 1950 to 2012. In this period there are 2.246.098 papers and 1.337.195 authors. However, in order to study the evolution of authors' interests it is necessary to observe some change in their semantic profile during time; therefore only authors that have published papers in, at least, two different years can be analysed. We named those authors "treatable". Only 519.886 authors out of 1.337.195 are treatable in the considered time period.

A first analysis was performed by natural language processing techniques [15]. By that means a preliminary set of 7632 topics was identified. Within the observation period, we indexed papers and in turn we assigned a semantic profile to each member by means of the relative frequencies of "expressions of interest" (publications):

$$\xi_{h_i}(c_k, t) = \frac{\nu_{h_i}(c_k, t)}{\sum_{c_k} \nu_{h_i}(c_k, t)}, \tag{12.5}$$

where $\nu_{h_i}(c_k, t)$ represents how many papers, written before the considered year, are indexed by the topic c_k. This function, by definition, spans the [0, 1] range; the unitary value represents a total interest in the subject while a null value means no interest at all. These semantic profiles represent our estimates of the probabilities $L_{h_i}(c_k, t)$ evolving according to (12.2); that is, one can estimate the likelihood of an author h_i to publish on a topic c_k through its share of interest ($L_{h_i}(c_k, t) \sim \xi_{h_i}(c_k, t)$).

The popularity of a topic ξ_s was estimated by its relative frequency over all published papers:

$$\xi_s(c_k, t) = \frac{\nu(c_k, t)}{\sum_{c_k} \nu(c_k, t)} \tag{12.6}$$

where $\nu(c_k, t)$ is the frequency of the topic c_k at time t. It can be regarded as the likelihood of a random person to be interested in the concept c_k at time t.

We assumed that interests propagate according to a diffusion-like process (12.2). The model contains free parameters (x_{ij}) that need to be specified. We formulated three different hypotheses on susceptibility with increasing level of complexity that we tested against the DBLP dataset. The parameters were fit by the maximum likelihood outcomes.

For the sake of simplicity (and to prevent possible overfitting), we assumed that x_i, x_{ij}, and x_{is} do not depend on the specific topic c_k. This means that a member influences her/his neighbours with the same intensity regardless of the subject.

In general, to estimate the susceptibility parameters, we constructed the mean square differences χ^2 between the predicted $L's$ and the observed ones:

$$\chi^2 = \sum_{t,h_i,c_k} \left[L^{th}_{h_i}(c_k.t + \Delta t) - L_{h_i}(c_k, t) - \delta\xi_{h_i}(c_k.t)\right]^2; \tag{12.7}$$

where the symbol δ indicates the variation of a quantity from one year to the next ($\delta\xi(c_k, t) = \xi(c_k, t + \Delta t) - \xi(c_k, t)$).

One performs the optimization using the χ^2 as an object function, that is minimizing the deviation of prediction from observed values.

Since the L's represent likelihoods, they must be confined to the [0, 1] range. This implies that also the x_{ij} and x_{is} belong to the same interval. Therefore the feasible solutions of the optimization process must respect these constraints:

$$\begin{cases} x_{is} \geq 0 \\ x_{ij} \geq 0 \\ \sum_j x_{ij} + x_{is} \leq 1 \end{cases} \tag{12.8}$$

The optimum values of the parameters are achieved analytically if the point at which the gradient of the χ^2 vanishes corresponds to a feasible solution:

$$\frac{\partial}{\partial \theta} \chi^2 = 0 \tag{12.9}$$

Table 12.1 A Summary overview of the different hypotheses

Hypothesis	Free parameters	Estimated values	χ^2/dof
HP_1	$x_{ij}=0$ $x_{is}=x_{s0}$	$x_{ij}=0$ $x_{is}=x_{s0}=0.084$	4.606×10^{-6}
HP_2	$x_{ij}=\bar{x}$ $x_{is}=\bar{x}_s$	$x_{ij}=\bar{x}=0.051$ $x_{is}=\bar{x}_s=0.053$	4.576×10^{-6}
HP_3	$x_{ij}=x_i$ x_{is}	$\bar{x}_=0.087$ $\bar{x}_s=0.059$	3.780×10^{-6}
$HP_{3\alpha}$	$x_{ij}=x_i>0$ $x_{is}>0$	$\bar{x}_=0.093$ $\bar{x}_s=0.071$	3.920×10^{-6}

HP_1 All people have the same susceptibility to trends and are not influenced by friends
HP_2 All people have the same susceptibility to trends and to neighbours
HP_3 People have individual susceptibility to trends and to neighbours
$HP_{3\alpha}$ People have individual susceptibility to trends and to neighbours. In case of negative values of x_i and x_{is} they are considered null
The χ^2 are normalized by the degrees of freedom (dof) for comparison

On the other side, when the analytical solution is unfeasible, we attribute to the parameters the closest value at boundary.

We considered the following three hypothesis. The first hypothesis (HP_1), that we took into account, states that all members have the same susceptibility to trends $(x_{is}=x_{s0})$ and are not influenced by neighbours $(x_{ij}=0)$. The second hypothesis (HP_2), that we took into account, states that all people have the same susceptibility to trends $(x_{is}=\bar{x}_s)$ and to the neighbours $(x_{ij}=\bar{x})$. The third hypothesis (HP_3), that we took into account, states that people have both individual susceptibility to trends (x_{is}) and neighbours $(x_{i,j}=x_i)$.

Table 12.1 presents a summary of the testing hypotheses of the *interest propagation theory* and the main numerical results.

The results presented in Table 12.1 show that the fitness of the hypothesis improves with the complexity of the model behind the *interest propagation theory*. In fact, taking into account the number of degrees of freedom (dof) and the value of the χ^2/dof function (representing a good index for the method), HP_3 fits the dataset better than HP_2 and HP_1.

Optimizing χ^2 resulted in some negative values of x_i and x_{is}. In such cases, they were considered null (case α in Table 12.1). Even if the χ^2/dof increases, it is still less than that of HP_2.

The hierarchical ranking of hypotheses supports the validity of the *interest propagation theory*.

The HP_3 hypothesis is more complex than the others and deserves some further discussion. The analysis of the determinant of the best fit equations (12.9) shows that there are 420290 cases where $det\mathcal{A}\neq0$ and 11627 cases where $det\mathcal{A}=0$. Hereby $\hat{x}_i$ and $\hat{x}_{is}$ indicate the solutions of those equations when they exist. Unfortunately in some cases those solutions are not feasible.

The average susceptibility under the $HP_{3\alpha}$ hypothesis due to neighbours is 9.3 %, whereas the contribution due to trends is 7.1 % for a total average susceptibility of

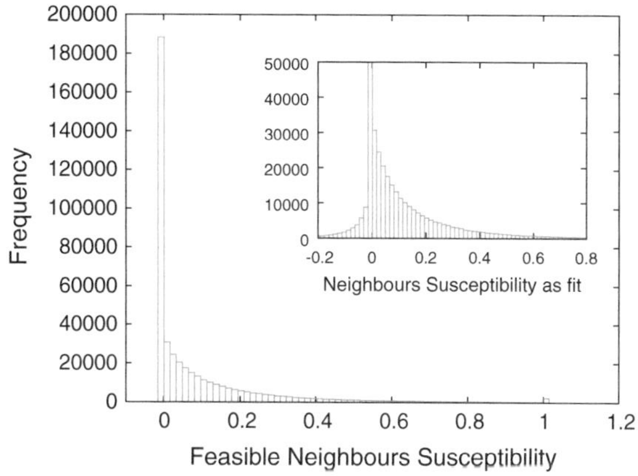

Fig. 12.1 Histogram of neighbours' susceptibilities: feasible solutions and solutions as fit (in the inset)

16.4 %. Roughly speaking, this means that about 85 % of the subjects of publications are along the line of the previous works while some 15 % do exhibit new topics due to the influence of collaborators and trends. The distribution profile of neighbours susceptibilities is presented in the Fig. 12.1. This shows a very pronounced peak at null susceptibility, while being smooth for other values. The existence of such peak may be an artifact of an insufficient semantic analysis. As a matter of fact, there is a large set of papers (1,215,200 out of 2,246,098) that can not be indexed by means of our selected set of topics. This is expected to result into spurious null susceptibilities. Similar considerations apply to trends susceptibilities.

The authority coefficients (authorities, shortly) spans the [0,52] range; their mean value is $\bar{a} = 0.44$ and its standard deviation is about twice that value (0.89). Figure 12.2 shows the distribution of authorities. As can be seen, there is a very long queue of few authors at high values. It is known that there exist different authors with the same full name; those people are very often treated as a single author in several

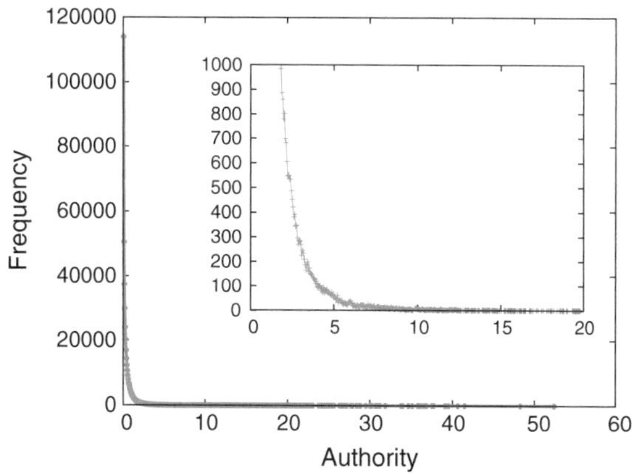

Fig. 12.2 Histogram of the authority coefficients

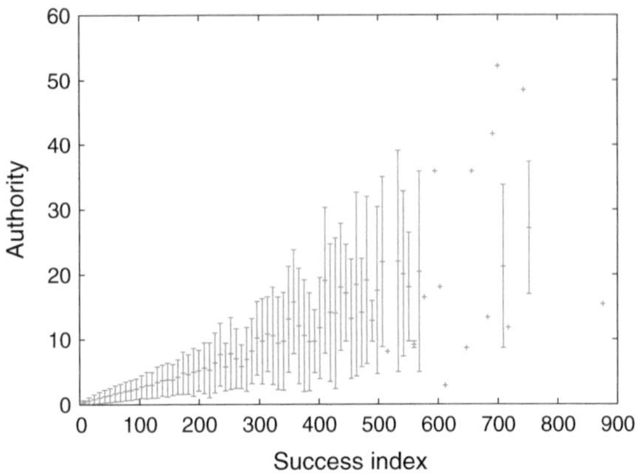

Fig. 12.3 Scatter plot of the authority of different treatable authors versus their success index

Table 12.2 Notorious authors in computer science

Name	$\hat{x}_i$	$\hat{x}_{is}$	Authority a_i
Wil M. P. van der Aalst	+0.111	+0.058	+12.809
Jack Dongarra	−0.019	+0.028	+10.259
John Mylopoulos	+0.021	+0.037	+8.852
Georg Gottlob	+0.055	+0.009	+5.081
Ian Horrocks	+0.198	−0.080	+4.835

datasets. This problem is known as "ambiguity" of the papers indexing; it results in gathering different authors into a single member of our social network.

We also tried to relate the success of an author with its authority. We employed the number of published papers as an index of success, however a more appropriate index should be the total number of citations [18, 25] which were not available. As shown in Fig. 12.3, the higher is the success index the higher is the authority.

Finally, Table 12.2 presents some individual features of some famous authors. As expected they all exhibit high levels of authority.

12.6 Conclusions and Future Work

This paper combines known methods in complexity science with the semantic analysis of natural language to provide insights in the propagation of people interests in social networks. We assumed that interests propagate according to a diffusion mechanism, while being continuously created. The human behaviour was described by means of two basic attitudinal characteristics (the susceptibility and the authority) that quantify the tendency to influence and to be influenced by "friends" and the environment.

We presented a very general model that we tested against the DBLP database. However the theory applies also to a broader set of social networks (including popular ones such as Facebook or Twitter). Application to those types of social networks will possible lead to member-tailored services and/or commercial exploitations. A wider treatment of the subject is available at [7].

Acknowledgments The authors are indebted to Fulvio D'Antonio for providing the preliminary set of topics. Salvatore Tucci, Emiliano Casalicchio, and Francesco Lo Presti are kindly acknowledged for stimulating discussions.

References

1. Aral, S., Muchnik, L., Sundararajan, A.: Distinguishing influence-based contagion from homophily-driven diffusion in dynamic networks. Proc. Nat. Acad. Sci. **106**(51), 21544–21549 (2009)
2. Aral, S., Walker, D.: Identifying influential and susceptible members of social networks. Science **337**(6092), 337–341 (2012)
3. Barabási, A.L., Albert, R.: Emergence of scaling in random networks. Science **286**(5439), 509–512 (1999)
4. Bojārs, U., Breslin, J., Finn, A., Decker, S.: Using the semantic web for linking and reusing data across web 2.0 communities. Web Seman.: Sci., Ser. Agents World Wide Web **6**(1), 21–28 (2008)
5. Castellano, C., Fortunato, S., Loreto, V.: Statistical physics of social dynamics. Rev. Mod. Phys. **81**(2), 591 (2009)
6. Cucchiarelli, A., D'Antonio, F., Velardi, P.: Semantically interconnected social networks. Soc. Netw. Anal. Min. **2**(1), 69–95 (2012)
7. D'Agostino, G., D'Antonio, F., De Nicola, A., Tucci, S.: Interests diffusion in social networks. Physica A **436**, 443–461 (2015)
8. De Nicola, A., Missikoff, M., Navigli, R.: A software engineering approach to ontology building. Inf. Syst. **34**(2), 258–275 (2009)
9. Galam, S.: Local dynamics vs. social mechanisms: A unifying frame. EPL (Europhys. Lett.) **70**(6), 705 (2005)
10. Gomez, S., Diaz-Guilera, A., Gomez-Gardeñes, J., Perez-Vicente, C.J., Moreno, Y., Arenas, A.: Diffusion dynamics on multiplex networks. Phys. Rev. Lett. **110**(2), 028701 (2013)
11. Goyal, A., Bonchi, F., Lakshmanan, L.V.: Learning influence probabilities in social networks. In: Proceedings of the 3rd ACM International Conference on Web Search and Data Mining, pp. 241–250. WSDM '10, ACM (2010)
12. Kempe, D., Kleinberg, J., Tardos, E.: Maximizing the spread of influence through a social network. In: Proceedings of the 9th ACM SIGKDD International Conference on Knowledge Discovery and Data Mining, pp. 137–146. KDD '03, ACM (2003)
13. Manski, C.F.: Identification Problems in the Social Sciences. Harvard University Press (1995)
14. Mika, P.: Ontologies are us: A unified model of social networks and semantics. Web Seman.: Sci., Serv. Agents World Wide Web **5**(1), 5–15 (2007)
15. Navigli, R., Velardi, P.: Learning domain ontologies from document warehouses and dedicated web sites. Comput. Linguist. **30**(2), 151–179 (2004)
16. Palla, G., Vicsek, T.: Statistical properties of community dynamics in large social networks. Int. J. Agent Technol. Syst. (IJATS) **1**(4), 1–16 (2009)
17. Pastor-Satorras, R., Vespignani, A.: Epidemic spreading in scale-free networks. Phys. Rev. Lett. **86**, 3200–3203 (2001)

18. Petersen, A.M., Fortunato, S., Pan, R.K., Kaski, K., Penner, O., Rungi, A., Riccaboni, M., Stanley, H.E., Pammolli, F.: Reputation and impact in academic careers. Proc. Nat. Acad. Sci. **111**(43), 15316–15321 (2014)
19. Quattrociocchi, W., Caldarelli, G., Scala, A.: Opinion dynamics on interacting networks: media competition and social influence. Sci. R. **4** (2014)
20. Resnik, P.: Semantic similarity in a taxonomy: An information-based measure and its application to problems of ambiguity in natural language. J. Artif. Intell. Res. **11**, 95–130 (1998)
21. Richardson, M., Domingos, P.: Mining knowledge-sharing sites for viral marketing. In: Proceedings of the 8th ACM SIGKDD International Conference on Knowledge Discovery and Data Mining, pp. 61–70. KDD '02, ACM (2002)
22. Shao, J., Havlin, S., Stanley, H.E.: Dynamic opinion model and invasion percolation. Phys. Rev. Lett. **103**(1), 018701 (2009)
23. Sowa, J.F.: Semantic networks. Encyclopedia of Cognitive Science (2006)
24. Vespignani, A.: Modelling dynamical processes in complex socio-technical systems. Nat. Phys. **8**(1), 32–39 (2012)
25. Wang, D., Song, C., Barabási, A.L.: Quantifying long-term scientific impact. Science **342**(6154), 127–132 (2013)
26. Wang, D., Wen, Z., Tong, H., Lin, C.Y., Song, C., Barabási, A.L.: Information spreading in context. In: Proceedings of the 20th International Conference on World Wide Web. pp. 735–744. WWW '11, ACM, New York, NY, USA (2011)

Chapter 13
Nonparametric Estimation of the Preferential Attachment Function in Complex Networks: Evidence of Deviations from Log Linearity

Thong Pham, Paul Sheridan and Hidetoshi Shimodaira

Abstract We introduce a statistically sound method called PAFit for the joint estimation of preferential attachment and node fitness in temporal complex networks. Together these mechanisms play a crucial role in shaping network topology by governing the way in which nodes acquire new edges over time. PAFit is an advance over previous methods in so far as it does not make any assumptions on the functional form of the preferential attachment function. We found that the application of PAFit to a publicly available Flickr social network dataset turned up clear evidence for a deviation of the preferential attachment function from the popularly assumed log-linear form. What is more, we were surprised to find that hubs are not always the nodes with the highest node fitnesses. PAFit is implemented in an R package of the same name.

13.1 Introduction

By the turn of the 20th century, complex networks in diverse domains were found to share some universal characteristics, such as the well-known scale-free property [21]. These intriguing universalitics have prompted researchers to hypothesize common mechanisms that can explain the observed network properties. Preferential attachment (PA) [1, 26] and node fitness [2, 3] combined with growth have been advanced as plausible mechanisms underlying the scale-free property in real-world networks.

T. Pham (✉) · H. Shimodaira
Osaka University, Osaka, Japan
e-mail: thongpham@sigmath.es.osaka-u.ac.jp

H. Shimodaira
e-mail: shimo@sigmath.es.osaka-u.ac.jp

P. Sheridan
The University of Tokyo, Tokyo, Japan
e-mail: sheridan@ims.u-tokyo.ac.jp

S. Battiston et al. (eds.), *Proceedings of ECCS 2014*, Springer Proceedings in Complexity, DOI 10.1007/978-3-319-29228-1_13

Temporal networks are closely connected to the concept of growth. Real-world networks are rarely static, in the sense that they gradually grow in both number of nodes and edges. Let G_t denote the network at time t. The growth of real-world temporal networks is often modelled as follows. Starting from some initial network G_0, at each time-step t, new edges and nodes are added to G_{t-1} to form G_t. A temporal network is in fact a time series of static networks observed at each time-step: $G_0, G_1, \ldots$. How new edges at a time-step t connect to nodes in G_{t-1} is determined by preferential attachment and fitness mechanisms.

The PA mechanism [1, 26] basically states that the probability a node v_i of degree k acquires a new edge is proportional to some function A_k:

$$Pr(v_i \text{ acquires a new edge}) \propto A_k. \tag{13.1}$$

A_k is often called the *attachment kernel*. The so-called preferential attachment then corresponds to the case when A_k is an increasing function on average. In this case, a node with high degree tends to acquire new edges more readily than a low degree node. This rich get richer phenomenon [29] has been observed in complex networks across heterogeneous domains. In social networks, for example, a person with many acquaintances tends to make more new acquaintance than does a less sociable person. Or in citation networks of research papers, to take another example, a paper with many citations may pick up new citations more readily than a comparatively lesser known paper. Similar examples abound in other domains. The true form of A_k underpins underlying network properties.

On the other hand, in the node fitness mechanism [2, 3], the probability node v_i receives a new edge is proportional to its fitness f_i:

$$Pr(v_i \text{acquires a new edge}) \propto f_i. \tag{13.2}$$

The fitness of a node can be interpreted as its intrinsic quality [12].

Taken in tandem, the PA and fitness mechanisms yield

$$Pr(v_i \text{ acquires a new edge}) \propto f_i A_k. \tag{13.3}$$

This important equation encompasses a large number of well-known network models [1–3]. This equation underlies the temporal network model in our paper.

The inverse problem of estimating A_k and f_i from observed data ought to be one of great importance to the working network scientist. Besides practical applications in link prediction algorithms [17], its solution gives valuable insights into the global characteristics of networks. There are several fundamental questions surrounding PA and node fitness in real-world temporal networks. For starters, does the signature of PA persist even after having accounted for the contribution of node fitness? If so, then what, if any, simple functional form does it take? For example, the log-linear form $A_k = k^\alpha$ has been shown to affect greatly the topology of the network. When $\alpha = 1$, the network is scale-free. On the other hand, an asymptotically sub-linear A_k ($\alpha < 1$) gives rise to a stretch-exponential degree distribution, while an asymptotically super-linear A_k ($\alpha > 1$) leads to a "winner take all" situation when a finite number of nodes takes all the edges [14]. The limiting case $\alpha = 0$ corresponds to the Erdös-Rényi model [7], in which the network is also not scale-free. The question then arises as to whether this widely accepted log-linear form true, or does the attachment kernel take other forms? Conversely, do we find non-uniform node fitnesses exist after having taken the effect of PA into account? At present answers to these fundamental questions remain obscure on account that no method has been proposed to jointly estimate both A_k and f_i.

The problem of estimating A_k while fixing $f_i = 1$ for all i has attracted the attention of many researchers and there are a number of estimation methods [8, 11, 18, 20, 28]. The primary drawback of most of these methods is that they explicitly assume the log-linear form $A_k = k^\alpha$ and focus only on estimating the attachment exponent α [8, 15, 18, 28]. Massen et al. [18] used a fixed point iterative algorithm, Sheridan et al. [28] a Markov Chain Monte Carlo method, Gomez et al. [8] a maximum likelihood estimation to estimate the value of α.

While the remaining attachment kernel estimation methods [11, 20] do not assume a functional form for A_k, they are not founded on any rigorous statistical framework. Firstly, Newman's method [20] estimates A_k by building a histogram of A_k over multiple time-steps. It has been used to estimate the attachment kernel a number of times [4, 9]. In simulated examples, Newman's method appears to work well in estimating A_k when k is small, but systematically underestimates A_k for large k [22]. This is thought to be an artifact of the method [9].

The method of Jeong et al. [11], by contrast, estimates A_k by observing the rates at which degree k nodes acquire new edges in a small time window. Jeong' s method chooses a small time window and create a histogram of the degrees of nodes to which new edges appeared in this time window connect. Note that the time window must be small enough so that all changes in the degrees of nodes can be ignored. Normalizing the histogram value of the degree k by the number of nodes with degree k at the on-start of the time window will give us the estimated value of A_k. Its simplicity makes it is the most frequently used method in practice [6, 13, 27]. One problem with Jeong's method, however, is that the time window should be kept small in comparison with the size of the network, leading to a lack of robust estimation of the attachment kernel.

Lastly, the Growth method of Kong et al. [12] is the only method we are aware of that estimates node fitness, albeit under the assumption that $A_k = k$. Their method estimates f_i by using an asymptotic formula related to the degree of a node with its fitness. This method is also ad hoc from a statistical point of view.

In contrast to the existing methods, we propose a maximum likelihood estimation (MLE) method called PAFit for estimating both A_k and f_i. PAFit is nonparametric in the sense that it does not assume any particular functional form for either A_k or f_i. The algorithm underlying PAFit is known as a Minorize-Maximization (MM) algorithm [10, 16]. We can prove that our algorithm increases the log-likelihood function monotonically, and that it moreover converges to a global maximizer of this function. We also calculate confidence intervals for the estimated values of A_k and f_i using standard statistical theory.

By using PAFit to analyze a publicly available Flickr social network dataset [19], we found that the estimated attachment kernel differed considerably from the functional form $A_k = k^\alpha$. This result suggests that it is important to look beyond the classical log-linear hypothesis in modelling the attachment kernel. With regard to node fitness, we found that hubs are not always the nodes with the greatest fitnesses.

In summary, PAFit is a statistically sound method for the joint estimation of PA and node fitness nonparametrically in temporal complex networks. We have demonstrated its potential in this paper to uncover interesting findings about real-world complex networks. This method is implemented in the `R` package `PAFit` [24].

13.2 The Network Model

The statistical estimation method we present in this paper is tailored for the following temporal model for directed networks. Note, however, that our methodology can be easily adapted to work for undirected networks. Starting from a seed network at time $t_0 = 0$, we grow the network by adding $n(t)$ nodes and $m(t)$ edges at every time-step t, for $t = 1, \ldots, T$. Our method allows $m(t)$ to be consisted of both new edges that emanate from the $n(t)$ new nodes and emergent edges between existing nodes. This is important since in a large portion of real-world complex networks, new edges do emerge between existing nodes. At time-step t, the probability that an existed node v_i with in-degree $d_{v_i(t)}$ acquires a new edge is

$$Pr(v_i \text{ acquires a new edge}) \propto f_i A_{d_{v_i}(t)} \tag{13.4}$$

where A_k is the attachment value of degree k and f_i is the fitness of node v_i.

The temporal model defined by (13.4) includes a number of important network models as special cases. When $A_k = k$, it reduces to the Bianconi-Barabási model [2]. When $A_k = 1$ for all k, the model corresponds to the model in [3]. When $f_i = 1$ for all i and $A_k = k^\alpha$, it corresponds to Price's model [25, 26] or Barabási-Albert (BA) model in the undirected case [1]. Note, however, that the original Price and BA models only focused on the case $\alpha = 1$. Furthermore, when $f_i = 1$ for all i and $A_k = 1$ for all k, then the model reduces to the classical Erdös-Rényi random network model [7].

Here we note an important remark regarding an assumption about the distribution of $m(t)$ and $n(t)$. Let $\boldsymbol{\theta}(t)$ denote the parameter vector governs the distribution of $m(t)$ and $n(t)$. The likelihood of the data at time-step t is the product of $P(m(t), n(t)|G_{t-1}, \boldsymbol{\theta}(t))$ and $P(G_t|G_{t-1}, m(t), n(t), A_k, f_i)$. In this paper, we assume that $\boldsymbol{\theta}(t)$ does not involve A_k and f_i. With just only this assumption, the term $P(m(t), n(t)|G_{t-1}, \boldsymbol{\theta}(t))$ can be safely ignored when calculating the maximum likelihood estimation for A_k and f_i. This very mild assumption still allows broad and realistic models for $m(t)$ and $n(t)$. For example, $m(t)$ and $n(t)$ can be random variables whose means depend on G_{t-1}.

13.3 Illustrative Examples

In this section we present simulated examples for the case of estimating (1) the attachment kernel when $f_i = 1$ for all i, and (2) the joint estimation of the attachment kernel and node fitness. The purpose being to demonstrate the workings of our proposed method, PAFit.

13.3.1 Attachment Kernel Estimation

In this first example, the network generative process follows (13.4) with $f_i = 1$ for all i and $A_k = 3(\log \max(k, 1))^2 + 1$ Starting from a seed network of 20 nodes, we add $m(t) = 5$ new edges and $n(t) = 1$ new node at each time-step t until a total of $N = 5000$ nodes is reached. For a quantitative assessment, we also measure the average relative error $e_A = \frac{1}{K}\sum_{k=0}^{K-1}\frac{\left(A_k - \hat{A}_k\right)^2}{A_k^2}$. Here K denotes the maximum degree in the observed data.

We first apply Jeong's method [11] to estimate A_k. Here we choose the time window between when the 4500th node and the 5000th node are added. From Fig. 13.1a, one can see that Jeong's method captured the shape of the attachment kernel, but the estimated function was sparse and fluctuated considerably. These are inherent drawbacks of the method that arise from using only a small time window to estimate the attachment kernel. In this case, e_A is 0.20.

Second, we apply Newman's method. In Fig. 13.1b, the estimated attachment kernel follows the true values very closely, but the estimated value of A_k starts to fall off when the degree k becomes large. This phenomenon has been observed by other researchers, and is thought to be an artifact of Newman's method [9]. In this case, e_A is 0.14.

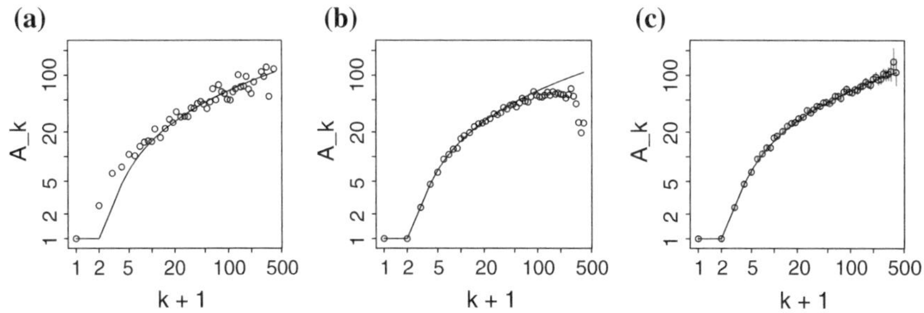

Fig. 13.1 Estimation of the attachment kernel $A_k = 3(\log \max(k, 1))^2 + 1$ in a simulated network of 5000 nodes. The *solid line* corresponds to the true A_k. In PAFit, the *vertical lines* correspond to confidence intervals of the estimated values of A_k. **a** Jeong's method. **b** Newman's method. **d** PAFit

Unlike the methods of Jeong and Newman, PAFit estimates the attachment kernel by maximum likelihood estimation using all the available data. It is clear from visual inspection alone of Fig. 13.1c that PAFit outperformed these two methods. The estimated attachment kernel follows the true values comparatively well, even in the high degree region. In this case, e_A is 0.01, which is the smallest error among three methods.

13.3.2 Joint Attachment Kernel and Node Fitness Estimation

Turning our attention to the case of estimating node fitness, we compare PAFit with the Growth method. The generative process follows (13.4) with true node fitnesses sampled from a gamma distribution with shape and scale parameters are both 5. As in the previous example a total of $m(t) = 5$ new edges and $n(t) = 1$ new node are added at each time-step t until the total number of nodes reached is $N = 5000$. Regarding the true attachment kernel, we consider two cases: $A_k = \max(k, 1)$ and $A_k = 3(\log \max(k, 1))^2 + 1$. As in the previous example, we also measure the performance of the two methods quantitatively. We use three numbers: average relative error e_A, correlation coefficient r between estimated fitnesses and true fitnesses, and finally the average relative error e_f, defined as $e_f = \frac{1}{N}\sum_{i=1}^{N} \frac{\left(f_i - \hat{f}_i\right)^2}{f_i^2}$.

Figures 13.2a and 13.3a show the estimated fitnesses of the nodes that acquired at least five edges by the Growth method when the true attachment kernels are $A_k = \max(k, 1)$ and $A_k = 3(\log \max(k, 1))^2 + 1$, respectively. The correlation coefficients in Figs. 13.2a and 13.3a are 0.51 and 0.55, respectively. For the two cases, the average relative error e_f are 0.93 and 0.94. One can say that the Growth method performed acceptably well in both cases.

PAFit simultaneously estimates the attachment kernel and node fitnesses by maximum likelihood estimation. When $A_k = \max(k, 1)$, the correlation coefficient r in Fig. 13.2b and the average relative error e_f are $r = 0.63$ and $e_f = 0.34$, while

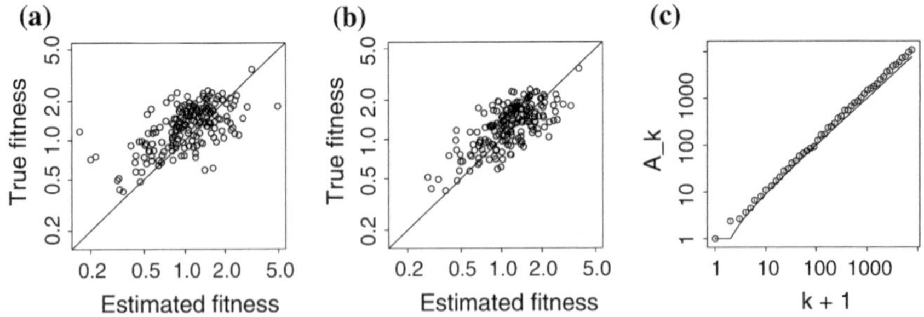

Fig. 13.2 Joint estimation of attachment kernel and node fitness when $A_k = \max(k, 1)$. **a** Estimated node fitnesses by Growth method. **b** Estimated node fitnesses by PAFit. **c** Estimated attachment kernel by PAFit

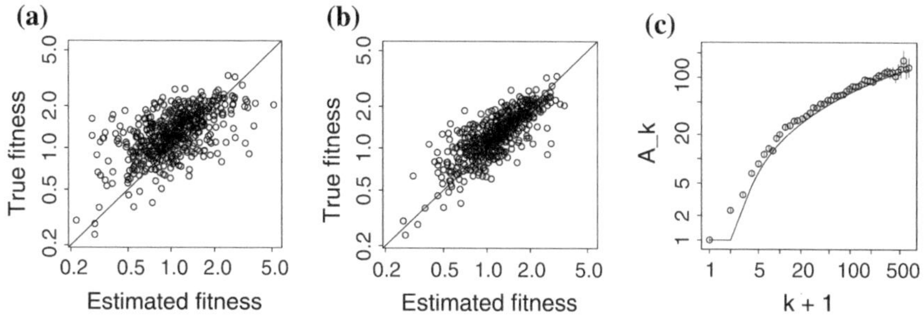

Fig. 13.3 Joint estimation of attachment kernel and node fitness when A_k=3 (log max $(k, 1))^2 + 1$. **a** Estimated node fitnesses by Growth method. **b** Estimated node fitnesses by PAFit. **c** Estimated attachment kernel by PAFit

when $A_k = 3\log^2 \max(k, 1) + 1$, these number are 0.78 and 0.38, respectively. In both cases, PAFit outperformed the Growth method. Not only did PAFit estimate the node fitnesses comparatively well, but it also succeeded in recovering the attachment kernel, as can be seen in Figs. 13.2c and 13.3c (the average relative error e_A are 0.004 and 0.009, respectively). To conclude, these two examples demonstrated that attachment kernel and node fitness can indeed be estimated simultaneously.

13.4 The PAFit Estimation Method

13.4.1 Attachment Kernel Estimation

It is instructive to first consider the simple yet important case of estimating the attachment kernel in isolation. In this case, we assume $f_i = 1$ for all i. Let $m_k(t)$ and $n_k(t)$ denote the number of new edges connect to nodes with degree k at time t, the number of existing nodes with degree k, respectively. Recall that K is the maximum degree in the observed data. The key observation is that given $m(t)$, the quantities $m_0(t), m_1(t), \ldots, m_K(t)$ follow a multinomial distribution with parameters $p_0(t), p_1(t), \ldots, p_K(t)$, where $p_k(t)$, the probability that a newly added edge at time t will link to a node with degree k, is

$$p_k(t) = \frac{n_k(t)A_k}{\sum_{j=1}^{K} n_j(t)A_j}. \tag{13.5}$$

This enables us to write down the log-likelihood function in this case:

$$l(\mathbf{A}) = \sum_{t=1}^{T}\sum_{k=1}^{K} m_k(t)\log A_k - \sum_{t=1}^{T} m(t)\log\left(\sum_{j=1}^{K} n_j(t)A_j\right) \tag{13.6}$$

where $\mathbf{A} = [A_0, A_1, \ldots]$ are parameters we want to estimate. Note that the A_k are only identifiable up to a multiplicative constant, as can be seen from (13.5).

We can maximize (13.6) by the following iterative algorithm. Define $\mathbf{A}^{(i)} = [A_0^{(i)} \cdots A_K^{(i)}]$ as the estimated parameter vector at iteration i. Starting from some initial value $\mathbf{A}^{(0)}$, we can update $A_k^{(i+1)}$, the value of A_k at iteration $i+1$, in parallel by

$$A_k^{(i+1)} = \frac{\sum_t m_k(t)}{\sum_t \frac{m(t)}{\sum_j n_j(t) A_j^{(i)}} n_k(t)} \tag{13.7}$$

until convergence. We can show that this algorithm is in fact a MM algorithm [23]. It follows from the theory of MM algorithms that the log-likelihood function is guaranteed to increase with number of iterations. It can also be shown that the algorithm converges to a global maximizer of (13.6) [23].

13.4.2 Joint Attachment Kernel and Node Fitness Estimation

In this section we proceed to the general case of jointly estimating A_k and f_i. Let $z_j(t)$ and N be the number of new edges that connect to node v_j at time t and the number of nodes in the final network, respectively. The log-likelihood function is then

$$l(\mathbf{A}, \mathbf{f}) = \sum_{t=1}^{T} \sum_{j} z_j(t) \log \left(f_j A_{d_j(t)} \right) - \sum_{t=1}^{T} m(t) \log \sum_{l} f_l A_{d_l(t)} \tag{13.8}$$

where $\mathbf{A} = [A_0, A_1, \ldots]$ and $\mathbf{f} = [f_1, f_2, \ldots]$ are parameters we want to estimate.

In order to maximize (13.8), we start from some initial value $\mathbf{A}^{(0)} = [A_0^{(0)} \cdots A_K^{(0)}]$ and $\mathbf{f}^{(0)} = [f_0^{(0)} \cdots f_N^{(0)}]$ for $\mathbf{A}$ and $\mathbf{f}$ at step $i = 0$ and then iteratively update:

$$A_k^{(i+1)} \leftarrow \frac{\sum_t m_k(t)}{\sum_t \frac{m(t)}{\sum_l f_l^{(i)} A_{d_l(t)}^{(i)}} \sum_{j:d_j(t)=k} f_j^{(i)}}, \tag{13.9}$$

$$f_j^{(i+1)} \leftarrow \frac{\sum_t z_j(t)}{\sum_t \frac{m(t)}{\sum_l f_l^{(i)} A_{d_l(t)}^{(i+1)}} A_{d_j(t)}^{(i+1)}}. \tag{13.10}$$

We can show that this algorithm is also a MM algorithm by a similar argument as in [23]. Thus the log-likelihood in this case is also guaranteed to increase with the number of iterations.

13.4.3 Confidence Intervals and Regularization

Here we make some remarks about the PAFit methodology. First, we can calculate confidence intervals for the estimated values $\hat{A}_k$ and $\hat{f}_i$ using standard statistical theory. Second, binning may be employed to achieve more robust estimation of the attachment kernel. In this paper, we use logarithmic binning. Together with binning, we may add regularization terms to (13.6) or (13.8) for more stable estimation when the data is sparse. For the attachment kernel, we can use the following regularization term:

$$-\frac{\lambda}{\sum_k w_k}\sum_k w_k(\log A_{k+1} + \log A_{k-1} - 2\log A_k)^2 \qquad (13.11)$$

with $w_k = \sum_t m_k(t)$. This regularization term will be approximately 0 if $A_k = k^\alpha$. Estimating A_k with small λ is then equivalent to making almost no assumptions on the functional form of A_k. When λ is large, we then estimate A_k with the assumption that its functional form is k^α. This is reasonable since $A_k = k^\alpha$ is the most frequently assumed functional form in the literature. For each fitness f_j, we can add the following regularization term that has the same effect as placing a gamma distribution prior on f_j:

$$(s-1)\sum_j \log f_j - s\sum_j f_j. \qquad (13.12)$$

The larger s is, the more f_i concentrates around 1. This will reduce the degree of freedom of the parameters, which in turn will help the estimation in the case of sparse data. We can still maximize the penalized log-likelihood functions using MM algorithms [23].

13.5 The Flickr Social Network

In this section we present the results from our analysis of a publicly available Flickr social network dataset [19]. It consists of a simple directed network of friendship relations between Flickr users. Table 13.1 shows some important summary statistics of the dataset.

Table 13.1 Summary statistics for the Flickr social network dataset

Dataset	Type	$\|V\|$	$\|E\|$	T	$\Delta\|V\|$	$\Delta\|E\|$	$\hat{\gamma}$
Flickr [19]	Directed simple	2,302,925	33,140,018	134	815,867	16,105,211	2.15

The numbers $|V|$ and $|E|$ are the total number of nodes and edges in the final network, respectively. Meanwhile, T is the number of observed time-steps, while $\Delta|V|$ and $\Delta|E|$ are the increments of nodes and edges after time $t = 0$, respectively. The value $\hat{\gamma}$ is the scaling exponent of the degree distribution of the final network [5]

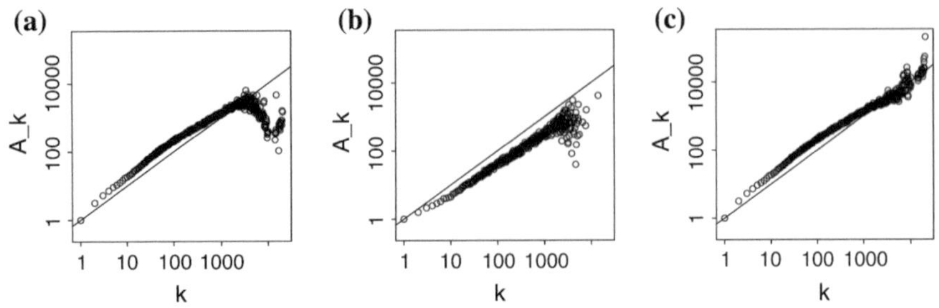

Fig. 13.4 Estimation of the attachment kernel in the Flickr social network dataset without regard to node fitness. The *solid line* corresponding to $A_k = k$ is plotted as a visual guide. **a** Newman's method. **b** Jeong's method. **c** PAFit

13.5.1 Flickr Attachment Kernel Estimation

First we estimate only the attachment kernel while fixing $f_i = 1$ for all i. For Jeong's method, we chose the time window from $T = 1$ to $T = 133$, i.e., using all available data.

In the estimated result of Newman's method (Fig. 13.4a), we once again spotted the falling off the estimated A_k when k is high. We note that when k is small, the estimated A_k of Newman's method are almost identical with those of the proposed PAFit. In this region of small k up to about 1000, while Jeong's method gave a sub-linear function (Fig. 13.4b), PAFit gave a super-linear function (Fig. 13.4c). It is worth noting that we spotted a clear signal of deviation from the log-linear model $A_k = k^\alpha$ (Fig. 13.4c).

13.5.2 Flickr Joint Attachment Kernel and Node Fitness Estimation

Here we use PAFit to estimate jointly the attachment kernel and node fitnesses. For more insights to the estimated fitnesses, we plot the estimated fitness of a node versus the number of new edges that node acquired during the growth of the network. The result is shown in Fig. 13.5. Regarding the attachment kernel, except for the high degree region, the estimated function appears to follow the log-linear model $A_k = k^\alpha$ well with the attachment exponent α is estimated to be 0.89. Regarding the estimated fitnesses, we notice a trend here: a node with high number of acquired edges tends to have high fitness value. This trend is expected since fitness of a node directly contributes to the probability that node will acquire new edges. But we also noticed that some hubs whose the number of acquired edges is large can have much lower fitness than that of a non-hub node. This intriguing fact suggests that in order to

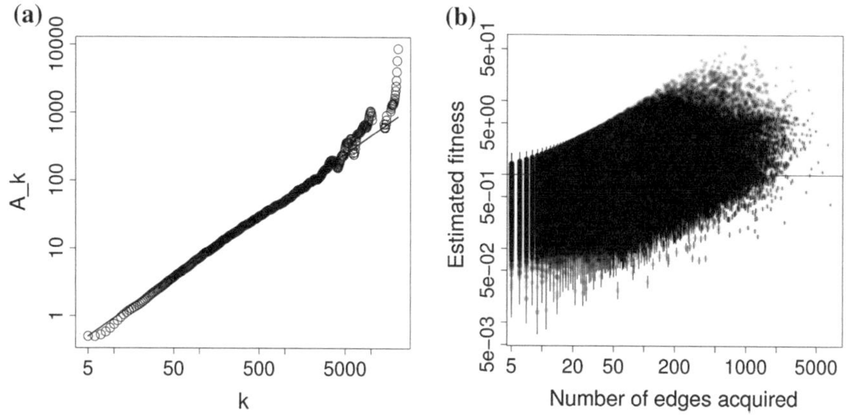

Fig. 13.5 Joint estimation of attachment kernel and node fitness in the Flickr social network dataset. **a** Estimated attachment kernel. **b** Estimated fitnesses

measure the attraction of a node, one must take into account the PA mechanism and the growth process.

13.6 Conclusion

We propose a statistically sound method, called PAFit, for estimating both the attachment kernel (A_k) and node fitnesses (f_i) in temporal networks by maximizing their joint log-likelihood function. Our methodology is nonparametric in the sense that it does not assume any particular functional form for either A_k or f_i, so that it is able to detect different types of functional forms. We report clear evidence for the presence of PA and fitness in the Flickr social network. We also found that the functional form of the attachment kernel differs from the classically assumed log-linear form, $A_k = k^{\alpha}$. What is more, we also observed that hubs do not necessarily have the highest node fitness values, and that even some low degree nodes can have high fitness. Given these interesting discoveries, we expect that PAFit will prove to be a useful tool in the analysis of temporal complex networks.

Acknowledgments This work was supported by grants from Japan Society for the Promotion of Science KAKENHI (Grant numbers 26120523 and 24300106 to Hidetoshi Shimodaira).

References

1. Albert, R., Barabási, A.: Emergence of scaling in random networks. Science **286**, 509–512 (1999)
2. Bianconni, G., Barabási, A.: Competition and multiscaling in evolving networks. Europhys. Lett. **54**, 436 (2001)
3. Caldarelli, G., Capocci, A., De Los Rios, P., Muñoz, M.A.: Scale-free networks from varying vertex intrinsic fitness. Phys. Rev. Lett. **89**, 258702 (2002). http://link.aps.org/doi/10.1103/PhysRevLett.89.258702
4. Capocci, A., Servedio, V., Colaiori, F., Buriol, L., Donato, D., Leonardi, S., Caldarelli, G.: Preferential attachment in the growth of social networks: The internet encyclopedia wikipedia. Phys. Rev. E **74**, 036116 (2006). http://link.aps.org/doi/10.1103/PhysRevE.74.036116
5. Clauset, A., Shalizi, C.R., Newman, M.E.J.: Power-law distributions in empirical data. SIAM Rev. **51**(4), 661–703 (2009). http://dx.doi.org/10.1137/070710111
6. Eom, Y.H., Jeon, C., Jeong, H., Kahng, B.: Evolution of weighted scale-free networks in empirical data. Phys. Rev. E **77**, 056105 (2008). http://link.aps.org/doi/10.1103/PhysRevE.77.056105
7. Erdös, P., Rényi, A.: On random graphs. Publicationes Math. Debrecen **6**, 290–297 (1959)
8. Gómez, V., Kappen, H.J., Kaltenbrunner, A.: Modeling the structure and evolution of discussion cascades. In: Proceedings of the 22Nd ACM Conference on Hypertext and Hypermedia, pp. 181–190. HT '11, ACM, New York, NY, USA (2011). http://doi.acm.org/10.1145/1995966.1995992
9. Herdagdelen, A., Aygn, E., Bingol, H.: A formal treatment of generalized preferential attachment and its empirical validation. EPL (Europhysics Letters) **78**(6), 60007 (2007). http://stacks.iop.org/0295-5075/78/i=6/a=60007
10. Hunter, D., Lange, K.: Quantile regression via an mm algorithm. J. Comput. Graph. Stat 60–77 (2000)
11. Jeong, H., Néda, Z., Barabási, A.: Measuring preferential attachment in evolving networks. Europhys. Lett. **61**(61), 567–572 (2003)
12. Kong, J., Sarshar, N., Roychowdhury, V.: Experience versus talent shapes the structure of the web. Proc. Nat. Acad. Sci. USA **37**, 105 (2008)
13. Kou, Z., Zhang, C.: Reply networks on a bulletin board system. Phys. Rev. E **67**, 036117 (2003)
14. Krapivsky, P., Rodgers, G., Redner, S.: Organization of growing networks. Phys. Rev. E 066123 (2001)
15. Kunegis, J., Blattner, M., Moser, C.: Preferential attachment in online networks: Measurement and explanations. In: WebSci'13. France (May 2013)
16. Lange, K.: Numerical Analysis for Statisticians. Springer, New York (2014)
17. Lu, L., Zhou, T.: Link prediction in complex networks: A survey. Phys. A: Stat. Mech. Appl. **390**(6), 1150– 170 (2011). http://www.sciencedirect.com/science/article/pii/S037843711000991X
18. Massen, C., Jonathan, P.: Preferential attachment during the evolution of a potential energy landscape. J. Chem. Phys. **127**, 114306 (2007)
19. Mislove, A., Koppula, H., Gummadi, K., Druschel, P., Bhattacharjee, B.: Growth of the flickr social network. In: Proceedings of the Workshop on Online Social Networks, pp. 25–30 (2008)
20. Newman, M.: Clustering and preferential attachment in growing networks. Phys. Rev. E **64**(2), 025102 (2001)
21. Newman, M.: Power laws, pareto distributions and zipf's law. Contemp. Phys. **46**, 323–351 (2005)
22. Onodera, T., Sheridan, P.: Maximum likelihood estimation of preferential attachment in growing networks. Topologica **3** (2014)
23. Pham, T., Sheridan, P., Shimodaira, H.: Pafit: A statistical method for measuring preferential attachment in temporal complex networks. PLoS ONE **10**(9), e0137796 (09 2015). http://dx.doi.org/10.1371/journal.pone.0137796

24. Pham, T., Sheridan, P., Shimodaira, H.: PAFit: Nonparametric Estimation of Preferential Attachment and Node Fitness in Temporal Complex Networks (2015). http://cran.r-project.org/package=PAFit (r package version 0.7.5)
25. Price, D.d.S.: Networks of scientific papers. Science **149**(3683), 510–515 (1965). http://www.sciencemag.org/content/149/3683/510.short
26. Price, D.d.S.: A general theory of bibliometric and other cumulative advantage processes. J. Am. Soc. Inf. Sci. **27**, 292–306 (1976)
27. Redner, S.: Citation statistics from 110 years of physical review. Phys. Today **58**(6), 49–54 (2005)
28. Sheridan, P., Yagahara, Y., Shimodaira, H.: Measuring preferential attachment in growing networks with missing-timelines using Markov chain Monte Carlo. Phys. A, Stat. Mech. Appl. **391**, 5031–5040 (2012)
29. Simon, H.A.: On a class of skew distribution functions. Biometrika **42**(3–4), 425–440 (1955). http://biomet.oxfordjournals.org/content/42/3-4/425.short

Chapter 14
N-gram Events for Analysis of Financial Time Series

Igor Borovikov and Michael Sadovsky

Abstract Discretization of time series and encoding it as a string in a finite alphabet allows application of information theory methods developed for discrete signals. Computing information values of n-grams extracted from such string leads to introduction of events as occurrences of n-grams that possess specific properties, e.g. abnormally high (or low) information value. We define information value of an n-gram via maximum entropy lifts over frequency dictionaries. We also look for correlation between market events and n-gram events. The paper shows that the proposed method of time series analysis when applied to events study may provide new insightful perspective.

14.1 Introduction

An analysis of string in finite alphabet produced from financial time series may offer additional benefits enhancing traditional analysis (e. g [1]) on the original real-valued time series. The methods we discuss here targeted to apply to financial time series were inspired by pioneering works in analysis of genetic texts [2–4]. The methods are based on maximum entropy principle. Generality of the underlying principle allows to extend these methods to any strings (texts) from finite alphabets. Genetic texts from four letter alphabet $\aleph = \{A, C, G, T\}$ or natural texts with the usual alphabet(s) are the examples of those sequences. In finance, such alphabet can consist of up A and down a tick: $\aleph = \{A, a\}$.

In this paper, following several previous works [5–7], we obtain string from adjusted close time series by quantizing returns. Next we extract n-grams from the text to build a frequency dictionary (a probability distribution over the n-grams encountered for the given value of n). This step is followed by calculating relative information of particular n-grams and cumulative relative information. It is done

I. Borovikov (✉)
Nekkar.net: Int. Labs, 51 Commons Ln., Foster City, CA 94404, USA
e-mail: igor.borovikov@gmail.com

M. Sadovsky
Institute of Computational Modelling SB RAS, Krasnoyarsk, Russia

S. Battiston et al. (eds.), *Proceedings of ECCS 2014*, Springer Proceedings in Complexity, DOI 10.1007/978-3-319-29228-1_14

by introducing max entropy lift (intuitively it corresponds to "extrapolation") of a dictionary of n-grams of length n to a dictionary of $(n + 1)$-grams. The difference (as Kullback-Leibler divergence) between the actual dictionary of $(n + 1)$-grams extracted from the text and the lifted dictionary is treated as the cumulative relative information. Computing difference between original and lifted frequencies (again as relative information) for individual $(n + 1)$-grams gives their information value (information capacity).

There are some challenges to address when using such approach in finance. The most notable one is noise, which is always present in the data and is part of the reality. The simplest model of time series for returns is Bernoulli process, which is pure noise with single parameter p—the probability of up tick against down tick. Yet it gives quite reasonable representation of reality and was used in historical works by Bachelier [8]. As such, we used Bernoulli process with equivalent parameter p to normalize the data we obtain for information value and cumulative divergence. The memory-less nature of Bernoulli process eliminates any related bias from normalization. Additionally, we aggregated information values for head n-grams (head n-gram is the one found at the end of the string in sliding window) within the limits of so called "noise limit" [5, 6].

The results obtained using these methods appear to be novel and quite insightful at least in the cases we studied. The n-grams techniques are widely used in text mining; in bio-informatics they are proven tool of analysis (references are too numerous to list here). But in the context of the time series, particularly financial ones, the authors couldn't point at directly comparable published works. Specifically, n-grams relative information derived from maximum entropy lifts, normalized and de-noised using aggregation offers new tool for the analysis of time series.

14.2 *N*-grams Dictionaries from Time Series

To avoid ambiguity we shall use the 'ticker' term when talking about securities like company shares, ETFs or indexes (e. g. GOOG or ^DJI). We reserve the terms 'letter' and 'symbols', which we use interchangeable, for the elements of the alphabets we are going to construct.

We consider the simplest case of a financial time series, namely Adjusted Close daily price on a ticker denoted by $z(t)$,[1] from which we calculate either log- or simple returns $p(t)$:

$$p(t) = \log(z(t)/z(t-1)) \approx \frac{z(t)}{z(t-1)} - 1.$$

[1]The source of the data used throughout this work is the publicly available financial data from *Yahoo!*Finance unless indicated otherwise.

Here Adjusted Close price $z(t)$ is a real number and t is (trading) day treated as an integer index. The choice of simple returns over log-returns is not critical for this work so we will not distinguish them.

14.2.1 Constructing Texts from Time Series

To apply the n-grams-based methods, we will represent the time series $p = p(t)$ as a string in an alphabet $\aleph$. We will call this string *input text*. The terms string and text will be used interchangeably. The letters of the alphabet encode quantized values of $p(t)$.

Definition 1 A finite alphabet $\aleph_N$ of size $2N > 0$ is called an output alphabet if it is ordered by bijective mapping to the set of integers $Z_N = \{-N, -(N-1), \ldots, -2, -1, 1, 2, \ldots, N-1, N\}$ (note the absence of 0). This mapping $X\colon Z_N \to \aleph_N$ will be called indexing.

The choice of mapping $R \to \aleph$ and the choice of the alphabet $\aleph$ are parameters of the method. Our main practical choice will be binary quantization (i. e. strings representing only up- and down-ticks) that produces strings in alphabet $\{A, a\}$. Here A corresponds to positive price change on the certain day and a to non-increasing price (i.e. no change or down tick). For the discussion of other alphabets and quantization mappings see [5, 6].

14.2.2 Dictionaries from the Input Text

Given an input text T of a finite length we build natural frequency dictionary $D(n)$ by first counting all n-grams occurrences C_w for each n-gram w in the text T. This results in a set of pairs (w, C_w). Let's denote by C_* the total number of n-grams in T. Obviously, $C_* = |\mathsf{T}| - n$, where $|\mathsf{T}|$ is the text length. Normalization by C_* gives the *frequency* of the n-gram w: $f_w = C_w/C_*$.

Definition 2 The (natural) frequency dictionary $D(n)$ of the text T is the set of all pairs $\{(w, f_w)\}$ where w are unique n-grams and f_w are the corresponding frequencies constructed as described above. The parameter n is called the thickness of the dictionary.

The set $\Omega = \{w\}$ is called the *support* of the dictionary.

A dictionary $D(n)$ of thickness n can be naturally projected to the frequency dictionary $D_1(n)$ of thickness $n-1$ consisting of $(n-1)$-grams and their induced frequencies. More generally, we can compute $D_k(n)$ that is the dictionary of $(n-k)$-grams with their induced frequencies. It is a straightforward procedure that calculates all $(n-k)$-grams and their frequencies not from the original text T but rather from $D(n)$ with proper counting of the corresponding frequencies. This procedure uniquely defines projection operator P_k: $D(n) \to D_k(n)$ for k in the range $0, \dots, n$.

The inverse lifting operator L_k: $D(n) \to D^k(n)$ reconstructs a frequency dictionary $D^k(n)$ of the thickness $n+k$ from the dictionary $D(n)$. It is easy to see that it is not uniquely defined. We can address this by using the maximum entropy principle, see [2–4, 9] for the details and proofs. A brief outline of these results follows in the Sect. 14.2.3.

Note that the original works [2–4, 9] considered circularly looped input texts for the dictionaries generation. Here we can not require any periodicity of the input text because it will create artificial connection between otherwise disconnected trading days at the beginning and at the end of the analyzed time interval. The absence of the loop will create a complication that we will discuss later but for now we will just ignore it. The approximation by the results from the looped texts improves as $|T| \to \infty$. We leave discussion of practical choice of text length outside of this article.

14.2.3 Reconstructed Dictionary and the Information Valued n-grams

Again consider an input text T defined over a finite alphabet $\aleph$. We can construct a sequence of dictionaries D_j of increasing thickness j:

$$D_1 \leftrightarrow D_2 \leftrightarrow \cdots \leftrightarrow D_j \leftrightarrow D_{j+1} \leftrightarrow \cdots \leftrightarrow D_L. \tag{14.1}$$

As we already pointed out, the projection operator (arrows pointing left in (14.1)), i.e., the transition $D_j \mapsto D_{j-1}$ is unique. On the contrary, the lift is not a unique transformation generally because an n-gram w may have multiple valid continuations (not more than the cardinality of the alphabet $|\aleph|$).

A *valid* 1*-lift* is a transformation L_1: $D_j \mapsto W_{j+1}$ such that W_{j+1} is a dictionary of thickness $n+1$ and $P_1(W_{j+1}) = D_j$. So, by definition, a valid 1-lift L_1 satisfies $P_1 \circ L_1 = I$ where I is the identity mapping of D_j. Thus, a lifted ("reconstructed") dictionary consists of n-grams $w \in D_j$ extended by adding prefix or suffix of length 1 and such a way that it projection yields the original frequency dictionary. Note that adding an infix to the original n-grams may not lead to a valid lift. In other words, each combined set $f^*_{\nu_1\nu_2\nu_3\dots\nu_{q-1}\nu_q\nu_{q+1}}$ of the extended n-grams must satisfy the constraint

$$\sum_{\nu_{q+1}} f^*_{\nu_1\nu_2\nu_3\dots\nu_{q-1}\nu_q\nu_{q+1}} = \sum_{\nu_{q+1}} f^*_{\nu_{q+1}\nu_1\nu_2\nu_3\dots\nu_{q-1}\nu_q} = f_{\nu_1\nu_2\nu_3\dots\nu_{q-1}\nu_q}, \tag{14.2}$$

where $f_{\nu_1\nu_2\nu_3\ldots\nu_{q-1}\nu_q}$ is the frequency of an n-gram $w \in D_j$ in the original frequency dictionary D_j. Linear constraints (14.2) eliminate some of the possible extensions for the original n-grams, but still do not define the lift uniquely.

As the final step to define the lift uniquely, we can use the maximum entropy requirement:

$$\max_j \left\{ -\sum_{w^*} f^{(j)}_{w^*} \ln f^{(j)}_{w^*} \right\} \tag{14.3}$$

Here $w^* = \nu_1\nu_2\nu_3\ldots\nu_{q-1}\nu_q\nu_{q+1}$ denotes an n-gram satisfying the linear constraint (14.2). The maximum-entropy dictionary $\widetilde{D}_{q+1}$ satisfying both (14.2) and (14.3) exists always, since the set of the dictionaries which could be constructed from the given one is finite.

The frequency of the n-grams in the max-entropy lift $\widetilde{w} \in \widetilde{D}_{q+1}$ could be computed explicitly using Lagrange multipliers method [2–4, 9]. Frequency of an n-gram $\widetilde{w} \in \widetilde{D}_{q+1}$ is determined by the expression

$$\widetilde{f}_{\nu_1\nu_2\nu_3\ldots\nu_{q-1}\nu_q\nu_{q+1}} = \frac{f_{\nu_1\nu_2\nu_3\ldots\nu_{q-1}\nu_q} f_{\nu_2\nu_3\ldots\nu_{q-1}\nu_q\nu_{q+1}}}{f_{\nu_2\nu_3\ldots\nu_{q-1}\nu_q}}. \tag{14.4}$$

The 1-lift to a thicker dictionary via (14.4) yields the dictionary that contains no "additional" information, external with respect to the one contained in the original dictionary. It consists of the n-grams of the length $q+1$ that are the most probable continuations of the strings of the length q. The lifted dictionary $\widetilde{D}_{q+1}$ contains all the strings that occur in the original dictionary D_{q+1} and, possibly, some other ones. For any $q \geq 1$ the following inequality of the entropy:

$$S\left[\widetilde{D}_{q+1}\right] \geq S\left[D_{q+1}\right]$$

holds true. This approach may be generalized for l-lifts with $l > 1$ also yielding a unique solution. In this paper we focus on 1-lifts. Also in the following we will consider only max-entropy lifts.

14.2.4 Information Capacity as KL Divergence

Here we outline the idea of the information valuable n-grams (see Sect. 14.2.3). Consider two sequences of the frequency dictionaries: the one of the dictionaries constructed directly from the input text (14.1), i.e. the natural dictionaries, and the the sequence

$$\widetilde{D}_2 \leftrightarrow \widetilde{D}_3 \leftrightarrow \cdots \leftrightarrow \widetilde{D}_j \leftrightarrow \widetilde{D}_{j+1} \leftrightarrow \cdots \leftrightarrow \widetilde{D}_L$$

of lifted dictionaries. Here we assume that $\widetilde{D}_j$ is lift of D_{j-1}, $j = 2, \ldots, L$.

Definition 3 Information capacity $\overline{S}_j$ of a natural dictionary D_j is the mutual entropy

$$\overline{S}_j = \sum_{w \in \Omega} f_w \ln \left(\frac{f_w}{\widetilde{f}_w} \right) \tag{14.5}$$

of the natural dictionary D_{j-1} calculated against its lift $\widetilde{D}_j$ from the dictionary D_{j-1}.

The expression (14.5) is also known as Kullback-Leibler divergence, or divergence for short. This definition is applicable to any valid lifts. For the case of (14.4) (max-entropy lift), the information capacity could be easily determined:

$$\overline{S}_j = 2S_{j-1} - S_j - S_{j-2} \quad \text{and} \quad \overline{S}_2 = 2S_1 - S_2, \tag{14.6}$$

where S_j is absolute entropy of the natural dictionary D_j.

14.2.5 Information Valuable (Divergent) n-Grams

Consider again the information capacity (14.5). Sufficiently close values of natural frequencies f_w and lifted frequencies $\widetilde{f}_w$ of the same n-gram w make smaller contribution (per n-gram) to the overall value of the sum. And the n-grams with the greatest deviation provide greater-than-average contribution. This observation motivates the following:

Definition 4 Information valuable n-gram $\widehat{w}$ (an element of the frequency dictionary D_j) is an n-gram satisfying the inequality

$$\left| \log f_{\widehat{w}} - \log \widetilde{f_{\widehat{w}}} \right| > |\log \alpha| ,$$

where $1 \geq \alpha > 0$ is the information value threshold.

We will also call such n-grams α-*divergent* n-grams, or *divergent* n-grams when parameter α is obvious from the context or its specific value is not important.

The subset of the divergent n-grams is complemented with the subset of α-*ordinary* n-grams (or just *ordinary* n-grams). The Definition (4) depends on the parameter α and its practical choice depends on the application. We connect information valuable n-grams with market behavior in Sect. 14.4.

14.2.6 Normalization of the Information Capacity for Finite Input Texts

There are several issues stemming from the finite length of the input texts. The first issue we have to deal with is that some of the formulas derived for looped or infinite input text will hold for finite texts only approximately. In particular, (14.6) would not be a valid formula for information capacity. It is easy to see from its derivation as it relies on the exact match of the frequencies of the n-grams of different length. This assumption holds only for looped or infinite input text. Hence for practical calculations we should use direct formula (14.5).

Another issue is related to the noise resulted from the finite length of the input text affecting the value of the information capacity. Consider for simplicity the binary alphabet with close probabilities for both letters. The total number of *different* n-grams of length n is 2^n. There are total $L - n + 1$ of *all* n-grams of the length n in the input text of the length L. For $L \gg n$ we can take the number of n-grams $\approx L$. If $L = 2^n$ then each occurrence of each n-gram is "critical" in a sense that every difference of lifted dictionary from the natural one will "amplify" the random nature of the input text. When dictionary thickness exceeds $\log_2 L$, some of the n-grams will not be present at all (go "extinct"). This leads to degeneration of the information capacity as many of the longer n-grams will be lifted from the shorter ones uniquely and other will not be present. The number of uniquely reconstructed n-grams will grow as the ratio L/n gets smaller. This results in the bell curve of information capacity discussed earlier. Its peak is located at the value of n close to $j_{\max} \approx \log_2 L$.

To separate the virtual signal from the noise in the information capacity $\overline{S}_n$ of the input text T we need to compare it to the expectation $\mathsf{E}(S'_n)$ of the information capacity S'_n calculated from the randomly generated texts T'. Random texts T' must have the same length and generated by the source with the same probabilities for the alphabet letters as T. The absolute difference of the values S_n and S'_n will not be as useful as the one normalized by the standard deviation $\sigma(S'_n)$ of the information capacity of the corresponding random input texts T'. Such normalization will give σ-distance from the purely random signal. For the binary alphabet the model of purely random signal is Bernoulli process (random walk) with parameter equal to the frequency of up-ticks versus down-ticks. More complicated statistical models can be introduced as the normalization basis: simple auto-regressive models, GARCH (Generalised Autoregressive Conditional Heteroscedasticity) and other [1]. Fixing a model for normalization basis leads to the following:

Definition 5 The normalized information capacity of the input text T is defined by the formula:

$$S_n^* = \frac{\overline{S}_n - \mathsf{E}(S'_n)}{\sigma(S'_n)}. \tag{14.7}$$

where $\overline{S}_n$ is the information capacity of the original input text; $\mathsf{E}(S'_n)$ and $\sigma(S'_n)$ are the expectation and the standard deviation correspondingly of the information capacity of the random input text T' such that $\mathsf{E}(D_1(T')) = D_1(T)$.

The last condition means that the letters in the source of the random texts T' are distributed in the same way as for the original text T. In practice the estimates for the values of $\mathsf{E}(S'_n)$ and $\sigma(S'_n)$ can be computed using Monte-Carlo method.

14.2.7 Sliding Window

Analysis of a single static text information properties obtained from financial data is limited in ability to discover interesting connections between abnormal divergence and information values with market events. In particular, n-grams with high information value did not show any correlation with market events in the case studies in [5]. One of the problems with that approach was exactly use the static nature of the text and the investigated n-grams were located (mostly) in the middle it. Such n-grams can be of interest in analysis of genetic texts but in application to time series they represent "old news", i. e. correspond to the events which market already reacted to and their information properties are not directly related to the current market. The most meaningful n-gram for such analysis is the one at the very end of the text and corresponds to the most recent dates. We call it head n-gram.

From that perspective, consideration of texts obtained form a sliding window seems more appropriate. It may discover connections between information properties of the text in sliding window and head n-gram in particular with the market events. For sliding window, we found that normalization defined in the previous section becomes more important for an obvious reason. The random walk parameter (or parameters of other normalization basis model) will inevitably change between sliding window locations. Unnormalized information properties may not be directly comparable between two neighboring window locations because of that. Normalization using plain Bernoulli model for the head n-gram will be called "naïve" and we will usually drop this adjective.

Thus, we follow the framework: the complete input text T is traversed by smaller sliding window of length m and normalized information value of the head n-gram is denoted as $h = h(t) = h(t;\ n, m)$ for the position t of the window; and the normalized divergence for the text in the window $H = H(t) = H_m(t;\ n, m)$, where parameters n and m will be dropped unless it brings an ambiguity.

Introduction of normalized values opens possibility to apply a trick motivated by processing data with very low signal-to-noise ratio (SNR), and intuitively markets are source of such low SNR signals. Under assumption of i.i.d. noise multiple instances of the input data can be averaged, which improves SNR logarithmically in the number of averaged instances. In finance, we have no luxury of running same experiment multiple times, but we may extract same quantities from the unique instance of the data using different parameters (e. g. n-gram length, number k days used to calculate returns, see below) and average their normalized values. The averaged quantities obtain using such approach will be called "aggregated".

14.3 Aggregated Divergence and Information Value

Normalization introduced earlier removes only some part of the inherent noise present in the computed divergence and information values. It also places the values into a common scale well defined by the mean and variance of the normalization base model. This opens possibility to combine values of $h(t;\ n, m)$ and $H(t;\ n, m)$ over some range of n. The range of aggregation is naturally defined by lower limit of $n_0 = 3$ and the noise limit being the upper limit $n_* \approxeq log(m)$. The following averages will be called an aggregated value of head h-gram:

$$\overline{h}(t; m) = \frac{1}{n_* - n_0 + 1} \sum_{n=n_0}^{n_*} h(t; n, m). \tag{14.8}$$

The aggregated value for divergence $H(t; n, m)$ is defined similarly.

We may take the idea of aggregation a step further: we can use returns for multiple days. Multiple k days returns are defined similarly:

$$p_k(t) = \log(z(t)/z(t-k)).$$

The resulting text is parameterized by k. Instead of single quantity $h(t; m)$ we get parameterized family $h(t; m, k)$ with k covering some "reasonable" range. We may consider averaging over parameter k as well, but for the purpose of this paper we leave this discussion out, see [7].

14.4 Extreme Divergence and Information Value n-Gram Events as Market Indicator

Recall the hinted earlier intuitive interpretation of the information value of n-gram. It tells how far probability to find it in the text diverges from the probability induced by the dictionary of thickness $n-1$. In Markov model approach information value is a measure of how far is the prediction produced by $n-1$ model from the n-order model constructed directly from the text. Note that appealing to Markovian nature of the financial time series is not really necessary as it was not used as an assumption anywhere in any of the derivations. Its only purpose is to illustrate the intuitive meaning of the notions we introduces. By stretching interpretations, aggregation described in the previous section is a cumulative measure of how unexpected is the current market behavior. For head n-gram this corresponds to how far the current moment from following the patterns expected by the participants. Unlike head n-gram information value, divergence is not local and characterizes of the entire text in sliding window.

Aggregated values computed in sliding window may display some notable local maxima and minima, which may provide an insight into inner workings of the market not accessible via conventional methods. They may indicate extreme processes happening in the market. Of course, the exact nature of such processes can't be induced from the quantities we calculate.

14.5 Case Studies

The following case studies use sliding window $m = 256$, that is close to the conventional length of the trading year (250 days). On the figures below the upper plot is adjusted close price of GSPC index (S&P 500), the middle one is aggregate 256 days divergence for 5 days returns, and the lower plot is aggregate head n gram information value with same parameters (window size 256 for 5 days returns). The value $k = 5$ is picked to avoid possible influence of day of the week (regular trading week lasts five day).

14.5.1 *Case Study: Russian Financial Crisis of 1998*

The Russian financial crisis of 1998 significantly affected markets worldwide and the US market in particular. Here we investigate how the US market reflects the events of 1998. We focus on the preceding events and consider only few that happened in the aftermath of Russian default. The key observation is that some, but not all, important events preceding the official default announcement on August 17, 1998 coincide with abnormally high information value head n-grams. The events (in the numbered list, following below) selected for the plot Fig. 14.1 are pulled from the Wikipedia page [10]. It remains to find what were the events that created other noticeable peaks in information value between labels 2 and 4 on the Fig. 14.1; these events are not necessarily tied to the unfolding crisis in Russia and hence are not covered by the cited Wikipedia article [10].

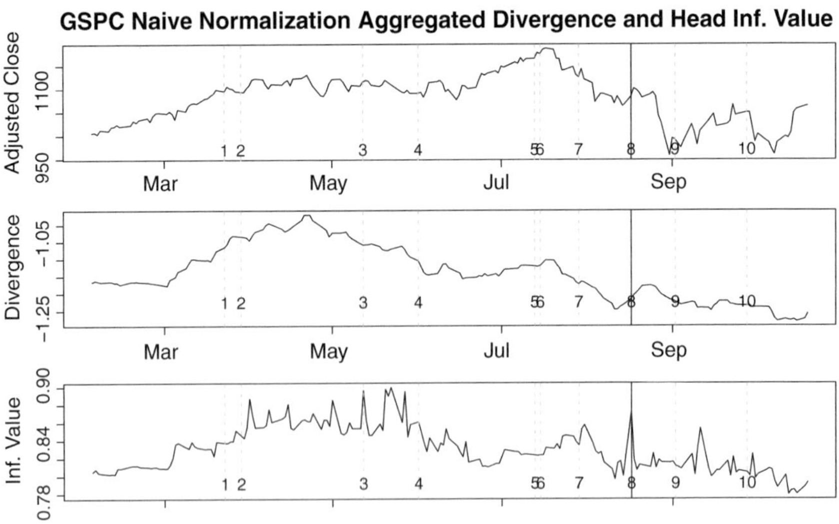

Fig. 14.1 *Solid vertical line* marks the official date of GKO default. *Dashed lines* mark other important events. See text below for details

Some of the Events of Russian GKO Crisis of 1998 cited by [10]:

(1) 1998-03-23 *event*: B. Yeltsin suddenly dismissed Prime Minister V. Chernomyrdin and his entire cabinet. *Comment*: Aggregate head information value starts uptrend, which may be viewed as increase of market breaking previous behavior patterns.
(2) 1998-03-29 *event*: Yeltsin appointed Boris Fyodorov as Head of the State Tax Service. *Comment*: An anticipated or not too important event? There is no obvious connection between several following peaks and any significant political or economical events.
(3) 1998-05-12 *event*: Coal miners went on strike over unpaid wages, blocking the Trans-Siberian Railway. *Comment*: A prominent peak in information value suggests that the event was considered to be both significant and unanticipated.
(4) 1998-06-01 *event*: Kiriyenko hiked GKO interest rates to 150 %. *Comment*: Possibly this action was either anticipated or, less likely, not treated as an important news.
(5) 1998-07-13 *event*: A $22.6 billion International Monetary Fund and World Bank financial package was approved to stabilize the Russian market. *Comment*: Apparently this was an anticipated event with only details being in flux.
(6) 1998-07-15 *event*: The State Duma dominated by left-wing parties refused to adopt most of the government anti-crisis plan. *Comment*: A zero-surprise event given political situation in Russia?
(7) 1998-07-29 *event*: Yeltsin interrupted his vacation and flew to Moscow; replaced Federal Security Service Chief N. Kovalyov with V. Putin *Comment*: Was it not anticipated or not deemed important? The following peak probably reflects post-factum digestion of Yeltsin's actions by market participants.
(8) **1998-08-17** *event*: **The official date of Russian default crisis** (solid vertical line in Fig. 14.1). Note, the previous date is Sunday, making the change in information value abrupt. *Comment*: An obvious peak suggests the event was not anticipated while very important. The market didn't move just yet.
(9) 1998-09-02 *event*: The Central Bank of the Russian Federation decided to abandon the "floating peg" policy and float the ruble freely. *Comment*: Probably not a very surprising event given the circumstance?
(10) 1998-09-28 *event*: Boris Fyodorov was discharged from the position of the Head of the State Tax Service. *Comment*: A minor event given the scale of the crisis?

14.5.2 Case Study: 9–11 Terrorist Act

We used the same methodology as in Sect. 14.5.1 to probe into the market behavior before 9–11 terrorist act. There are no obvious preceding events that could be revealed by high information value or cumulative divergence, but there is a steady growth of head n-grams information value and even less disputable growth of aggregate

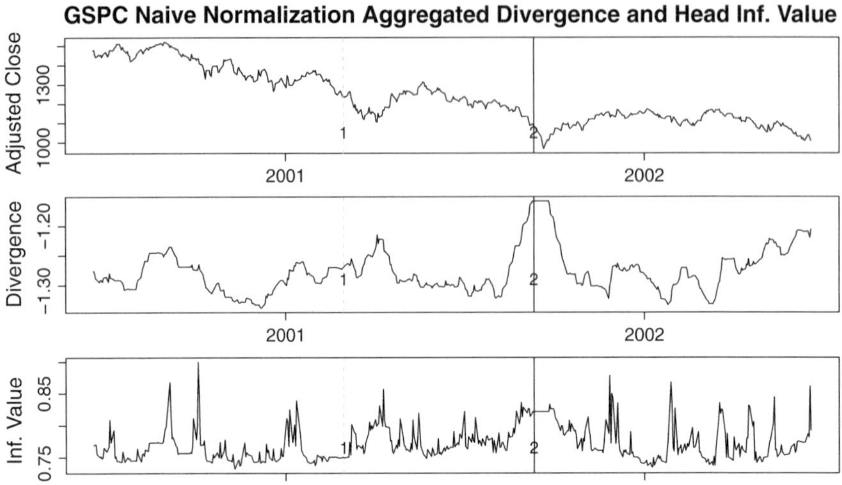

Fig. 14.2 *Solid vertical line* marks the date of 9–11 terrorist act. An approximate date of the peak of the US business activity marking the beginning of the recession is marked as *dashed line 1* [11]. Note steady growth of aggregate divergence and head information value in the preceding days

divergence in the days preceding the terrorist act, see Fig. 14.2. These observations suggest that some market participants were either aware of or were affected by the trades placed by fully informed of the upcoming events traders. Since the number of informed participants was unlikely large it suggests that the fully or partially informed participants were sufficiently influential ones and acting with high caution (i. e. avoiding direct price shift). The absence of abnormally high information value of head n-gram also suggest that the event was not a total surprise, at least in terms of market behavior of some of its participants. Of course, all observed effects were amplified by the recession on which background 9–11 occurred. These proposed interpretations are not intended to support any of the popular conspiracy theories built around 9–11. However they offer a circumstantial evidence that the tragic event didn't occur in total information vacuum.

14.6 Discussion and Conclusion

One of the challenges of the proposed approach is the choice of meaningful normalization. It may be argued that Bernoulli process, while being attractive for its simplicity, provides only very crude model of the background noise. As better alternative, we may contemplate using boundary conditions for the bootstrapped series to enforce the same leading n-gram, or use Markov process instead, or both. We leave investigation of these options to future publications.

Additional applications of the approach could be in the area of general change (or fault) detection in various time series. One specific practical example that we consider is EEG analysis for early detection of onsetting epileptic seizure.

The goal of this research (which is still work in progress) is to evaluate if n-grams information theory offers new insights into market behavior. We described the developed methodology and some of the preliminary results. In the presented case study we found some correlation between market events and abnormal behavior of computed information values. One possible interpretation is the following: the abnormal values observed before an event correspond to the market participants processing the event in its anticipation; the events corresponding to the abnormal information values in the immediate wake of the event suggest that the event was not anticipated and was unexpected.

References

1. Tsay, R.S.: Analysis of Financial Time Series, p. 448. Inc, Financial Econometrics. Wiley & Sons (2002)
2. Bugaenko, N.N., Gorban, A.N., Sadovsky, M.G.: Towards the definition of information content of nucleotide sequences. Mol. Biol. Moscow **30**(5), 529–541 (1996)
3. Bugaenko, N.N., Gorban, A.N., Sadovsky, M.G.: The information capacity of nucleotide sequences and their fragments. Biophysics **5**, 1063–1069 (1997)
4. Bugaenko, N.N., Gorban, A.N., Sadovsky, M.G.: Maximum entropy method in analysis of genetic text and measurement of its information content. Open Syst. Inf. Dyn. **5**(2), 265–278 (1998)
5. Borovikov, I., Sadovsky, M.: A relative information approach to financial time series analysis using binary N-grams dictionaries; arXiv:1308.2732 [q-fin.ST] (2013) 13 pp
6. Sadovsky, M.G., Borovikov, I.: Analysis of financial time series with binary n-grams frequency dictionaries. J. Siberian Fed. Univ., Math. Phys. **7**(1), 112–123 (2014)
7. Borovikov, I., Sadovsky, M.: Sliding Window Analysis of Binary n-Grams Relative Information for Financial Time Series, LLNL CASIS proceedings (2014). https://casis.llnl.gov/content/pages/casis-2014/docs/poster/Borovikov-CASIS-2014.pdf
8. Bachelier, L., Théorie de la spéculation. Annales Scientifiques de l'École Normale Supérieure **3**(17), 21–86
9. Hu, R., Bin, W.: Statistically significant strings are related to regulatory elements in the promoter regions of Saccharomyces cerevisiae. Physica A **290**, 464–474 (2001)
10. 1998 Russian financial crisis, Wikipedia, the free online encyclopedia. http://en.wikipedia.org/wiki/1998_Russian_financial_crisis
11. Early 2000s recession, Wikipedia, the free online encyclopedia. http://en.wikipedia.org/wiki/Early_2000s_recession

Chapter 15
Human Mobility and the Dynamics of Measles in Large Geographical Areas

Ramona Marguta and Andrea Parisi

Abstract In recent years the global nature of epidemic spread has become a well established fact, however there have been limited studies on the detailed propagation of infectious diseases on regional scales. We have recently introduced a simulation program that explores disease propagation on such scales: the model uses a gridded geographical description of human settlements on top of which mobility is implemented using the Radiation Model. Parallel computation permits unlimited complexity. Both individual and equation based simulations of epidemiological models can be performed, thus permitting the exploration of the effects of mobility locally and globally. Using a SIR model parametrized for measles, we perform simulations for the area of British Isles, which we assume isolated. Exploring how the dynamics is influenced by human mobility, we show that mobility influences the dynamics globally and locally. In particular, the interplay of mobility and city size, enhances or reduces the contribution of the different mechanisms involved.

15.1 Introduction

Several publications in recent years have been devoted to the study of the geographical spread of infectious diseases: contact networks are used at many levels of the description of several epidemic models [1–4]; for instance Colizza et al. use a sophisticated geographical model in which they use the airline transportation network to make prediction on the worldwide spread of flu epidemics [4–6]. Simulations on regional level however require data on mobility at lower scales, and have been developed more recently thanks to data coming from mobile phone operators [7]. Here we describe a simulation model [8] that is based on a recently developed description of human mobility [9] and uses it on a regional scale to analyze the spread of measles.

R. Marguta · A. Parisi (✉)
BioISI-Biosystems Integrative Sciences Institute and Departamento de Física, Faculdade de Ciências da Universidade de Lisboa, 1749-016 Campo Grande, Lisbon, Portugal
e-mail: andreaparisi.physics@gmail.com

S. Battiston et al. (eds.), *Proceedings of ECCS 2014*, Springer Proceedings in Complexity, DOI 10.1007/978-3-319-29228-1_15

Measles is a well studied disease: several publications have been devoted to its dynamics and the mechanisms that control its periodic outbreaks [10–13], but the *human mobility* factor has not been thoroughly explored so far. However, some key mechanisms depend on mobility: for instance mobility is a key ingredient in explaining the recurrent outbreaks occurring in low populated areas [14, 15]. By considering a detailed geographical description of human settlements as well as human mobility, we have been able to uncover the link between human mobility, the frequency of measles outbreaks and the mechanisms that control its dynamics. Specifically we see that different mechanisms are at work depending of the intensity of human mobility and the size of the local population. We also see that, when we consider the whole British Isles, the resulting sequence of outbreaks might have characteristics that differ from what is observed locally: in other words, what is observed at the global level has a complex relation with what occurs in the multiplicity of cities and low populated areas that constitute the British Isles.

15.2 The SIR Epidemiological Model

The basic compartmental models used to describe disease dynamics divide the population into different classes according to their epidemiological status. The SIR model, describes many infectious diseases including measles, rubella and mumps. It uses three classes: susceptible to the disease (S), infected (I) and recovered (R). Infected individuals can transmit the disease to other susceptibles, whereas recovered individuals have a lifelong immunity to the disease. Three processes alter the status of an individual: transmission of the disease from an infective to a susceptible individual, recovery of an infected individual and demography that acts by replacing individuals regardless of their epidemic status with new susceptible individuals. The three processes are thus:

$$\begin{array}{rl} SI \xrightarrow{\beta} II & \text{transmission} \\ I \xrightarrow{\gamma} R & \text{recovery} \\ \{S, I, R\} \xrightarrow{\mu} S & \text{demographic death/birth} \end{array}$$

where the three parameters β, γ and μ are the rates at which these events occur. The analytic description of these models, known as deterministic description, is provided by a set of coupled differential equations describing the time evolution of the average number of individuals that belong to each class and can be formally derived as the mean field limit of the master equation describing the above processes [16]. In this case the SIR model has the form:

$$\begin{cases} \dot{S} = -\beta SI/N + \mu(N - S) \\ \dot{I} = \beta SI/N - (\gamma + \mu)I \end{cases}$$

where S, I, R represent the number of individuals in the corresponding class: the population size $N = S + I + R$ is constant, and the equation for R is hence easily obtained. This formulation uses a single infective class: this corresponds to a distribution of recovery times that is exponential. Instead of a single infective class, it is possible to use $L > 1$ infective classes: this has the effect of altering the distribution of recovery times making it gamma distributed [17], which provides a more realistic description of recovery from a disease. Seasonal effects, including climate signals or school opening and closing, are simulated using a time dependent transmission rate.

15.3 A Model for Geographical Spread

The underlying geographical description uses a gridded map of the human geographical distribution based on the Gridded Population of the World database (GPW) [18] which provides estimates of resident population with a resolution of 2.5 arc-minutes. The world is described as a collection of cells, each corresponding to a square in which an estimate of the resident population is provided. When a cell has a fractional number of individuals, we assign to the cell a number of individuals corresponding to the integer part, plus one individual with probability equal to the fractional part. This means that if four cells have an estimated number of individuals equal to 1.25 each, we assign 1 individual per cell plus one individual in each cell with probability 0.25. This insures that on average the four cells will have 5 individuals.

Individuals are moved among cells according to the fluxes predicted by the radiation model [9]. This model gives the flux of commuters between two cells i and j with populations m_i and n_j as:

$$T_{ij} = T_i \frac{m_i n_j}{(m_i + s_{ij})(m_i + n_j + s_{ij})}$$

where s_{ij} is the number of individuals living in an area of radius r_{ij} (the distance between the i and j cells) with the exclusion of n_i and m_j, and T_i is the total flux of commuters leaving location i and is given by:

$$T_i = m_i \frac{N_c}{N}$$

with N_c/N being the fraction of the total number of commuters with respect to the total population. This ratio is the only parameter of the mobility description and can be tuned to simulate different levels of mobility.

Finally, each cell is described as a well mixed population and a standard epidemiological SIR model is implemented. Its specific implementation depends on whether we wish to perform individual based or equation based simulations.

15.3.1 Individual Based Implementation

In the individual based implementation [8], the population of each cell is described as a set of distinct individuals, each storing information on its epidemiological status, current location and home location. Each individual is able to move from its home location to other locations, and the probability of moving depends on the fluxes between cells. Each individual has a set of preferred locations which is built by sampling the probability distribution of moving to different cells according to the fluxes predicted by the radiation model. The sampling procedure also determines the probability of visit to each preferred location.

Each day, each individual chooses one of his preferred locations and, if this preferred location is not his home cell, he is moved there. There he participates to the local dynamics. At the end of the day he is moved back to his home location and a new day elapses. Given that in each cell individuals leave for other locations and new individuals arrive from other locations, all individuals ending up in the same location will participate to the dynamics of the cell for that day. The dynamics follows the SIR model and is implemented using a Gillespie algorithm for one day [19].

15.3.2 Equation Based Model

The equation based version of the simulation model uses the same geographical description, with the difference that in each cell we do not generate a set of distinct individuals. Rather, in each cell we use a compartmental description of the resident population by storing the number of resident individuals belonging to each epidemiological class. Each individual is supposed to have his own set of preferred locations, however since in this version the notion of distinct individuals is lost, we build a common set of preferred location for the individuals of each cell following the observation that most individuals in a cell will have similar preferred locations. The set is built as follows: for each cell we produce a set of preferred locations for each individual following the same procedure implemented for the individual based implementation. These locations are then merged into a single set with corresponding probability of visit which is supposed to be valid at the level of the cell. It is this set that will be used to move fractions of the population belonging to each epidemiological class to different cells.

Mobility occurs once per day. Movement of individuals is simulated by moving fractions of the number of individuals among cells according to the probabilities of visit of the various preferred locations. When individuals move to a new cell they do not fully mix: the idea here is that the same group of individual will be moved back to their home cell, therefore in the new cell we keep track of where the various fractions came from. The result of this mobility phase is that in each cell, the population will be composed of individuals belonging to different epidemiological classes and coming from different cells.

Let us consider a SIR model and a specific cell i. Each individual of the cell belongs to a specific epidemiological class $K_i \in \{S_i, I_i, R_i\}$. We call p_{ij} the probability of an individual to move to cell j, with p_{ii} the probability of remaining in cell i. Thus the number of individuals K_{ij} in class K moving from cell i to cell j is given by $K_{ij} = K_i p_{ij}$. Thus the number of individuals in class K in cell i during the mobility phase will be given by the population of the cell that did not move to other cells, augmented by individuals coming from other cells:

$$N_i = \sum_j \sum_{K \in \{S,I,R\}} K_{ji}$$

Using this formulation, individuals of each class will interact with all other individuals within the cell; hence the SIR model takes the form:

$$\begin{aligned} S_{ji} &= -\beta S_{ji} \frac{I_i}{N_i} + \mu(N_i - S_{ji}) \\ I_{ji} &= \beta S_{ji} \frac{I_i}{N_i} - (\gamma + \mu) I_{ji} \end{aligned}$$

where $I_i = \sum_j I_{ji}$. The equations are integrated for one day using a fourth order Runge-Kutta algorithm, and then individuals are moved and mixed back to their home locations.

15.4 Parallelization

The parallelization of the simulation program allows unlimited complexity to be included in the simulation in the form of large geographical areas or complex epidemiological modelling. Parallelization of the simulation program is based on the observation that individuals move mostly to nearby locations, while long-distance trips are rare: the radiation model reflects this characteristic of the mobility of individuals. As a result, to parallelize the simulation it is sufficient to find an efficient partitioning of the gridded map into regions that can be fed to different computing units. The algorithm that performs this partitioning procedure is based on simulated annealing [20] but uses a coarsening technique that permits high efficiency [8]. In practice, the original map is coarsed by merging neighbouring grid cells in groups of four into larger square grid cells. This coarsening is repeated multiple times until a coarsed version of the original map consisting only of a few hundred coarsed cells is obtained. The map is partitioned into a set of connected regions and a simulated annealing algorithm modifies the position of the frontier between regions in order to find the optimal partition. Since the map consists only of a few hundred grid cells, the optimal partition is rapidly found. At this point, the map is decoarsened one level, by recovering the constituting four grid cells and the final part of a simulated annealing is repeated to adapt the optimal solution to the new map resolution. This cycle is repeated several times, until the original map is recovered.

The constraints under which simulated annealing is performed depend on the specific implementation: for individual based simulations, an optimal solution that partitions the map into regions with similar population size and compact shape is sought. For equation based simulations, the optimal solution must support a similar number of populated cells (i.e. with non-zero population) and must be compact in shape. The difference is due to the fact that for equation based simulations we integrate a set of differential equations only in populated cells, and this integration does not depend on the number of individuals.

The two algorithms are not optimal because they do not directly minimize transmission between neighbouring regions, however they provide a good approximation of an optimal partitioning.

15.5 Results

We consider the seasonal forced SIR epidemiological model, with multiple infective classes ($L = 2$). Our analysis is limited to individual based simulations: we have not at the moment exploited the equation based description, except for verification of the endemic equilibria. Seasonality is included using a term time forcing [10], $\beta(t) = \beta_0\,[1 + \beta_1\,\mathrm{Term}(t)]$, where $\mathrm{Term}(t)$ takes value $+1$ during school opening and -1 during school vacations. Parameters for measles are taken from literature [11] and shown in Table 15.1.

In Fig. 15.1 we show the global behavior for the British Isles, for different values of N_c/N. For a mobility ratio of $N_c/N = 0.001$ the outbreaks are multi-annual, with an interval of several years between the major peaks. As we increase the fraction of commuters, the dynamics change to annual cycles for $N_c/N = 0.05$ and shows biennial cycles for $N_c/N = 0.2$. These results point out that the global behavior is strongly influenced by the mobility of individuals.

At the level of cities, the dynamics is different for high populated and low populated areas. For the city of London the mobility of individuals has less impact on the dynamics as shown in Fig. 15.2 for (a) $N_c/N = 0.05$ and (b) $N_c/N = 0.2$: in both cases we observe biennial cycles. In the case of a non-seasonal simulation (c) the resulting time series presents a substantially different behaviour from the typical time series observed for London [10]: this shows that seasonality is a key factor for the dynamics of high populated areas.

Table 15.1 Parameters for SIR model

β_0	γ	μ	β_1
$1.175\,\mathrm{days}^{-1}$	$1/13\,\mathrm{days}^{-1}$	$5.5 \times 10^{-5}\,\mathrm{days}^{-1}$	0.25

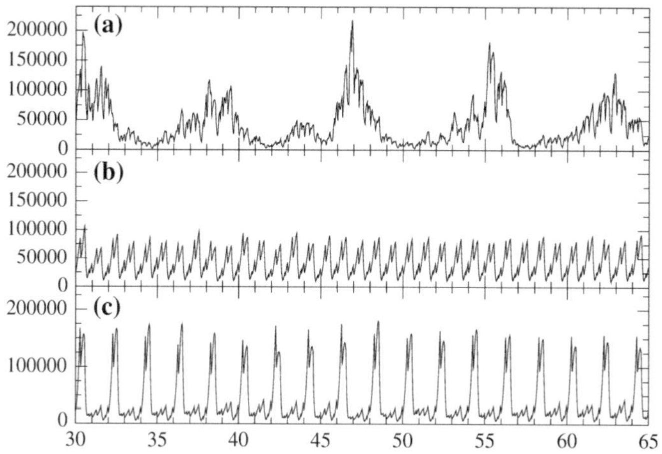

Fig. 15.1 Infective incidence for the whole British Isles as a function of time. From *top* to *bottom*: **a** $N_c/N = 0.001$, **b** $N_c/N = 0.05$ and **c** $N_c/N = 0.2$

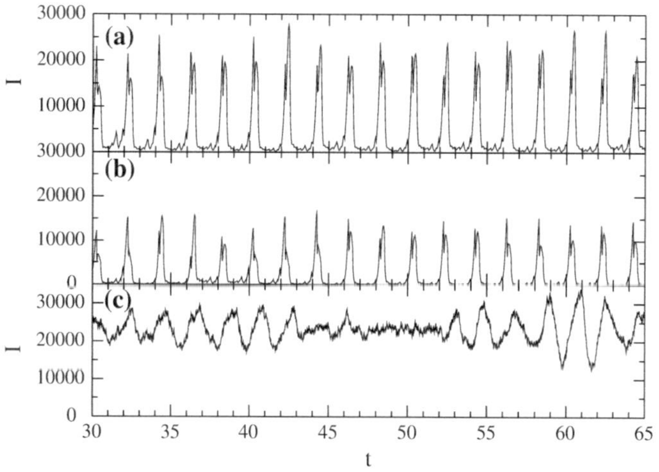

Fig. 15.2 Infective incidence as a function of time for the city of London, for different mobility ratios: **a** $N_c/N = 0.05$; **b** $N_c/N = 0.2$. **c** Non-seasonal ($\beta_1 = 0$) simulation for $N_c/N = 0.2$

For low populated areas, like in the case of Chester, mobility of individuals influences the dynamics due to extinction and reinfection as shown in Fig. 15.3: the top plot (a) corresponding to $N_c/N = 0.05$ shows outbreaks roughly triennial, while the periodicity is slightly reduced to 2.5 years for $N_c/N = 0.1$ (b). The last plot shows non-seasonal ($\beta_1 = 0$) simulations for $N_c/N = 0.05$: these simulations are very similar to case (a) in term of behaviour and frequency of outbreaks: this suggests that seasonality here has essentially no effect.

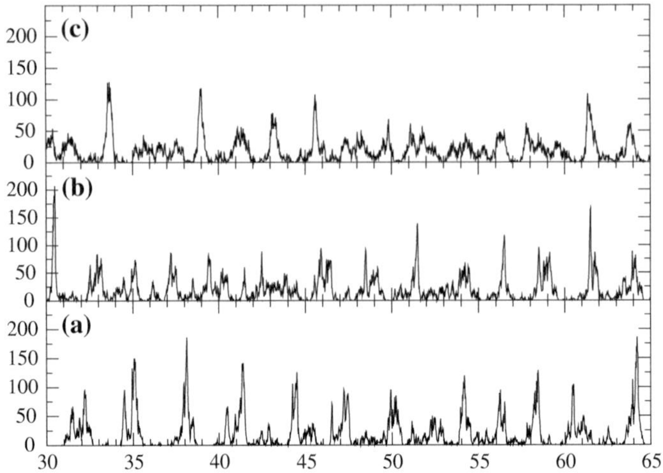

Fig. 15.3 Infective incidence as a function of time for the city of Chester: **a** $N_c/N = 0.05$; **b** $N_c/N = 0.1$. **c** Non-seasonal ($\beta_1 = 0$) simulation for $N_c/N = 0.1$

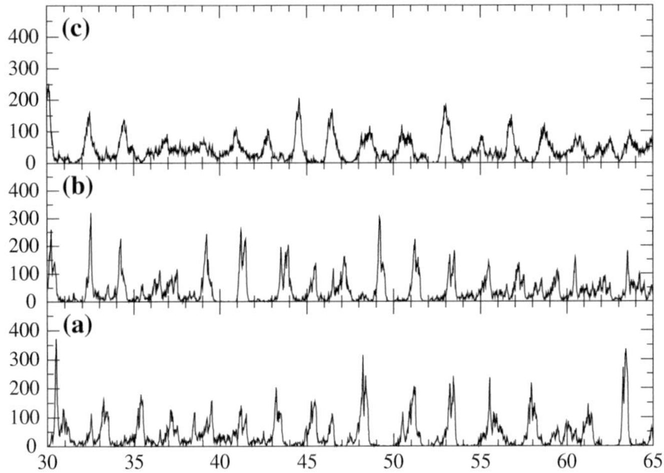

Fig. 15.4 Infective incidence as a function of time for the city of York: **a** $N_c/N = 0.05$; **b** $N_c/N = 0.1$. **c** Non-seasonal ($\beta_1 = 0$) simulation for $N_c/N = 0.1$

Intermediate sized cities are subjected to the amplification of stochastic fluctuations: in Fig. 15.4 we show the time series for the seasonal case for two values of mobility: (a) $N_c/N = 0.05$ and (b) $N_c/N = 0.1$. Non-seasonal ($\beta_1 = 0$) time series are shown in (c) for $N_c/N = 0.1$: the two time series in (b) and (c) show similar periodicity. An analysis using power spectra shows that indeed the location of the stochastic peak is not influenced by the seasonal forcing (Fig. 15.5).

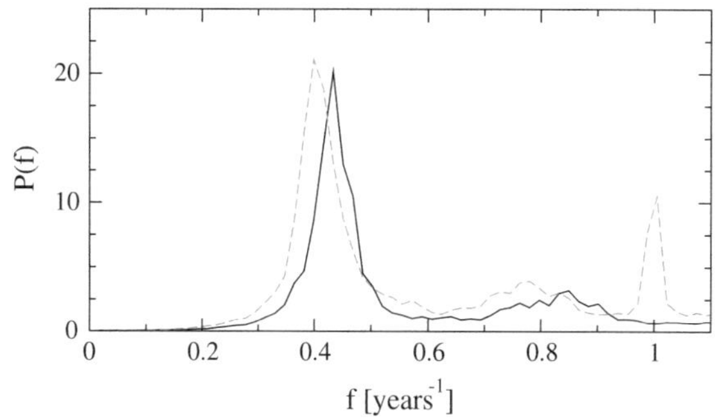

Fig. 15.5 Power spectrum averaged over 100 simulations for the time series of the city of York, for $N_c/N = 0.05$

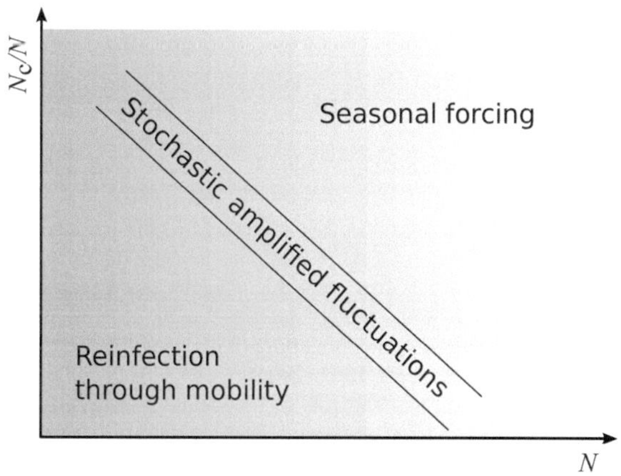

Fig. 15.6 The interplay between human mobility and population size

The results discussed can be summed up into a schematic representation of the interplay of population size and mobility ratio on the mechanisms involved in the propagation of measles (Fig. 15.6). For a given mobility level the three mechanisms can be found at work, with seasonality being the major mechanism in highly populated locations, and mobility mediated reinfection, after extinction, in low populated areas. In between these two regimes, stochastic amplification is found for locations with populations of intermediate sizes. Mobility however changes the importance of the mechanism by shifting the population sizes at which these mechanisms are predominant.

15.6 Conclusions

Our computer model describes the geographical distribution of human population using gridded maps, with each grid element representing a well mixed population: the disease evolves according to a given epidemiological model. The simulations can be individual based or equation based: in both cases individuals commute to different grid elements where they can eventually be infected or transmit the disease, thus participating in the long-range transmission of the disease.

We have analyzed individual based simulations of measles spread on the British Isles in conditions corresponding to those of the pre-vaccination period. Our results show that the dynamics of the disease is influenced by the intensity of human mobility both globally and locally. In particular, different mechanisms are at work depending on the level of human mobility and the local population size.

Acknowledgments The authors acknowledge funding from the Fundação para a Ciência e a Tecnologia (FCT) under contract no. PTDC/SAU-EPI/112179/2009, and centre grant (to BioISI, centre reference: UID/MULTI/04046/2013), obtained from FCT/MCTES/PIDDAC, Portugal.

References

1. Dunyak, J., Martin, C., Lampe, R.: Analysis of the influence of social structure on a measles epidemic. Appl. Math. Comput. **92**, 283 (1988)
2. Sander, L.M., Warren, C.P., Sokolov, I.M.: Epidemics, disorder and percolation. Phys. A **325**, 1 (2003)
3. Eubank, S., et al.: Structure of social contact networks and their impact on epidemics. DIMACS Ser. Discrete Math. Theor. Comput. Sci. **70**, 181 (2010)
4. Colizza, V., et al.: The modelling of global epidemics: stochastic dynamics and predictability. Bull. Math. Biol. **68**, 1893 (2006)
5. Colizza, V., et al.: The role of the airline transportation network in the prediction and predictability of global epidemics. PNAS **103**, 2015 (2006)
6. Colizza, V., et al.: Predictability and epidemic pathways in global outbreaks of infectious diseases: the SARS case study. BMC Med. **5**, 34 (2007)
7. Gonzalez, M.C., Hidalgo, C.A., Barabasi, A.L.: Understanding individual human mobility patterns. Nature **453**, 779 (2008)
8. Marguta, R., Parisi, A.: Impact of human mobility on the periodicities and mechanisms underlying measles dynamics. J. R. Soc. Interface **7**, 20141317 (2015). doi:10.1098/rsif.2014.1317
9. Simini, F., Gonzalez, M.C., Maritan, A., Barabasi, A.L.: A universal model for mobility and migration patterns. Nature **484**, 96 (2012)
10. Keeling, M.J., Rohani, P., Grenfell, B.T.: Seasonally forced disease dynamics explored as switching between attractors. Phys. D **148**, 317 (2001)
11. Alonso, D., McKane, A.J., Pascual, M.: Stochastic amplification in epidemics. J. R. Soc. Interface **4**, 575 (2007)
12. Black, A.J., McKane, A.J., Nunes, A., Parisi, A.: Stochastic fluctuations in the susceptible-infective-recovered model with distributed infectious periods. Phys. Rev. E **80**, 021922 (2009)
13. Bjornstad, O.N., Finkenstädt, B.F., Grenfell, B.T.: Endemic and epidemic dynamics of measles: Estimating epidemiological scaling with a time series SIR model. Ecol. Monogr. **72**, 169 (2002)
14. Cliff, A.D., Haggett, P.: Changes in the seasonal incidence ofmeasles in Iceland, 18961974. J. Hyg. Camb. **85**, 451 (1980)
15. Cliff, A.D.: Spatial Diffusion: an Historical Geography of Epidemics in an Island Community. Cambridge University Press, Cambridge (1981)
16. Alonso, D., McKane, A.J., Pascual, M.: Stochastic amplification in epidemics. J. R. Soc. Interface **4**, 575 (2007)
17. Lloyd., A.: Destabilization of epidemic models with the inclusion of realistic distributions of infectious periods. Proc. R. Soc. Lond. B **268**, 985 (2001)

18. Center for International Earth Science Information Network (CIESIN)/Columbia University, United Nations Food and Agriculture Programme (FAO), and Centro Internacional de Agricultura Tropical (CIAT).: Gridded Population of the World, Version 3 (GPWv3): Population Count Grid. Palisades, NY: NASA Socioeconomic Data and Applications Center (SEDAC) (2005)
19. Gillespie, D.T.: A general method for numerically simulating the stochastic time evolution of coupled chemical reactions. J. Comput. Phys. **22**, 403 (1976)
20. Kirkpatrick, S., Gelatt, C.D., Vecchi, M.P.: Optimization by Simulated Annealing. Science **220**, 671 (1983)

Chapter 16
Does Training Lead to the Formation of Modules in Threshold Networks?

D. Nicolay, A. Roli and T. Carletti

Abstract This paper addresses the question to determine the necessary conditions for the emergence of modules in the framework of artificial evolution. In particular, threshold networks are trained as controllers for robots able to perform two different tasks at the same time. It is shown that modules do not emerge under a wide set of conditions in our experimental framework. This finding supports the hypothesis that the emergence of modularity indeed depends upon the algorithm used for artificial evolution and the characteristics of the tasks.

16.1 Introduction

Modularity is a widespread feature of biological and artificial networks such as animal brains, protein interactions and robot controllers. This feature makes networks more easily evolvable, i.e. capable of rapidly adapting to new environments and offers computational advantages. Indeed, an intuitive idea is that it is easier and less costly to rewire functional subunits in modular networks. Despite its advantages, modularity remains a controversial issue, with disagreement concerning the nature of the modules that exist as well as over the reason of their appearance in real networks [16]. Moreover, there is no consensus concerning the conditions for their emergence.

Whereas most hypotheses assume indirect selection for evolvability, Bullinaria [3] suggests that the emergence of modularity might depend on different external factors such as the learning algorithm, the effect of physical constraints and the tasks to learn. Clune et al. [4] also claimed that the pressure to reduce the cost of connections between network nodes causes the emergence of modular networks. On their side, Kashtan and Alon [8] found that switching between several goals leads to the spontaneous evolution of modular networks.

D. Nicolay · T. Carletti (✉)
Department of Mathematics and naXys, University of Namur, Namur, Belgium
e-mail: timoteo.carletti@unamur.be

A. Roli
Department of Computer Science and Engineering (DISI), University of Bologna, Campus of Cesena, Bologna, Italy

S. Battiston et al. (eds.), *Proceedings of ECCS 2014*, Springer Proceedings in Complexity, DOI 10.1007/978-3-319-29228-1_16

In this work, we study the emergence of modularity in the field of evolutionary robotics. First of all we remark that there is not a uniquely accepted definition of modularity; moreover, the existence of many different definitions, each one appropriate for different levels of abstraction [6, 13], makes this question even more intricated. We thus decided to analyse our results by considering two kinds of modularity, namely *topological modularity*, which is a measure of the density of links inside modules as compared to links between modules, and *functional modularity*, which groups together neurons that have similar dynamic behaviours.

The case under study consists in learning conflicting tasks where robots controllers are realised as neural networks. The learning phase is performed using a genetic algorithm that optimises both network structure and weights. Our starting working assumption is that only two conditions are needed for the emergence of modularity. Firstly, at least two tasks should be learnt. Secondly, the learning must be incremental, i.e. the modifications in both topology, weights and activation thresholds must be gradual and the structure of the networks can not be too strongly modified in one step. Let us also remark that the outcome of the learning process is path dependent as the learning algorithm is heuristic. Because the results obtained from the first assumption were unsatisfactory we decided to improve our working assumption by considering: switching between the tasks learning, cost of the connections and cost of the nodes, and thus to study their impact on the networks evolution.

Because we were not able to detect any kind of modules in all the performed experiments, our findings support the hypothesis that the emergence of modularity is not exclusively conditioned by the learning conditions but also depends upon the algorithm used for artificial evolution and the characteristics of the tasks. Furthermore, we conjecture that the computational nature of the tasks, namely combinatorial or sequential—i.e. requiring memory to be accomplished—may also play a role in the emergence of modularity.

The paper is organised as follows. In Sect. 16.2, we present our experimental settings, namely our model of networks, the tasks the robot has to perform and our learning algorithm. Experiments and results are described in Sect. 16.3. Section 16.4 concludes the contribution with a summary of our results.

16.2 Model and Tasks Description

The abstract application we focussed on is based on the experimental framework introduced by Beaumont in [1]. Virtual robots are trained to achieve two different tasks in a virtual arena. This arena is a discretised grid with a 2 dimensional torus topology on which robots are allowed to move into any neighbouring cell at distance 1 at each displacement. Assuming the arena possesses one global maximum, the aim of the first task, task A in the following, is to reach and to stay on this global maximum. The second task, task B, consists in moving incessantly by avoiding zones where robots lose energy, called "dangerous zones". Let us observe that such tasks are conflicting, because the first task will imply the robot to reach the peak and stay there, while the second would rather make the robot to wander around the arena.

The robots we considered have 17 sensors and 4 motors. The 9 first sensors check the local slope, that is the heights of the cell on which the robot is located and the cells around it. The 8 others are used to detect the presence of "dangerous zones" on the cells surrounding the robots. The 4 motors control the movements of robots, i.e. moving to the north-south-east and west and their combinations.

We used neural networks [11, 12] as robot controllers. They are made of 43 nodes: 17 inputs, 4 outputs and 22 hidden neurons. This number is large enough to let sufficient possibilities of connections to achieve the tasks, but it is still low to use reasonably level of CPU resources. The topology of these networks is completely unconstrained, except for self-loops that are prohibited. The weights and thresholds are real values between -1 and 1. The states of the neurons are binary and the updates are performed following the perceptron rule:

$$\forall j \in \{1,\ldots,N\}: \quad x_j^{t+1} = \begin{cases} 1 & \text{if } \sum_{i=1}^{k_j^{in}} w_{ji}^t x_i^t - \theta \geq 0 \\ 0 & \text{otherwise} \end{cases}$$

where k_j^{in} is the number of incoming links in the jth neuron, θ the threshold of the neuron, and w_{ji}^t is the weight, at time t, of the synapse linking i to j.

The robot controllers, i.e., the neural networks, are trained to achieve both tasks. This training consists in finding the suitable topology and the appropriate weights and thresholds to obtain neural networks responsible for good robot's behaviour. We resort to genetic algorithm [5, 7], for short GA in the following, to perform this optimisation heuristically. Let us observe that we can not use here a standard backpropagation algorithm, looking at the computed output and the required one, to fix the weights to minimise such difference. In fact the behaviour we wish to optimise depends on the full path followed by the robot, so act on the weights aimed at minimising the right solution—that we yet don't know—with the followed path would result in an optimisation problem per se. This GA is real-valued, as genotypes encode weights and thresholds of the networks. The selection is performed by a roulette wheel selection. The operators are the classical 1-point crossover and 1-inversion mutation. Their respective rates are 0.9 and 0.005, while the population size is 100 and the maximum number of generations is 50,000. To ensure the legacy of best individuals, the population of parents and offsprings are compared at each generation before keeping the best individuals among both populations. New random individuals (one-tenth of the population size) are also introduced at each generation by replacing worst individuals to avoid premature convergence.

Further details on the application and the model have already been presented in previous works [9, 10].

16.3 Experiments and Results

At the first stage of the experimental set up, we designed experiments that respect the two conditions that we assumed to be necessary for the emergence of modularity. These conditions are to learn more than one task and to modify the networks gradually during the training. As our optimisation process is stochastic, we performed 10 independent replicas of each experiment and in the rest of the paper we report results in terms of such averages.

For each experiment, we analysed two kinds of modularity. The first one is topological modularity, a measure of the density of links inside modules as compared to links between modules. We use the Louvain method [2] to study this modularity. The second one is functional modularity, which groups together neurons that have similar dynamic behaviours. The functional modularity is analysed by using the *dynamical cluster index* [14, 15], which makes it possible to identify subsets of variables that are integrated among themselves and segregated with the rest of the system. The analysis is made by collecting the multidimensional time series composed of network's variables values during the execution of the task. For details on this method, we refer the interested reader to [15].

16.3.1 *Initial Conditions*

We originally considered three ways to train the robot. First, it is trained on the two tasks at the same time (*i*). In this case, the fitness is obtained by averaging the fitness of each task using equal weights. Second, the robot is trained first on one task and then on both (*ii*). The GA is first applied with one fitness function—the one related to the task under scrutiny—and then once again with the weighted sum of the two. The third possibility consists in training two smaller networks so as to accomplish each task separately, then combine the networks by adding a small extra network and train the new larger network (called *juxtaposition* in the following), using as fitness once again the weighted sum of the fitness for each task, as in case (*i*). In terms of robot's performance, these three learning ways are equivalent with final fitness value around 0.75 (observe that we work with a normalised fitness in [0, 1]).

To study the topological modularity in case (*i*) and (*ii*), we analysed the modularity and the number of modules got by the Louvain method when we kept strong enough connections, i.e. connections whose absolute values of the weights are superior to a fixed threshold value. The method thus returns a modularity value in [0, 1], where 0 means that there are no modules and 1 that there are no interactions between modules. We compared the results of our evolved networks with those of random networks used as null hypothesis. These random networks have the same features of our networks, i.e. the same number of nodes, the same density of connections and the same distribution of weights. Figure 16.1 presents this comparison in case (*i*) for

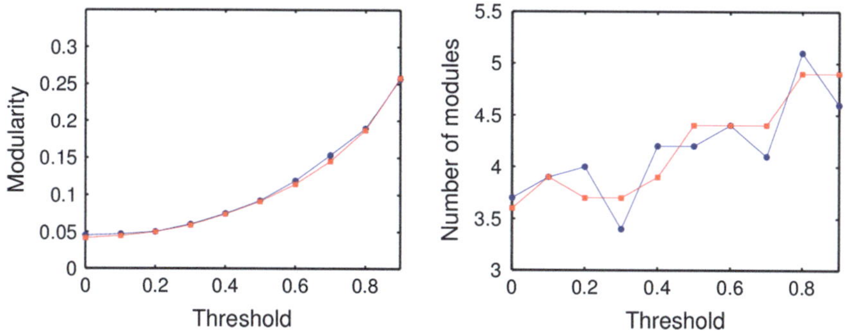

Fig. 16.1 Comparison of the topological modularity between evolved networks in case (*i*) and random networks. *Blue* (on *line*) *circles* represent the evolved networks, while *red* (on *line*) *squares* represent the random networks used as comparison. *Left panel* Modularity. *Right panel* Number of modules. In both cases, the modularity is computed on networks whose weights have been set equal to 0 if their absolute value is below a given threshold. Results are similar for evolved and random networks, which leads to the conclusion that the evolved networks are not topologically modular

different threshold values. We can observe that evolved and random networks behave similarly and hence we can conclude that the evolved networks are not modular. We obtained analogous results in case (*ii*) (data not shown).

The case *juxtaposition* is analysed in a different way. Indeed, initial networks are modular as they consist in the combination of two smaller networks, each trained to accomplish one task, and a small extra network. Thus, we decided to observe the evolution of topological modularity as a function of the increase of robot's performance. The evolution of trained networks is presented in Fig. 16.2 for the modularity (left panel) and for the number of modules (right panel). Results are also compared with those of random networks with the same features as previously. We can notice that the results obtained for evolved and random networks are significantly different. We also found that, in 9 simulations out of 10, the number of modules doesn't change during the optimisation (data not shown). Each module is made by one initial small network (able to perform a given task), while the nodes of the extra small network are shared between the two main modules. The size of these modules is sometimes slightly modified when one node of the extra network jumps from one module to the other. Although the number of modules is almost constant, we can observe a strong decrease of modularity along the optimisation process.

Following these results, we concluded that none of the training schemes leads to topologically modular networks. In the case *juxtaposition*, it even seems to make disappear the initial modularity, as long as the modularity score is considered.

Regarding functional modularity, no clear modules are found in all the analysed cases. The search for functional modules returned either a subset composed of all but a few nodes—i.e. almost all nodes are involved in the processing—or few small subsets with no statistical significance. For this reason we decided not to show data. When the robot is trained according to scheme (*ii*), i.e. sequential learning, naive modules form in the first phase of the training, as only one part of the sensors is stim-

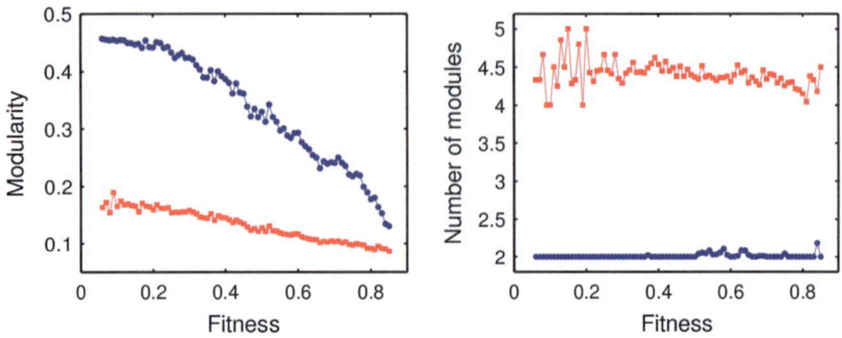

Fig. 16.2 Evolution of the topological modularity as a function of the fitness increase for the evolved networks in the case *juxtaposition* and for random networks. *Blue circles* represent the evolved networks while *red squares* represent the random networks. *Left panel* Modularity. *Right panel* Number of modules. We clearly observe that modularity decreases along the training process even if the number of modules is almost steady

ulated. Nevertheless, these modules disappear when the robot is subsequently trained to accomplish both the tasks. Furthermore, the same results as for the topological analysis are observed in the case *juxtaposition*, showing that the initial modules tend to be blended together in the final training phase.

The results returned by the analysis of functional modularity strengthen those on the topology, as they show that not only the networks have no apparent modules, but that they do not even show clusters of nodes which work in coordinated way and corresponding to either of the two tasks.

16.3.2 Improving the Experimental Setting

Results presented in the previous section do not support the presence of modules. To check if this is due to our main assumptions, we consider additional conditions that could be important for their emergence according to the literature. These conditions are the alternance between different goals, the penalty on the number of connections and the decrease of the number of hidden nodes.

Switching Between Goals

We first followed the suggestion of Kashtan and Alon that switching between different goals is important for the emergence of modules. We considered a fourth training scheme (*iv*) in which the target task is alternated every 100 generations. In preliminary tests, we also considered to alternate every 20 or 50 generations but the simulations with 100 gave us the best robot performance. Even if one particular task is trained in each epoch, all sensors are stimulated. Otherwise, robots can accomplish both tasks but not simultaneously. The results obtained by the simulations of this fourth scheme are similar to those of previous schemes. Indeed, the robot performance is also around 0.75 and the analysis of the topological modularity leads us to similar results to those presented in Fig. 16.1. The same also holds true for functional modularity, with analogous results to the previous cases.

Cost of Connections

Bullinaria [3] as well as Clune et al. [4] claimed that the penalty on the number of connections is essential for the formation of modules. So, we added this penalty for each of the four training schemes considered previously. The penalty contributes to 3/10 of the average fitness in each experiment. In case (*i*), the weight of each task is reduced to 0.35 instead of 0.5. For the second scheme, robot is trained first on one task with the penalty on the number of connections. Then robot is trained on both tasks with the penalty as described for scheme (*i*). For the *juxtaposition* scheme, the penalty is only added for the last phase of learning because if the penalty is also used while training the two smaller networks, the resulting fitness is too low (0.35 which is smaller than the half fitness of other experiments). For the case (*iv*), the penalty on the number of connections is considered during the training of the two alternated task. Let us observe that the fitness described in this paragraph are only used for the training phase. Results are then analysed using robot's performance corresponding to the fitness of the two tasks summed using equal weights, in this way we can compare them with the former ones.

When we analysed topological modularity, we obtained similar results for scheme (*i*), scheme (*ii*) if the learning procedure starts with task B (avoid dangerous zones) and scheme (*iv*). Indeed, in these cases, the fitness is nearly the same than without the penalty on the number of connections. Moreover, the modularity is low (close to 0.1) while random networks with the same density of links and the same distribution of weights have comparable values $\sim$0.15. The difference appears in the number of modules that is slightly higher in evolved networks as shown in the left panel of Fig. 16.3, which shows the distribution of modules according to their size. Indeed, in trained networks, some modules consist of isolated nodes, i.e. nodes without any link with the rest of the network. Let us observe that this never happens in random networks.

As the cost on the number of connections did not lead to the emergence of modules, we might suspect that the penalty that we had fixed was not strong enough to involve modularity. Nevertheless, we obtained similar results with a larger penalty of 0.5, which is a quite high value representing half of the fitness during the optimisation process.

If we consider scheme (*ii*) when the first trained task is task A (reach the peak), the value of the fitness decreases slightly with an average value of 0.67. Moreover if we compare the modularity between these networks and random networks with the same features by keeping connections whose absolute values of weights are larger than a given threshold value (see Fig. 16.4), we can observe a significantly different behaviour. The value of the topological modularity is lower for evolved networks while their number of modules is higher. If we consider evolved networks without eliminating any connections (threshold of 0), we observe that the number

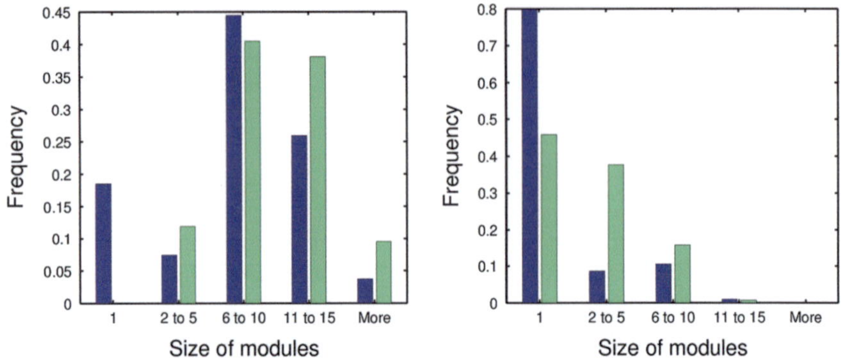

Fig. 16.3 Comparison of the distribution of modules according to their size between evolved networks with penalty and random networks with the same density of links and distribution of weights. *Blue bars* represent the results for evolved networks while *green bars* represent those for the random networks. *Left panel* Scheme (*i*). *Right panel* Scheme (*ii*) when training starts with task A. The number of isolated nodes is significantly higher in evolved networks

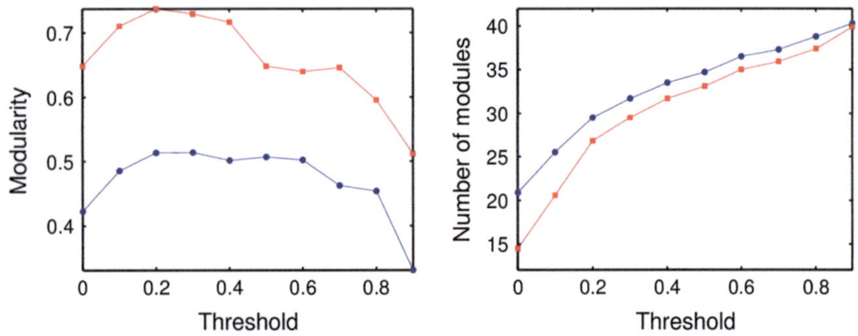

Fig. 16.4 Comparison of topological modularity between evolved networks in case (*ii*), when the training starts with task A and random networks with the same features. *Blue* (on *line*) *circles* represent the evolved networks and *red* (on *line*) *squares* the random networks used as comparison. *Left panel* Modularity. *Right panel* Number of modules. Evolved networks have a slower rate of modularity than random networks but they contain more modules

of modules is ∼21, out of which ∼17 are isolated nodes for evolved networks, for random networks we got respectively ∼15 and ∼7. This high frequency of isolated nodes is also clearly apparent in the right panel of Fig. 16.3. We can explain such results by the simplicity of task A, which indeed requires few connections to be accomplished. When this task is trained alone with the penalty on the number of connections, we got networks with a high level of modularity and a high number of modules, most of which are isolated nodes, comprising non-stimulated inputs. As the penalty cost is always active in the second phase of learning, useless hidden nodes remain isolated, which leads to a higher modularity than for previously analysed training schemes.

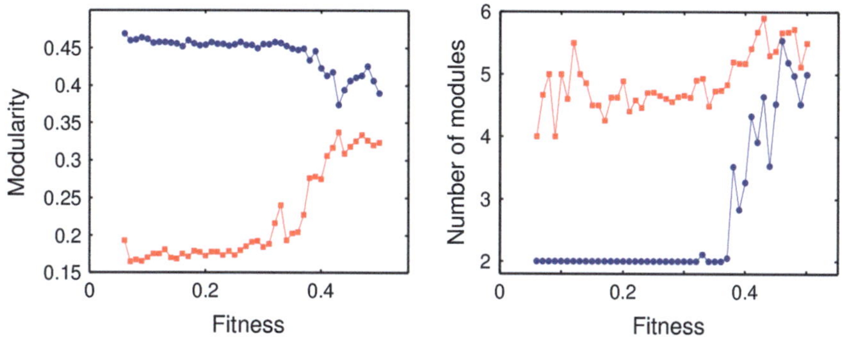

Fig. 16.5 Evolution of the topological modularity as a function of the fitness increase in the case *juxtaposition* with penalty. *Blue circles* represent the evolved networks while *red squares* represent the random networks. *Left panel* Modularity. *Right panel* Number of modules. The decrease of modularity is less important than in the case without penalty. Contrarily, the number of modules rises with the appearance of isolated nodes

Once we eliminated these isolated nodes, the analysis of evolved networks gives us similar results to those of our initial assumption.

In the case *juxtaposition*, the mean fitness of simulations is 0.44, which is considerably smaller than for other experiments. Figure 16.5 shows the evolution of modularity according to the increase in robot's performance. The decrease of modularity seems to be less important than without the penalty but the final fitness is lower. The difference that exists between the modularity of evolved networks and random networks significantly decreases during the learning process.

Following this experiment, we conclude that the penalty on the number of links allows to keep a level of modularity close to the initial one. Indeed, the decline in the number of connections leads to the emergence of isolated nodes that increases the modularity but to the detriment of robot's performance. As for functional modularity, no significant groups of nodes are identified with coordinated behaviour.

Number of Hidden Nodes

Another argument by Bullinaria [3] is that the number of hidden nodes plays a role in the emergence of modules. Indeed, modularity has more possibilities to appear if the number of hidden nodes is small. Thus, our last experimental settings consisted in decreasing the number of hidden nodes and testing if this can lead to the formation of modules in evolved networks. This case can be considered as a very strong implementation of the previous analysis where, instead of removing one link we remove several links, i.e. all the ones connected to a given node. With this aim, we only considered the first training scheme (*i*).

We checked the dependence on the number of hidden nodes on robot performance and modularity. One would expect the performance to be very poor for very small number of hidden nodes—i.e. not enough to perform the required computation—then the performance should increase as long as the number of hidden nodes increases

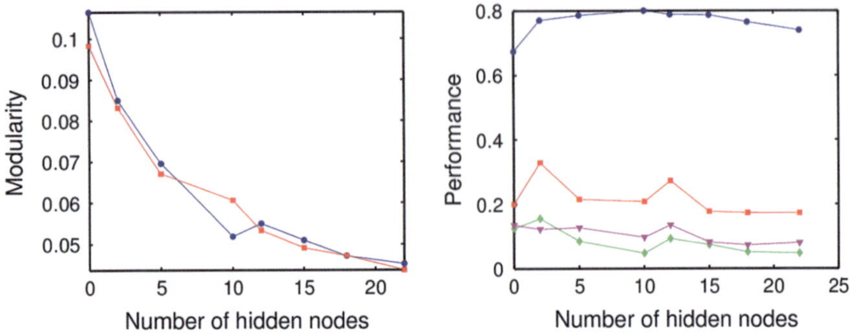

Fig. 16.6 Evolution of robot's performance and modularity according to the number of hidden nodes. *Left panel* Modularity compared to the one of random networks with the same features. Modularity decreases when the number of hidden nodes rises. *Right panel* Robot's performance on the training set and on three validation sets. *Blue* (on *line*) *circles* represent the performance on the training set while *red squares* and *green diamonds* show respectively the performance on an arena whose the shape of the surface and the zones where robots lose energy have been modified. The *magenta triangles* represent the fitness when both modifications are performed. We can not observe significant differences of performance

up to some number, beyond which no improvement is found. The robot performance is tested on the training set and on three different validation sets, i.e. scenarios robots never seen before. In the first two cases, we modify respectively the location of loss of energy zones and the shape of the surface to climb (position of the peak and slope of the surface). The last third case takes into account both modifications.

Results are presented in Fig. 16.6. The left and right panel respectively present the modularity and the robot performance according to the number of hidden nodes. Modularity decreases when the number of hidden nodes increases and no functional modules have been detected. Regarding robot performance, we cannot observe significant decrease when the number of hidden nodes is small. Even more, the performance seems to be better in the validation phase for networks without any hidden node. This result may be explained by the fact that memory is not needed to solve the problem and thus hidden nodes, responsible for information storage, are not relevant to accomplish the task. Observations are the same if we add the penality on the number of connections to the learning process.

In conclusion, also in this experiment no topological modularity is observed. Likewise, the analysis of functional modularity does not support the emergence of modules in these smaller networks.

16.4 Conclusion

Modularity is a major factor of evolvability in biological and artificial networks. Nevertheless, it remains a controversial issue with disagreement over the sufficient conditions for the appearance of modules. This paper analyses the emergence of modularity in the context of evolutionary robotics by taking into account some of the most frequently used conditions in the literature. With this aim, two kinds of modularity are considered, namely the topological modularity and the functional modularity.

We assume that robot controllers, made of neural networks, are trained to fulfil two conflicting tasks. The learning process is a GA that modifies both structure and weights of the controllers. Our initial working assumption is that the emergence of modularity only requires two conditions, namely the learning of at least two tasks and an incremental optimisation process. Faced with the unsatisfying results obtained by this first assumption, we considered supplementary conditions such as switching between the tasks learning, penalising the cost of connections and decreasing the number of hidden nodes. However, contrary to results obtained in previous studies, we can not observe the emergence of topological and functional modularity whatever the conditions that we consider.

Even more, under our initial assumptions, it seems that the learning phase leads to the disappearance of the initial modules. Our results suggest that tasks switching doesn't modify our former ones. When we introduce a penalisation to the density of connections, the level of modularity is higher, but associated to the appearance of isolated nodes in the evolved networks. The reduction of the number of hidden nodes doesn't lead to the emergence of modularity but brings us interesting results. Indeed, we can observe that hidden nodes do not seem to be needed to learn the tasks.

This fact suggests us a possible explanation of the absence of modularity and even a reduction of modularity as learning proceeds. In fact, neural networks composed of only input and output nodes cannot be modular. Indeed, outputs are shared among the tasks and not splitted as in the experiments of Clune. Thus, modular networks would imply that some inputs are disconnected from some outputs and this assumption seems hard to be satisfied because input signals are not correlated. Therefore, once we initialise the neural network with hidden nodes and links, we are adding an "unnecessary" structure resulting in some detectable modularity, which will be slowly removed by the learning phase (creating isolated nodes or making all the nodes to work together) and so finally decrease the network modularity. As a consequence, a possible clue to have modular structures to emerge because of a learning phase with (at least) two tasks is that they require memory to be accomplished.

Even if it provides more questions than answers, the conclusion of this study is promising as it extends the study of the emergence of modularity to another context. Modularity has already been studied in the field of evolutionary robotics but our research differs from previous studies in the choice of the learning algorithm and of the application. Indeed, we trained both structure and weights by a GA whereas weights are usually trained by a backpropagation algorithm. Likewise, our conflicting tasks are computationally more complex than tasks generally considered in other studies (classification tasks or what-where tasks). The absence of modularity in our case strengthens the claim of Bullinaria [3] that the emergence of modularity might depend on external factors such as the learning algorithm and the tasks to learn. Further work will address this issue in more depth.

Acknowledgments This research used computational resources of the "Plateforme Technologique de Calcul Intensif (PTCI)" located at the University of Namur, Belgium, which is supported by the F.R.S.-FNRS. This paper presents research results of the Belgian Network DYSCO (Dynamical Systems, Control, and Optimisation), funded by the Interuniversity Attraction Poles Programme, initiated by the Belgian State, Science Policy Office.

References

1. Beaumont, M.A.: Evolution of optimal behaviour in networks of boolean automata. J. Theor. Biol. **165**, 455–476 (1993)
2. Blondel, V.D., Guillaume, J.-L., Lambiotte, R., Lefebvre, E.: Fast unfolding of communities in large networks. J. Stat. Mech: Theory Exp. **2008**(10), P10008 (2008)
3. Bullinaria, J.A.: Understanding the emergence of modularity in neural systems. Cogn. Sci. **31**(4), 673–695 (2007)
4. Clune, J., Mouret, J.-B., Lipson, H.: The evolutionary origins of modularity. Proc. R. Soc. B: Biol. Sci. **280**(1755), 20122863 (2013)
5. Deb, K.: Multi-Objective Optimization using Evolutionary Algorithms. John Wiley and Sons Ltd, Chichester (2008)
6. Geary, D.C., Huffman, K.J.: Brain and cognitive evolution: forms of modularity and functions of mind. Psychol. Bull, **128**(5), 667 (2002)
7. Goldberg, D.E.: Genetic Algorithms in Search, Optimization, and Machine Learning. Addison-Wesley (1989)
8. Kashtan, N., Alon, U.: Spontaneous evolution of modularity and network motifs. Proc. Nat. Acad. Sci. USA **102**(39), 13773–13778 (2005)
9. Nicolay, D., Carletti, T.: Neural networks learning: Some preliminary results on heuristic methods and applications. In: Perotti, A., Di Caro, L. (eds.) DWAI@AI*IA, volume 1126 of CEUR Workshop Proceedings, pp. 30–40. CEUR-WS.org (2013)
10. Nicolay, D., Roli, A., Carletti, T.: Learning multiple conflicting tasks with artificial evolution. In Advances in Artificial Life and Evolutionary Computation, volume 445 of Communications in Computer and Information Science, pp. 127–139. Springer International Publishing (2014)
11. Peretto, P.: An Introduction to the Modeling of Neural Networks. Alea Saclay. Cambridge University Press, Cambridge (1992)
12. Rojas, R.: Neural Networks: A Systematic Introduction. Springer, Berlin (1996)
13. Seok, B.: Diversity and unity of modularity. Cogn. Sci. **30**(2), 347–380 (2006)
14. Villani, M., et al.: The detection of intermediate-level emergent structures and patterns. Adv. Artif. Life, ECAL **12**, 372–378 (2013)
15. Villani, M. et al.: The search for candidate relevant subsets of variables in complex systems. Artificial Life, 2015. Accepted. Also available as arXiv:1502.01734
16. Wagner, G.P., Pavlicev, M., Cheverud, J.M.: The road to modularity. Nat. Rev. Genet. **8**(12), 921–931 (2007)

Chapter 17
Understanding Financial News with Multi-layer Network Analysis

Borut Sluban, Jasmina Smailović and Igor Mozetič

Abstract What is in the news? We address this question by constructing and comparing multi-layer networks from different sources. The layers consist of the same nodes (hence multiplex networks), but links are constructed from textual news on one hand, and empirical data on the other hand. Nodes represent entities of interest, recognized in the news. From the news, links are extracted from significant co-occurrences of entities, and from strong positive and negative sentiment associated with the co-occurrences. In a case study, the observed entities are 50 countries, extracted from more than 1.3 million financial news acquired over a period of 2 years. The empirical network layers are constructed from the geographical proximity, the trade connections, and from correlations between financial indicators of the same countries. Different network comparison metrics are used to explore the similarity between the news and the empirical networks. We examine the overlap of the most important links in the constructed networks, and compare their structural similarity by node centrality and main k-cores. The comparative analysis reveals that the co-occurrences of countries in the news most closely match their geographical proximity, while positive sentiment links most closely match the trade connections between the countries. Correlations between financial indicators have the lowest similarity to financial news.

17.1 Introduction

Methods developed in the fields of mathematics, computer science and statistical physics have contributed to the emergence of the theory of complex networks. The theory mathematically characterizes systems in the form of entities (nodes) connected by some interactions (links) [2]. The analyses of complex networks strongly influenced and advanced research in social media, biology, and economics [4, 10]. A special type of networks extracted from the data are co-occurrence networks, used in diverse fields, such as linguistics [7], bioinformatics [6, 18, 22], ecology [8],

B. Sluban (✉) · J. Smailović · I. Mozetič
Jožef Stefan Institute, Ljubljana, Slovenia
e-mail: borut.sluban@ijs.si

S. Battiston et al. (eds.), *Proceedings of ECCS 2014*, Springer Proceedings in Complexity, DOI 10.1007/978-3-319-29228-1_17

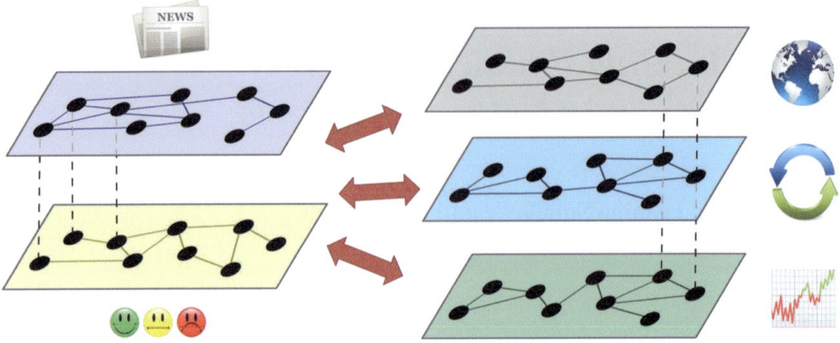

Fig. 17.1 Comparison of multiplex networks representing different types of relations between the same entities: significant co-occurrences and sentiment extracted from news (*at left*), versus geographical proximity, high trade, and high correlation of financial indicators (*at right*, *top to bottom*)

scientometry [12, 19], and socio-technological networks [5, 9, 23]. Co-occurrence networks are loosely defined as networks in which nodes represent some entities (for example persons, companies, genes, etc.), and links represent the fact that these entities exist together in some collection (for example database, article, etc.). For textual sources, it is of high importance to extract the links between the entities that represent a real relationship and are not created by chance [15].

In this paper we investigate the relation between the networks extracted from online texts and the networks drawn from empirical data. Both networks, the news network and the empirical network, are multi-layer, but with the same nodes, hence multiplex networks. In our case study, we analyze 2 years of financial news from 170 major English-language web sites. The nodes in all network layers are 50 countries recognized in the news. The news network has a co-occurrence layer and the sentiment (positive or negative) layer (associated with the co-occurrence links between the countries). The empirical network has three layers: a geo layer (corresponds to geographical proximity between the countries), a trade layer (volume of trade between the countries), and a financial layer (correlations between the Credit Default Swaps of the countries). See Fig. 17.1. Most of the layers, except geo and trade, vary in time. For the period of 2 years, we constructed monthly snapshots of the time-varying layers.

The goal of this research is to shed some light on the contents of online news. The research question we attempt to answer is: What are the similarities between the news and the empirical network layers? The results indicate that the co-occurrences of countries in financial news most closely match their geographical proximity. When we take the news sentiment into account, the positive sentiment layer is most similar to the trade layer. Somehow surprisingly for financial news, the financial layer of the empirical network has the lowest similarity to any news layer.

The chapter is organized as follows. In Sect. 17.2 we describe entity recognition in news, and the construction of the news network layers in terms of co-occurrences, and by sentiment analysis. The empirical network layers are constructed from the geographical proximity, the trade volumes, and the correlations between financial indicators. We then propose several metrics to compare layers of multiplex networks. In Sect. 17.3, we first describe the acquisition and processing of textual data used in our case study. We then present the most interesting results of comparison between the news and empirical network layers. We conclude in Sect. 17.4.

17.2 Methods

In general, when processing financial news, one first has to identify different entities, such as financial institutions, countries, or persons. Then one can construct layers of the news network, consisting of significant entity co-occurrences and their associated sentiment. We compare the news network to the empirical network, in order to discover connections between entity relations in different layers of both networks.

17.2.1 Entity Recognition

Financial news are about events related to companies, stocks, countries or persons, which we call financial entities. The process of identifying financial entities in textual documents requires three components: an ontology of financial entities and terms, gazetteers of the possible appearances of entities in the text, and a semantic annotation procedure that finds and labels the entities. We describe the entity identification approach as implemented in our NewsStream portal [11].[1]

The ontology we use for information extraction constitutes of three main categories: financial entities, financial terms, and geographical entities. The ontology also includes a dictionary of positive and negative words for dictionary-based sentiment analysis.[2] Most of the ontology is automatically induced from various data sources. The geographical entities (continents, countries, cities, organizations) were extracted from GeoNames.[3] The IDMS database and MSN Money[4] were used to organize stock indices and link them to the companies that issue these stocks. The hierarchy of financial terms related to the financial crisis was developed in collaboration with experts in economics. It includes the main European politicians, Central banks and other financial institutions, rating agencies, fiscal and monetary policy terms.

[1]http://newsstream.ijs.si/.

[2]Harvard-IV-4 sentiment dictionary [20, 21].

[3]GeoNames: http://www.geonames.org/.

[4]MSN Money: http://money.msn.com/.

Each entity in the ontology has associated gazetteers, which are sets of rules that specify the lexicographic information about the possible appearances of entities in text. For example, 'The United States of America' can appear in text as 'USA', 'US', 'the United States', etc. The rules include capitalization, lemmatization, POS tag constraints, must-contain constraints (i.e., another gazetteer must be detected in the document or in the sentence) and followed-by constraints.

Finally, the so-called semantic annotation procedure annotates the entities of interest. It traverses each document and searches for entities from the financial ontology. The gazetteers of the entities in the ontology provide information required for the disambiguation of different appearances of the observed entities, resulting in the correct uniform annotation of entities.

17.2.2 Network Construction

For a particular set of entities $E = \{e_1, \dots, e_n\}$ we construct networks that are obtained from different data sources. We distinguish two networks: *News network*—constructed from entities and relations appearing in financial news, and *Empirical network*—with the same entities, but linked by relations extracted from other information sources or databases. Each network consists of several layers, and each layer is constructed for a particular time period, resulting in a series of snapshots for each network.

17.2.2.1 News Network

Financial entities identified in a single news document can be connected with various types of relations. One of the simplest is their common appearance in the document, referred to as the co-occurrence of entities. Hence, for a selected set of entities $E = \{e_1, \dots, e_n\}$ we construct a layer of entity co-occurrences within a particular time frame—the *Co-occurrence layer*. Each link in such a layer represents a significant co-occurrence relation between two entities. We use the Significance algorithm proposed in [15] to assess whether the co-occurrence of two entities is significant.

The number of all documents with at least two entities from E is N. Let A and B be two entities that occurred with at least one other entity from E in N_A and N_B documents, respectively. Let N_{AB} denote the number of actual A and B co-occurrences. Then the expected number of co-occurrences is given by

$$\mathbb{E}(N_{AB}) = \frac{N_A N_B}{N} . \quad (17.1)$$

According to [15], the standard deviation is

$$\sigma_{AB} = \sqrt{\frac{N_A N_B}{N}\left(\frac{N^2 - N(N_A + N_B) + N_A N_B}{N(N-1)}\right)} \quad (17.2)$$

and hence the standard significance score of the co-occurrence N_{AB} from the data is

$$Z_{AB} = \frac{N_{AB} - \mathbb{E}(N_{AB})}{\sigma_{AB}} . \tag{17.3}$$

For a selected threshold Z_0, one can distinguish significant $Z_{AB} > Z_0$ and non-significant $Z_{AB} < Z_0$ co-occurrence relations between the two entities.

A set of entities can be linked also by other types of relations, e.g., based on expressed sentiment in documents which discuss the entities. We construct a *Sentiment layer* of the news network by detecting sentiment orientation and strength of financial news articles which mention pairs of entities in a specific time period. A sentiment link between two entities in the layer exists if its sentiment value is higher than a predefined threshold.

In order to calculate sentiment between entities, we use the sentiment analysis implementation of the NEWSSTREAM portal. The implementation is dictionary-based, meaning that sentiment polarity of a document is based on the count of predefined sentiment terms (positive and negative) in the document. The implementation relies on the Harvard-IV-4 sentiment dictionary [20, 21]. For each document it calculates the overall sentiment polarity by applying the following formula:

$$polarity = \frac{pos - neg}{pos + neg} \tag{17.4}$$

where *pos* is the number of positive and *neg* is the number of negative dictionary terms found in the document.

Using the NEWSSTREAM portal we obtain the sentiment results calculated on the level of a document and aggregated by summing the results for each day. Moreover, since we are interested in making a snapshot of a network over a longer time period of T days, we further aggregate the obtained results for T days. Based on the analysis of the sentiment distribution, we determine the thresholds p_0 and n_0 for the creation of positive and negative sentiment links.

17.2.2.2 Empirical Network

We observe the same set of entities E as in the 'News network', but the information regarding their mutual interactions is not acquired from the news. In particular, we explore three data sources to construct the empirical network layers: the geographical proximity of the financial entities, correlations between their financial indicators, and their direct interaction in terms of mutual trade. We use these layers as the underlying empirical representation of the complex relations between financial entities, from which we try to understand the dynamics of entity appearance in financial news.

The simplest among the 'empirical network' layers is the geographical proximity, or short the *Geo layer*. Each financial entity has a predominant geographical location, place of residence, headquarters address, area, country, continent of trade or market influence. A financial entity is assigned a geographical entity and hence a link between two entities is established if a certain proximity measures is above a given geographically feasible threshold. Examples of proximity measures include geographical distance $d(A, B)$, inverse distance $\frac{1}{d(A,B)}$, or inverse squared distance $\frac{1}{d(A,B)^2}$.

Constructing a layer from trade interaction data is also fairly straightforward. Consider the interaction between financial entities as the amount of mutual trade. Each entity $e_i \in E$ engages in $c(e_i)$ of trade with all other entities, therefore

$$c(e_i) = \sum_{e_j \in E \setminus \{e_i\}} c(e_i, e_j) \tag{17.5}$$

where $c(e_i, e_j)$ is how much e_i trades to e_j. Notice the implied direction of the trade. In our experiment we use an undirected *Trade layer*, and therefore define $t(e_i, e_j) = c(e_i, e_j) + c(e_j, e_i)$ as the cumulative trade exchange between e_i and e_j. A trade link between two entities e_i and e_j is established if any of the relative amounts $\frac{t(e_i,e_j)}{c(e_i)}$ or $\frac{t(e_i,e_j)}{c(e_j)}$ is above a given threshold t_0.

Important financial entities have also an associated time-varying financial indicator (e.g., price, trade volume, confidence index), which is represented as a time series. A basic approach to measure similar trends in the movement of financial indicators is the Pearson correlation [14] between time series s_i and s_j of the entities e_i and e_j, over a period of K time points:

$$\rho_{i,j} = \frac{\sum_{k=1}^{K} (s_{i,k} - \bar{s_i})(s_{j,k} - \bar{s_j})}{\sqrt{\sum_{k=1}^{K} (s_{i,k} - \bar{s_i})^2 \sum_{k=1}^{K} (s_{j,k} - \bar{s_j})^2}}, \tag{17.6}$$

where $\bar{s_i}$ and $\bar{s_j}$ stand for the average (arithmetic mean) value of the respective series. A *Financial layer* can hence be constructed using a threshold value c_0, which determines whether the indicator time-series of two entities are sufficiently correlated $(\rho_{i,j} > c_0)$ to form a link between them.

17.2.3 Network Comparison

We described the construction of the two-layer *News network* and the three-layer *Empirical network*. Both networks share the same set of nodes, i.e., entities $E = \{e_1, \dots e_n\}$, which are in each layer connected by a different type of relation. As a whole, we are considering a *multi-layer network* of the same nodes, also called a *multiplex network*.

We try to understand the 'news network' by comparison to the 'empirical network'. In the 'empirical network' we construct a link between two entities when there is a 'strong' empirical relation between the entities. Although all layers contain the same entities, in the comparison isolated nodes are not considered. The layers that we compare are constructed from the sets of strongest links for a particular relation type.

The most straightforward comparison of the network layers $\mathcal{L} = \{L_1, \dots, L_m\}$ is done by measuring the size of link overlap between the layers. Let $l(L_i)$ and $l(L_j)$ be the sets of links in layers L_i and L_j, where a link is defined as a pair of nodes it connects, e.g., (e_u, e_v), then

$$o(L_i, L_j) = \frac{|l(L_i) \cap l(L_j)|}{|l(L_j)|} \tag{17.7}$$

is the size of their link overlap relative to layer L_j.

Considering for each layer not only the links that indicate the strength of a relation above a certain threshold, but also their weight–strength of the relation, then a comparison of top strongest links in each layer can be performed. Let $sl(L_i)$ and $sl(L_j)$ be lists of links from layers L_i and L_j, ordered descending by their weights, and let $sl_k(L)$ denote the list of first k element of list $sl(L)$, then *precision-at-k* [16] is defined as:

$$r_k(L_i, L_j) = \frac{|sl_k(L_i) \cap sl_k(L_j)|}{k} . \tag{17.8}$$

If for all pairs of layers L_i and L_j, $i, j \in \{1, \dots, m\}$, the same k is selected, then a meta-network can be constructed with nodes representing layers L_i, $i \in \{1, \dots, m\}$ and links representing the relation between layers, where $r_k(L_i, L_j)$ values are weights of the links, indicating the magnitude of the relationship.

Other comparisons of the network layers induced on the 'strongest' links for a particular relation type, are based on the most important nodes in each layer. In one approach, we measure the importance of nodes in terms of their *centrality*, as denoted by the *eigenvector centrality measure* [3]. Let A be the adjacency matrix of nodes $e_1, \dots, e_n$ in the network, then the components of the eigenvector of the largest eigenvalue λ solving the equation $A\mathbf{x} = \lambda\mathbf{x}$ hold the centrality values of the corresponding nodes. Nodes connected to better-connected nodes get higher centrality values. This measure is used to compare which are the most central nodes between pairs of layers.

Another approach to identify the most important nodes of a network is the *k-core decomposition* [17]. This is an iterative process pruning all nodes with degree smaller than k, and the remaining part of the network which holds only nodes with degree greater or equal to k is called the k-core. The core with the largest k is called the main core of the network. Comparing the main cores of different network layers will be used to assess the similarity between layers.

17.3 Experiments

Financial news cover a wide range of topics, they include numerous entities of interest, and are influenced by different factors. We describe the data we use to show the presence and interaction of entities in the news, and the empirical data to model the real-world context that may shape the news. Next, we present the comparison results between the network layers.

17.3.1 Data

We acquire news articles and blogs from 2,503 RSS feeds from 170 English language web sites (14,567 domains), covering the majority of web news in English and focusing on financial news and blog sources. We collect data from the main news providers and aggregators (like yahoo.com, dailymail.co.uk, nytimes.com, bbc.co.uk, wsj.com) and also from the main financial blogs (like zerohedge.com). The fifty most productive web sites account for 80 % of the collected documents.

The documents used in our experiment cover the period from November 1st, 2011 until December 31st 2013. A total of 18 million documents were filtered for strictly financial news, resulting in 1.3 million documents. From these documents we extract relevant entities, and construct the 'news network' layers in monthly time windows. For our analyses, we select 50 countries as entities of interest. Snapshots of the resulting co-occurrence layer, positive sentiment layer, and negative sentiment layer are shown in Figs. 17.2, 17.3 and 17.4, respectively.

Fig. 17.2 Snapshot of the country co-occurrence network layer (October 2012)

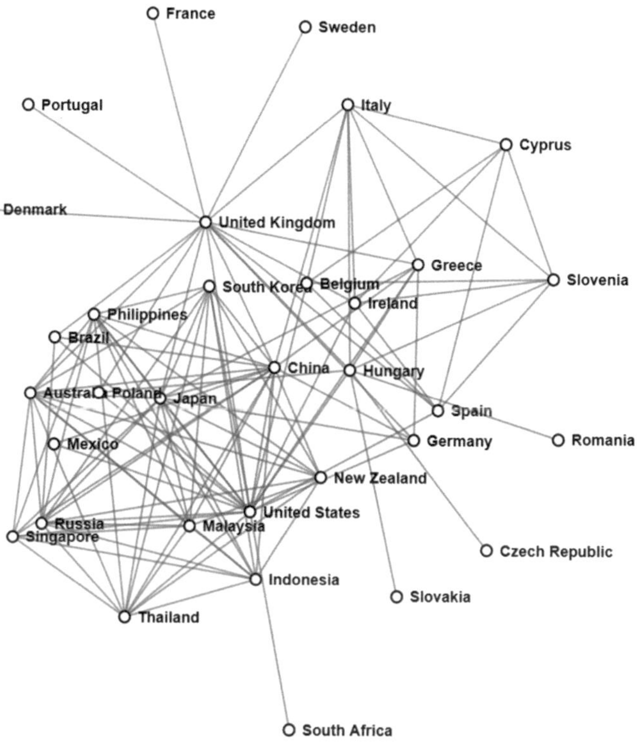

Fig. 17.3 Positive sentiment network layer (Jan 2012)

On the other hand, the construction of the empirical network, which should reflect the real-world context of the news, was done using data from different sources. For the 'Geo layer' we simply used the is-a-neighbour-of relation to link the selected countries. To the links representing common terrestrial borders we added also a few links between countries that are considered relatively adjacent in the local geographical context, such as Australia and New Zealand, South Korea and Japan, or Italy and Malta.

Trading relations between the countries were obtained from the *UNCTAD website*,[5] the United Nations statistics data center, providing yearly aggregations of trade data. Our 'Trade layer' was constructed from trade links that present relatively important trade relations (greater that 10 %, i.e. $t_0 = 0.1$) for at least one of the connected countries.

In our experiments we consider 50 countries that issue sovereign bonds, and which are insured by Credit Default Swaps (CDS), i.e., an insurance for the case when the bond issuer 'defaults' and is unable to repay the debt. To construct the 'financial layer' we used the time series of the countries' CDS prices, which are

[5]http://unctadstat.unctad.org.

Fig. 17.4 Negative sentiment network layer (May 2012)

often considered a good proxy for the risk of default of a financial institution issuing bonds [1, 13]. We create links between countries whose correlation between their CDS time series is above 0.9 ($c_0 = 0.9$). In order to ensure enough data for reliable correlation results, we use a 3-months time window for each snapshot, which is assigned to the last month (e.g., Nov-Dec-Jan for the 'Jan' snapshot). A snapshot of the 'Financial/CDS layer' and the 'Trade layer' are presented in Figs. 17.5 and 17.6, whereas the presentation of the 'Geo layer' is well known and therefore omitted.

17.3.2 Results

The results are presented for a multiplex network with 50 country nodes for the time period of 2 years in monthly steps. The Geo layer is static, as well as the Trade layer—we used the yearly aggregated trade data from 2012 also for 2013.

First, we present the analysis of overlapping links between the network layers L_{CO}, L_{Geo}, L_{Tr}, and L_{CDS}. We are interested in the number of links from the 'empirical network' that appear in the financial news as country co-occurrences over time. The relative overlaps $o(L_{CO}, L^*)$ for $L^* \in \{L_{Geo}, L_{Tr}, L_{CDS}\}$ are presented in Fig. 17.7. We see that most of the Geo layer links coincide with the country co-occurrences in the news, whereas on average less than half of the links between the countries in the Trade and CDS layers also appear in the co-occurrence layer.

Fig. 17.5 Trade layer (2012–2013)

Fig. 17.6 Financial/CDS layer (October 2012)

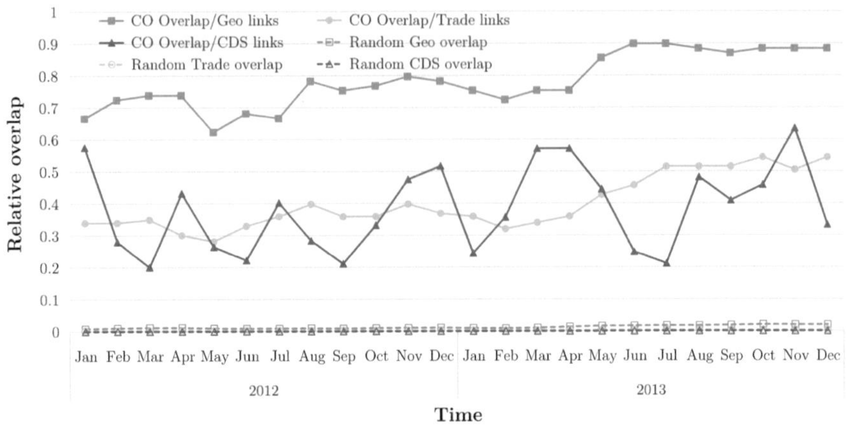

Fig. 17.7 Relative size of the empirical layer links present in the co-occurrence layer

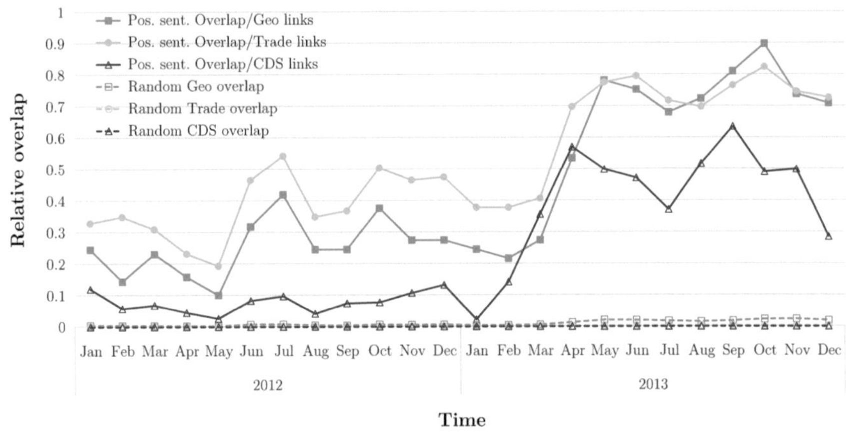

Fig. 17.8 Relative size of the empirical layers' links present in the positive sentiment layer

Next, we investigate how is the sentiment associated with the country co-occurrences related to the empirical network. Using the sentiment analysis approach presented in Sect. 17.2.2.1 we find that there is a strong bias towards positive sentiment in the financial news. We set thresholds n_0 and p_0 to two standard deviations apart from the average sentiment polarity in the documents, thus selecting only links that reflect the most negative and most positive sentiment in the context of two countries. The negative sentiment layer turns out to be predominantly small, even for a slightly less restrictive threshold n_0 (at 90 % st. dev. from the average) and therefore has mostly low overlap with the empirical layers. On the other hand, the comparison of the positive sentiment layer L_P with the empirical layers results in a larger number of common links, as shown in Fig. 17.8. Positive sentiment between the countries has the largest overlap with the trade relations, followed by geographical proximity and to the smallest extent by the correlation between the CDS time series.

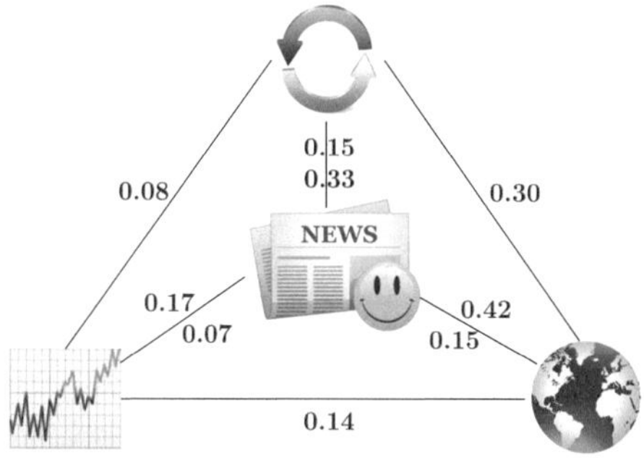

Fig. 17.9 A meta network between the news and empirical network layers

Comparison of the most important nodes in each layer shows similar results. The comparison of main k-cores results in the largest overlap between the co-occurrence and Geo layer cores, and positive sentiment and the Trade layer cores. The co-occurrence layer cores overlap with the Geo layer cores in central European countries, and with the Trade layer cores in western European countries. The overlaps between the co-occurrence and CDS layers show common presence of some eastern European countries in 2012, but no regular presence in 2013. Several countries regularly appear in the core overlap between the positive sentiment and the Trade layers (CN, DE, US, UK, JP, BR, FR, and AU). Germany is also almost all the time (23 months) in the core overlap of the positive sentiment and Geo layers.

Most central nodes of the co-occurrence layer coincide with the Geo layer in central European countries (AT, CZ, HU, SK, SI), with the CDS layer in few eastern European countries, and with the Trade layer only Finland appears often among the top ten most central nodes. For the positive sentiment layer, the common most central nodes are Germany and Russia for the Geo layer, and some of the largest economies (CN, DE, FR, JP, RU, US) for the Trade layer.

Finally, we use the *precision-at-k* method to measure the link overlap of the strongest relations in each layer. Limited by the number of links in the Geo layer, k was set to 69. Figure 17.9 illustrates the relations between layers weighted by the precision-at-'69' values.

17.4 Conclusions

In the chapter we present methods to extract nodes and layers of multiplex networks. The emphasis is on the network construction from large textual streams, where entity recognition, co-occurrence extraction, and sentiment analysis are performed. We propose several metrics to compare different layers of the same network. In the case study, we compare the news layers to the empirical layers, and conclude that financial news mostly reflect geographical proximity and trade relations, and much less financial relations. These results provide the background, long-term characterization of the news. We speculate that the underlying reason is the method to calculate the significant co-occurrences, which computes the expected co-occurrences from individual frequencies. Alternatively, one could compare actual co-occurrences in a short time window to a longer time window, and so detect potentially interesting deviations. In our future work, we intend to investigate this alternative approach and evaluate how the news layers constructed in such a way compare to different empirical layers.

Acknowledgments This work was supported in part by the EC FP7 projects SIMPOL (no. 610704) and MULTIPLEX (no. 317532), and by the Slovenian ARRS programme Knowledge Technologies (no. P2-103).

References

1. Aizenman, J., Hutchison, M., Jinjarak, Y.: What is the risk of European sovereign debt defaults? Fiscal space, CDS spreads and market pricing of risk. J. Int. Money Finan. (2012)
2. Albert, R., Barabási, A.L.: Statistical mechanics of complex networks. Rev. Mod. Phys. **74**(1), 47 (2002)
3. Bonacic, P.: Factoring and weighting approaches to status scores and clique identification. J. Math. Soc. **2**, 113–120 (1972)
4. Caldarelli, G.: Scale-free networks: complex webs in nature and technology. Oxford University Press (2007)
5. Cattuto, C., Schmitz, C., Baldassarri, A., Servedio, V.D., Loreto, V., Hotho, A., Grahl, M., Stumme, G.: Network properties of folksonomies. AI Commun. **20**(4), 245–262 (2007)
6. Cohen, A.M., Hersh, W.R., Dubay, C., Spackman, K.: Using co-occurrence network structure to extract synonymous gene and protein names from medline abstracts. BMC Bioinform. **6**(1), 103 (2005)
7. Edmonds, P.: Choosing the word most typical in context using a lexical co-occurrence network. In: Proceedings 35th Annual meeting of ACL, pp. 507–509. Association for Computational Linguistics (1997)
8. Freilich, S., Kreimer, A., Meilijson, I., Gophna, U., Sharan, R., Ruppin, E.: The large-scale organization of the bacterial network of ecological co-occurrence interactions. Nucleic Acids Res. **38**(12), 3857–3868 (2010)
9. Ghoshal, G., Zlatić, V., Caldarelli, G., Newman, M.: Random hypergraphs and their applications. Phys. Rev. E **79**(6), 066118 (2009)
10. Jackson, M.O.: Social and economic networks. Princeton University Press (2010)
11. Kralj Novak, P., Grčar, M., Sluban, B., Mozetič, I.: Analysis of financial news with newsstream. Tech. Rep. IJS-DP-11965, Jožef Stefan Institute, Ljubljana. http://arxiv.org/abs/1508.00027 (2015)
12. Mane, K.K., Börner, K.: Mapping topics and topic bursts in PNAS. Proc. Natl. Acad. Sci. **101**(Suppl 1), 5287–5290 (2004)
13. Pan, J., Singleton, K.J.: Default and recovery implicit in the term structure of sovereign CDS spreads. J. Financ. **63**(5), 2345–2384 (2008)
14. Pearson, K.: Note on regression and inheritance in the case of two parents. Proc. R. Soc. Lond. **58**, 240–242 (1895)
15. Popović, M., Štefančić, H., Sluban, B.: Kralj Novak, P., Grčar, M., Puliga, M., Mozetič, I., Zlatić, V.: Extraction of temporal networks from term co-occurrences in online textual sources. PLoS ONE **9**(12), e99515 (2014)
16. Raghavan, V., Bollmann, P., Jung, G.S.: A critical investigation of recall and precision as measures of retrieval system performance. ACM Trans. Inform. Syst. **7**(3), 205–229 (1989)

17. Seidman, S.B.: Network structure and minimum degree. Soc. Netw. **5**(3), 269–287 (1983)
18. Shalgi, R., Lieber, D., Oren, M., Pilpel, Y.: Global and local architecture of the mammalian microrna-transcription factor regulatory network. PLoS Comput. Biol. **3**(7), e131 (2007)
19. Su, H.N., Lee, P.C.: Mapping knowledge structure by keyword co-occurrence: a first look at journal papers in technology foresight. Scientometrics **85**(1), 65–79 (2010)
20. Tetlock, P.C.: Giving content to investor sentiment: the role of media in the stock market. J. Financ. **62**(3), 1139–1168 (2007)
21. Tetlock, P.C., Saar-Tsechansky, M., Macskassy, S.: More than words: quantifying language to measure firms' fundamentals. J. Financ. **63**(3), 1437–1467 (2008)
22. Wilkinson, D.M., Huberman, B.A.: A method for finding communities of related genes. Proc. Natl. Acad. Sci. **101**(Suppl 1), 5241–5248 (2004)
23. Zlatić, V., Ghoshal, G., Caldarelli, G.: Hypergraph topological quantities for tagged social networks. Phys. Rev. E **80**(3), 036118 (2009)

Chapter 18
Channel-Specific Daily Patterns in Mobile Phone Communication

Talayeh Aledavood, Eduardo López, Sam G.B. Roberts, Felix Reed-Tsochas, Esteban Moro, Robin I.M. Dunbar and Jari Saramäki

Abstract Humans follow circadian rhythms, visible in their activity levels as well as physiological and psychological factors. Such rhythms are also visible in electronic communication records, where the aggregated activity levels of e.g. mobile telephone calls or Wikipedia edits are known to follow their own daily patterns. Here, we study the daily communication patterns of 24 individuals over 18 months, and show each individual has a different, persistent communication pattern. These patterns may differ for calls and text messages, which points towards calls and texts serving a different role in communication. For both calls and texts, evenings play a special role. There are also differences in the daily patterns of males and females both for calls and texts, both in how they communicate with individuals of the same gender versus opposite gender, and also in how communication is allocated at social ties of different nature (kin ties vs. non-kin ties). Taken together, our results show that there is an unexpected richness to the daily communication patterns, from different types of ties being activated at different times of day to different roles of channels and gender differences.

T. Aledavood (✉) · J. Saramäki
Aalto University School of Science, P.O. Box 12200, 00076 Espoo, Finland
e-mail: talayeh.aledavood@aalto.fi

E. López · F. Reed-Tsochas
CABDyN Complexity Center, Saïd Business School, University of Oxford, Oxford OX1 1HP, UK

S.G.B. Roberts
Department of Psychology, University of Chester, Chester CH1 4BJ, UK

F. Reed-Tsochas
Department of Sociology, University of Oxford, Oxford OX1 3UQ, UK

E. Moro
Departamento de Matemáticas & GISC, Universidad Carlos III de Madrid, 28911 Leganés, Spain

R.I.M. Dunbar
Department of Experimental Psychology, University of Oxford, Oxford OX1 3UD, UK

S. Battiston et al. (eds.), *Proceedings of ECCS 2014*, Springer Proceedings in Complexity, DOI 10.1007/978-3-319-29228-1_18

18.1 Introduction

The human body is equipped with a circadian pacemaker that gives rise to 24-h rhythms in biological processes within the body, as well as in behavioural patterns [1–3]. Studies of human circadian rhythms have traditionally been small-scale studies that involve direct monitoring of human subjects. However, for more than a decade now, automated electronic records of human behaviour have given researchers the ability to study human dynamics and behavioural patterns in unprecedented ways. Circadian rhythms are clearly visible e.g. in records of Wikipedia and OpenStreetMap editing [4, 5], mobile telephone calls [6, 7] and on Twitter [8]. While it is well-known that there is a lot of individual variation in circadian rhythms, these and most other studies of electronic records have focused on aggregate-level phenomena. In [9], the authors of the present work studied the daily mobile telephone call patterns of individuals and the persistence of such patterns. Here, we expand on this work and also consider another communication channel: text messages.

In [9], we showed that individuals have their own distinct daily call patterns, and that these patterns are persistent for each individual even when their social networks undergo turnover. Further, these patterns were seen to have a social dimension: calls at late hours were often associated with close relationships. Because text messages may serve a different purpose in maintaining social relationships than calls (see, e.g., [10, 11]), we address the question of whether the daily patterns of text messaging are similar to those of calls, and whether individuals have their distinct, persistent text messaging patterns. Also, because significant differences were seen in call patterns to the same versus the opposite gender, as well as kin versus friendship ties, we study the daily text messaging patterns from this point of view.

We use the same longitudinal data set of time-stamped text communication records of 24 individuals as in [9] (for details, see Sect. 18.2). Our results are summarised as follows: first, each individual's text messaging frequency is seen to exhibit distinct daily patterns that are persistent over time, similarly to calls. However, the text messaging and call patterns may differ significantly for a given individual, and on average, text messages are sent more frequently at later hours of the day. Since there is a high level of social network turnover in the studied data set, the persistence of daily patterns for both communication channels indicates that these patterns are not explained in terms of preferred communication timings or channels with specific alters, but rather they contain a component intrinsic to each individual. Further, the difference between the channels is exemplified by daily entropy patterns: even though both calls and texts are targeted at a less diverse set of alters at the late hours, the clear correlation between calls to closest alters and the least diverse times of day is missing for text messages.

Regarding gender differences, we observed that the total number of text messages is about 1.5 times higher for females than males (for calls, the numbers are practically the same). At the same time, both genders have similar daily trends, sending out the largest numbers of texts in the evening. Calls to kin and family are overall much less frequent than to calls to friends and acquaintances, and text messages to kin are even more infrequent. However, females communicate with their kin by text around 3 times more frequently than males do.

18.2 Data

For this study, we have used a longitudinal data set of 18 months of auto-recorded, time-stamped phone calls and text communication records of 24 individuals ("egos" in the following). This data set has been used earlier in [9, 12, 13]. Altogether, this data set consists of 74,124 calls and 273,501 text messages, with time stamps at a resolution of one second. The large number of texts compared to calls may have to do with the young age of participants ($\sim$18 years at the beginning of the study), as well as with conversations via text messages generating a large number of messages, whereas a conversation via a phone call leaves one record only. As the original purpose of collecting this data was to study turnover in social networks of individuals, the setting was chosen such that all participants were in their last year of high school at the beginning of the study, and later went either to work or university (often in another city), after about 6 months of data collection. The participants also took part in 3 surveys, separated by 9 months, designed to provide complementary information about members of their communication network ("alters" in the following). Information on gender and kinship as well as data on how emotionally close egos felt to their alters were collected with these surveys. For further details, please see [12].

18.3 Results

18.3.1 Channel-Specific Daily Patterns and Their Persistence

In order to compute the daily patterns of texting for each individual, we begin by segmenting our data temporally. We make two temporal divisions: first, at the level of days, we divide each day to 24 one-hour bins and compute the number of calls and text messages inside each bin. In order to address the persistence of the observed patterns, we divide the 18-month time span of our data into three 6-month intervals, I_1, I_2, and I_3. The end of the first time interval I_1 coincides with early autumn, where the participants move on in their lives and begin work or studies at university. The second time interval, I_2, then spans a time range where a major change has taken place in the participants' lives and they are settling in a new environment with major turnover in their social networks.

For each ego, we aggregate all events (calls or texts) within each time interval (I_1, I_2 or I_3) to the hourly bins. To arrive at the daily text or call patterns measuring frequency as function of time, we then sum up and normalise the numbers of

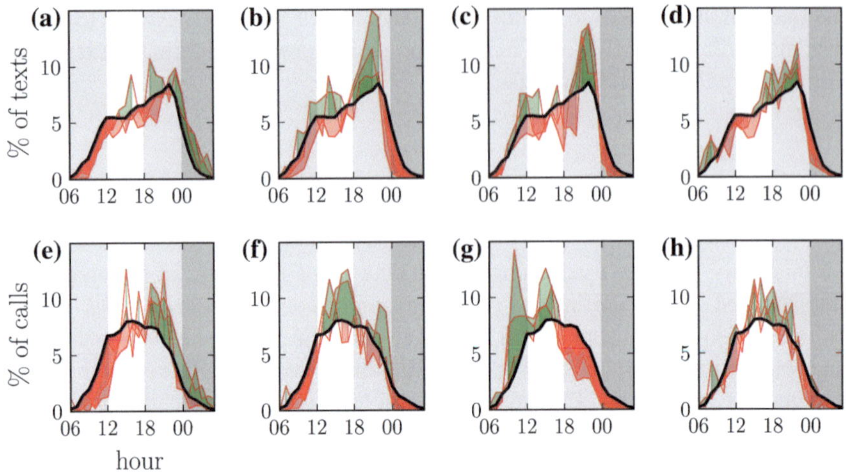

Fig. 18.1 **a–d** The daily text patterns of 4 individuals, **e–h** the daily call patterns of the same individuals. The average fraction of calls/texts at each hour of day is denoted by *red lines*. These have been computed for each of the three 6-month intervals, I_1, I_2, and I_3. The average call and text patterns, averaged over the patterns of all 24 individuals, are shown as *black lines*. *Green shading* indicates where an individual's call/text fraction is above average, whereas *red shading* indicates the opposite. *Note* The persistence of individual patterns (overlap of *green*/*red* areas for an individual) as well as the differences between call and text patterns

respective events at each hour of day. This is repeated separately for each 6-month interval, and each ego. Daily text and call patterns calculated in this manner are displayed in Fig. 18.1 for four different individuals, together with averages over all 24 individuals. It can be seen that each individual has their own distinct text and call pattern, and both patterns appear fairly persistent over time. It is also evident that the call and text patterns may significantly differ for a given ego. Likewise, there is a clear difference between the average daily patterns of calls and texts: the frequency of text messaging peaks at later hours of the day, whereas the majority of calls are made in the afternoon. This supports the notion of calls and text messages serving different social and communication functions.

For quantifying the persistence of each individual's daily text patterns, we use the Jensen-Shannon Divergence (JSD) as a measure of distance between two patterns, similarly to previous works [9, 13]. The JSD is a measure of the dissimilarity of two probability distributions. It is an extension of the Kullback-Leibler divergence (KLD), with the important difference that it can be used for discrete probability distributions with zero-valued elements. For two discrete probability distributions P_1 and P_2, the JSD is defined as

$$\mathrm{JSD}(P_1, P_2) = H\left(\frac{1}{2}P_1 + \frac{1}{2}P_2\right) - \frac{1}{2}[H(P_1) - H(P_2)], \qquad (18.1)$$

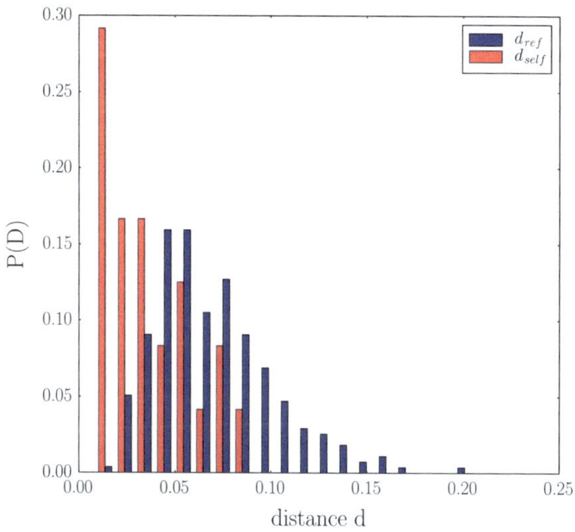

Fig. 18.2 Distributions of the values of the Jensen-Shannon divergence, measured between each individual ego's daily text patterns in different 6-month intervals (d_{self}) and between patterns of different egos (d_{ref}). Self-distances d_{self} are mostly lower than the reference distances, indicating that each individual's daily text patterns preserve their shape through the 6-month intervals

where H is the Shannon entropy, $H(P) = -\sum p(t) \log p(t)$. Here, we set $P_i = \{p_i(t)\}$, where t indicates the (binned) time of day, and $i = 1, 2$ denotes the two distributions to be compared (e.g. the two distributions corresponding to I_1 and I_2 for one ego). We calculate the self-distance d_{self} for each ego as the average of the JSDs between daily patterns for intervals I_1 and I_2 and for intervals I_2 and I_3. For a reference distance d_{ref} with which to compare these values, we calculate the JSD between the daily patterns of each ego and each of the other egos (within the same time interval), repeating this for all pairs of individuals and all time intervals. The result can be seen in Fig. 18.2. It is evident that on average the self-distances d_{self} are smaller than the reference distances d_{ref}, indicating that each individual's daily patterns are fairly persistent. The same was observed for calls in [9], but here the differences between self and reference distances are even more evident, i.e. daily text patterns appear to retain their shape even better than call patterns.

18.3.2 Specifity in Communication: Who Is Contacted and When?

Studies of call records with the data set at hand have revealed a social dimension within the daily patterns: for calls, the diversity of called individuals is on average lower in the evenings and especially at night [9]. When the called alters are ranked on the basis of the number of calls, it is seen that the fraction of calls to top-ranked alters is often high when the diversity is low; typically, evenings and nights are "reserved" for top-ranked alters.

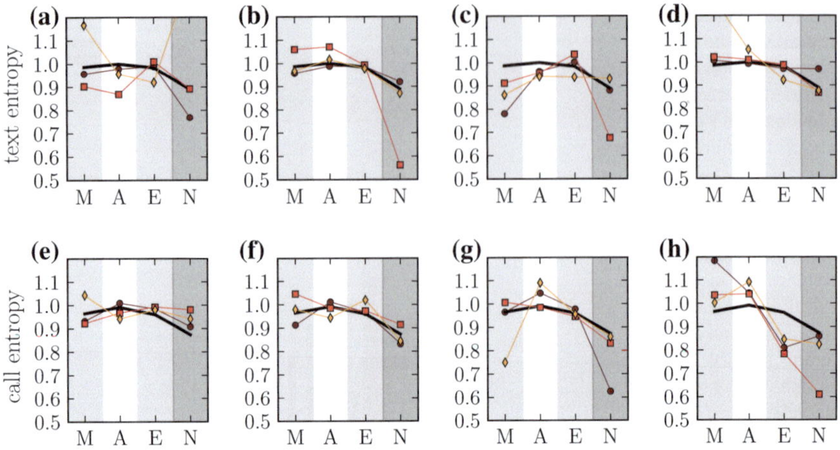

Fig. 18.3 The relative entropies for the same 4 individuals as in Fig. 18.1. **a–d** Relative entropies for text messages, **e–h** relative entropies for calls. The *black line* denotes average over all 24 individuals; the *coloured lines* correspond to the different 6-month intervals for each individual. Time periods are 6–12 AM (Morning), 12 AM–6 PM (Afternoon), 6–12 PM (Evening) and 12 PM–6 AM (Night)

Here, we set out to study whether similar effects can be detected with text messages (note that as seen in Fig. 18.1, calls and texts may follow different daily cycles). We approach the problem using relative entropies as in [9]: first, we measure the diversity of called/texted alters in the 6-h bins (6–12 AM (morning), 12 AM–6 PM (afternoon), 6–12 PM (evening) and 12 PM–6 AM (night)), by computing bin-wise call/text entropies for each ego and interval. These are then normalised by the average entropies computed with a null model, where all called alters are randomly shuffled among calls for one ego (see Methods for details). This null model corresponds to the hypothesis that given the cumulative numbers of calls/texts to each alter and the overall daily pattern, there are no preferred times of calling/texting.

As seen in Fig. 18.3, the average relative entropies for texting follow a similar pattern as the call entropies, the only difference being that the pattern is slightly more flat. Thus, similarly to calls, text messages in the afternoon are targeted at a more diverse subset of alters—the relative entropy close to unity indicates that there are no specific preferences. To the contrary, at night and to a smaller extent in the evening, text messages are frequently sent to a specific subset of alters. Note that there is a lot of variation in the individual entropy patterns.

In [9], it was seen that low entropy is often associated with calls to top-ranked alters. We computed the correlation coefficients between relative entropy and fraction of texts to top 3 alters separately for each ego (to avoid the ecological fallacy problem). Unlike for calls, only 7 out of 24 correlation coefficients had a p-value less than 0.05, and out of those, 6 coefficients displayed negative correlations. Hence, unlike for calls, communication focused at top-ranked alters do not necessarily explain the low-entropy time ranges. This may have to do with text messages serving a different role in communication, as also seen in the daily frequency patterns. However, 6 out of 7 statistically significant correlation coefficients were still clearly negative, averaging at $r \approx -0.75$, so for certain individuals, a high level of communication to top-ranked alters at certain hours explains the entropy variation.

18.3.3 Gender Differences in Communication Patterns

Next, we turn to gender differences in the daily communication patterns. Overall, when comparing the total numbers of calls and texts, we observe that there is a considerably higher total number of texts than calls, both for males and females. This may have to do with the study participants being about 18 years of age, as heavy users of text messaging are often found in the younger age groups. Further, carrying out a single conversation via text messages may involve a large number of texts.

We also see that calls and texts have a different average daily pattern, with calls peaking in the afternoon and texts in the evening (see Fig. 18.4). Frequent texts in the evening may be related to youth culture and communication conventions, and may also have to do with the intrusiveness of the channel; as seen in [9], calls at late hours are often targeted at a small subset of closest alters. Comparing the communication patterns of males and females, we see that the total numbers of calls made by males and females during the 18 months of data collection are almost equal, whereas females tend to text much more frequently than males. This difference is largest in the evening (see Fig. 18.4). Overall, the total number of text messages is about 1.5 times higher for females than males.

Focusing on different types of social ties, we see that even though the number of calls to friends and kin are similar, for texts they are very different.[1] This shows that calling is the dominant channel for communicating with kin. Despite the low numbers of texts to kin both by male and female egos, females in this study have texted their kin about 3 times more often than males, which agrees with other studies that males and females indeed make use of mobile telephones differently [14–16].

18.4 Summary and Conclusions

We have studied patterns of communication via mobile telephone calls and text messages, and have shown that like many other types of human activity, these patterns follow daily rhythms. Interestingly, the daily patterns of texts and calls appear

---

[1]Note that in reality the total numbers of calls to friends might be much higher, because for the majority of alters it is unknown whether they are friends, acquaintances, or social ties of some different type. Here, we call those alters for whom an emotional closeness score is available in the surveys "friends". However, we can still compare the ratio of calls to friends versus kin with texts to friends versus kin, because the set identified as friends is the same both for texts and calls.

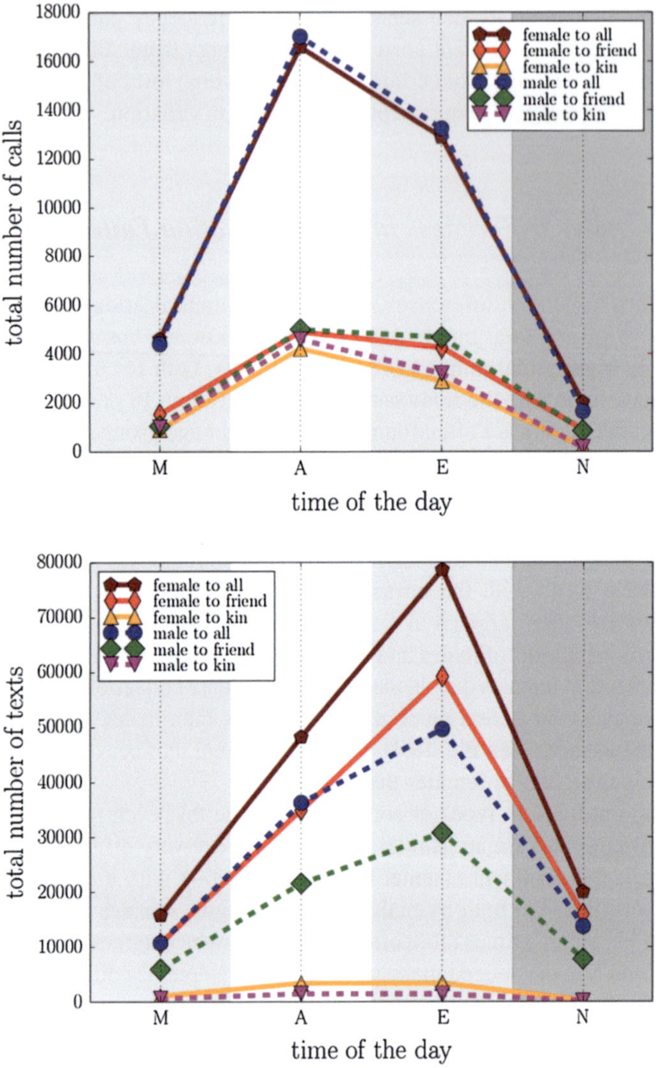

Fig. 18.4 Distribution of the total number of calls and texts (*top* and *bottom panels*, respectively) at different times of day, for calls/texts by female and male egos to alters of different types. The "Female to all" and "Male to all" categories contain those unknown alters who have not been recalled in surveys, and for whom no personal data is available

different and persistent for each individual. Furthermore, the two patterns may significantly differ for a given individual, pointing out that calls and texts may serve different functions.

One way of interpreting the observed patterns is that they are a superposition of common and unique patterns. First, humans naturally follow the day-night cycle, which is reflected in communication frequency. Second, on top, there may be social conventions and age-group-related effects that individuals typically follow: e.g. it is OK to text someone late in the evening, but calls can be made only to one's closest alters. Third, we have individual differences in personality, communication habits and social habits that give rise to each individual's distinct pattern: note that because of the high level of social network turnover in our data, these cannot be explained by communication conventions with specific alters.

Returning to the differences between calls and texts, our results point out that when studying social networks, data comprising communication along one channel only does not provide a full picture of the network, especially when using the temporal networks framework [17]: different channels play different roles, at different times of day.

Acknowledgments TA and JS acknowledge support from the Academy of Finland, project "Temporal networks of human interactions" (no. 260427), and computational resources by Aalto Science IT. RD's research is supported by an ERC Advanced grant. SGBR and RD acknowledge support from the UK EPSRC and ESRC research councils for collecting the data.

References

1. Kerkhof, G.A.: Inter-individual differences in the human circadian system: a review. Biol. Psychol. **20**, 83–112 (1985)
2. Czeisler, C.A., et al.: Stability, precision, and near-24-hour period of the human circadian pacemaker. Science **284**, 2177–2181 (1999)
3. Panda, S., Hogenesch, J.B., Kay, S.A.: Circadian rhythms from flies to humans. Nature **417**, 329–335 (2002)
4. Yasseri, T., Sumi, R., Kertész, J.: Circadian patterns of wikipedia editorial activity: a demographic analysis. PLoS One **7**, e30091 (2012)
5. Yasseri, T., Quattrone, G., Mashhadi, A.: Temporal analysis of activity patterns of editors in collaborative mapping project of openstreetmap. In: Proceedings of the 9th International Symposium on Open Collaboration, p. 13. ACM (2013)
6. Jo, H.-H., Karsai, M., Kertész, J., Kaski, K.: Circadian pattern and burstiness in mobile phone communication. New J. Phys. **14**, 013055 (2012)
7. Louail, T., et al.: From mobile phone data to the spatial structure of cities. Sci. Rep. **4** (2014)
8. ten Thij, M., Bhulai, S., Kampstra, P.: Circadian patterns in twitter. In: Data Analytics 2014, The Third International Conference on Data Analytics, pp. 12–17 (2014)
9. Aledavood, T., López, E., Roberts, S.G.B., Reed-Tsochas, F., Moro, E., Dunbar, R.I.M., et al.: Daily rhythms in mobile telephone communication. PLoS ONE **10**(9), e0138098 (2015). doi:10.1371/journal.pone.0138098
10. Stopczynski, A., et al.: Measuring large-scale social networks with high resolution. PLoS One **9**, e95978 (2014)
11. Ling, R.: The Mobile Connection: The Cell Phone's Impact on Society. Morgan Kaufmann (2004)
12. Roberts, S.G.B., Dunbar, R.I.M.: The costs of family and friends: an 18-month longitudinal study of relationship maintenance and decay. Evol. Human Behav. **32**, 186–197 (2011)
13. Saramäki, J., et al.: Persistence of social signatures in human communication. Proc. Natl. Acad. Sci. USA **111**, 942–947 (2014)
14. Zainudeen, A., Iqbal, T., Samarajiva, R.: Who's got the phone? gender and the use of the telephone at the bottom of the pyramid. New Media Soc. **12**, 549–566 (2010)

15. DeBaillon, L., Rockwell, P.: Gender and student-status differences in cellular telephone use. Int. J. Mobile Commun. **3**, 82–98 (2005)
16. Palchykov, V., Kaski, K., Kertész, J., Barabási, A.-L., Dunbar, R.: Sex differences in intimate relationships. Sci. Rep. **2** (2012)
17. Holme, P., Saramäki, J.: Temporal networks. Phys. Rep. **519**, 97–125 (2011)

Chapter 19
Investigating the Phonetic Organisation of the English Language via Phonological Networks, Percolation and Markov Models

Massimo Stella and Markus Brede

Abstract Applying tools from network science and statistical mechanics, this paper represents an interdisciplinary analysis of the phonetic organisation of the English language. By using open datasets, we build phonological networks, where nodes are the phonetic pronunciations of words and edges connect words differing by the addition, deletion, or substitution of exactly one phoneme. We present an investigation of whether the topological features of this phonological network reflect only lower or also higher order correlations in phoneme organisation. We address this question by exploring artificially constructed repertoires of words, constructing phonological networks for these repertoires, and comparing them to the network constructed from the real data. Artificial repertoires of words are built to reflect increasingly higher order statistics of the English corpus. Hence, we start with percolation-type experiments in which phonemes are sampled uniformly at random to construct words, then sample from the real phoneme frequency distribution, and finally we consider repertoires resulting from Markov processes of first, second, and third order. As expected, we find that percolation-type experiments constitute a poor null model for the real data. However, some network features, such as the relatively high assortative mixing by degree and the clustering coefficient of the English PN, can be retrieved by Markov models for word construction. Nevertheless, even Markov processes up to third order cannot fully reproduce other patterns of the empirical network, such as link densities and component sizes. We conjecture that this difference is related to the combinatorial space the real and the artificial phonological networks are embedded into and that the connectivity properties of phonological networks reflect additional patterns in word organisation in the English language which cannot be captured by lower order phoneme correlations.

M. Stella (✉) · M. Brede
Institute for Complex Systems Simulation, University of Southampton,
University Rd, Southampton SO17 1BJ, UK
e-mail: massimo.stella@inbox.com

S. Battiston et al. (eds.), *Proceedings of ECCS 2014*, Springer Proceedings in Complexity, DOI 10.1007/978-3-319-29228-1_19

19.1 Introduction

Human language relies on a hierarchical, multi-level combination of relatively simple meaningful components (i.e. graphemes, phonemes, words, periods) engaging in phonological, semantic, lexical and ortographic sophisticated interactions that ultimately allow for human communication [1–3]. An educated adult English speaker can retain up to $N_w \sim 50{,}000$ different words [4] composed of a small number of roughly 36 distinct phonological elements (i.e. phonemes).

Recently, complex networks have been applied to the investigation of human language in cognitive science, with nodes representing words and links indicating the presence of certain semantic, morphologic, orthographical or phonological relationships [1, 2, 5–13]. More in detail, at a high representation level, the cognitive process behind human language organisation can be analysed in terms of the so-called *mental lexicon* (ML) [4, 6, 14, 15]. This idealisation abstracts the details of the mental repository of words, that are stored and correlated according to multi-layered relationships (i.e. grammatical, semantic, phonological, syntactic, orthographical, etc.).

One area of psycholinguistics where complex networks have been successfully applied is phonology [8, 10–12, 16]. In [12], Vitevitch built a phonological network (PN), based on the Merriam Webster Pocket Dictionary, with nodes representing phonetic word transcriptions and with links connecting phonologically similar words (i.e. words differing in the addition, deletion or substitution of one phoneme [17]). For instance, the two words "cat" and "rat" would be connected in such a linguistic network. This operational definition of phonological similarity was based on empirical findings of spoken word production and recognition [17, 18]. Furthermore, adopting such metric for the network construction led to a convenient equivalence between node degree and the so-called *phonological neighbourhood density* (PND) of a given word, i.e. the number of its phonologically similar words [12, 17, 18]. Empirical studies indicate that a high PND/degree promotes speech errors such as malapropism[1] in high frequency words [18, 19].

This paper aims to add to this research by investigating phonetic correlations in the English language. We build a larger phonological network, which reveals interesting properties, quite uncommon in other real-world social, biological or technological networks, such as a very high assortative mixing by degree and global clustering coefficient and a dense giant component, surrounded by many smaller size connected components called linguistic islands. A theoretical framework for the understanding of these topological features has been recently suggested in [20]. Within the same context, the main research question of this work focuses on the extent up to which patterns in network organisation can be attributed to short or long range correlations between phonemes in words. We explore this issue by developing percolation- or Markov process-based models for word assembly. We test the validity of these processes as suitable models for investigating the structural patterns regarding the organisation of the English language that are reflected in the English PN.

[1] A malapropism is a type of word speech error where a target word is erroneously substituted by a phonologically similar word but from a different semantic context.

All in all, the work implements five different null models of randomly generated word repertoires that satisfy some constraints of the empirical data (i.e. phoneme frequency, word length distribution, phoneme correlations). These models are used to build "artificial" phonological networks which are then analysed and compared to the English PN. We conclude with a discussion of the results of the comparisons.

19.2 Structural Analysis of the Phonological Network in English

We based our construction on the English database WordData from Wolfram Research, a curated repository developed by the Princeton University Cognitive Science Laboratory "WordNet 3.0." (2006) and by the Oxford University Computing Service, British National Corpus, version 3 (2007).

The English corpus we used was composed of 29,750 phonological transcriptions of words, containing 36 different phonetic symbols and constituting a reasonable approximation of the mental lexicon at the phonological level. The network statistics are reported in Table 19.1. The English phonological network (PN) is disconnected and it exhibits relatively high values of the global clustering coefficient and assortativity coefficient, as defined in [21]. Our results are compatible with the previous analysis of Vitevitch [12], which was performed on a smaller dataset.

19.3 Five Null Models for One Hypothesis

Our null network models are based on artificial pseudolexica composed of:

1. random uncorrelated sequences of phonemes drawn uniformly at random by using the empirical word length distribution as the only constraint (**UR** ensemble);
2. random uncorrelated sequences of phonemes drawn at random by using the empirical word length distribution and the empirical phoneme frequency (**FR** ensemble);
3. random correlated sequences of phonemes drawn according to a Markov chain or a 2nd or a 3rd Order Markov processes trained on the empirical data, with no additional external constraint (**MC**, **2MP**, **3MP** ensembles, respectively).

All of our network ensembles are composed of 10 artificial network realisations. The random experiments **UR** and **FR** incorporate different constraints (i.e. word length distribution and empirical phoneme frequency) and they loosely relate to the WLO ensemble from [22]. These models constitute refined percolation experiments on the original discrete space phonological networks are embedded into [20]. The Markov process-based ensembles represent a different approach, since **MC**, **2MP** and **3MP** approximate the short range correlations present among phonemes.

In the following, we are going to explain each process, before analysing and comparing the results against the real data of the English phonological network.
Correlation Level 0—Sampling phonemes uniformly at random (UR) We sample the same number of words in the real data according to the empirical word length distribution $\mathcal{L}$. Subsequently, we structure each word of length l as a collection of l phonemes, drawn uniformly at random from the empirical alphabet of 36 phonemes.

Correlation Level 0—Sampling phonemes with empirical phoneme frequencies (FR) This ensemble is equivalent to the **UR** one, except for the fact that phonemes are drawn according to their relative frequencies in the empirical data.

Correlation Level 1—Markov Chain (MC) Different to Gruenenfelder and Pisoni's approach [22], we did not mimic consonant-vowel patterns but rather tried to infer the full bivariate correlation structures of phonemes in English. For this purpose, we adopted discrete time homogenous Markov chains [23] trained on the empirical data. Using this approach, we interpreted each word of the pseudolexicon $w_l = (s_i, ..., s_j)$ as a random walk among $|\mathcal{A}| = 36$ plus 2 states (including a "START" = 0 and "END" = 1 fictitious symbols). All the random walks emanate from the original state "START", then at each time step a state was visited and the relative phoneme added to the generated word, until the random walk reached the absorbing "END" state. Independent of the time step, the transition probability of going from state/phoneme s_i to state/phoneme s_j corresponds to the entry $W_{ij} = P(phon_{t+1} = s_i | phon_t = s_j)$ of a transition matrix **W**, obtained from the real data. Any time homogenous finite state Markov chain can be represented as a weighted graph [23]. For this end, one can represent states as nodes and associate the transition probabilities W_{ij} between them with weighted directed links. Figure 19.1 illustrates the structure of the transition matrix we calculated from the phonetic transcriptions in the English PN and a modularity based [21] graph community plot on the Markov chain. In the latter, phonemes are states and arrows represent accessible couples of phonemes. The fictitious starting phoneme 0 is clustered together with the most frequent first position phonemes in our database. Self-loops represent the possibility for phoneme repetitions. The final state 1 is an absorbing state, i.e. there is no way to transition to other states once it is reached. Interestingly, in the matrix plot of the transition matrix **W** a few preferential patterns are evident for given couples of phonemes. For instance, the all white entries in the row corresponding to 1 imply that it is impossible to transition away from 1 once it is reached. Simply put, all the generated artificial words never have 1 followed by another phoneme. Similarly, the phoneme "e" (as in "dress") is followed only by phoneme "ɪ" (as in "kit") and so on.

Correlation Levels 2 and 3—2nd and 3rd Order Markov Processes (2MP and 3MP) Sequences of phonemes representing real words contain higher than first order correlations and therefore the process of real word formation cannot be memoryless [4]. Hence, we generalised our approach to second and third order Markov processes. A

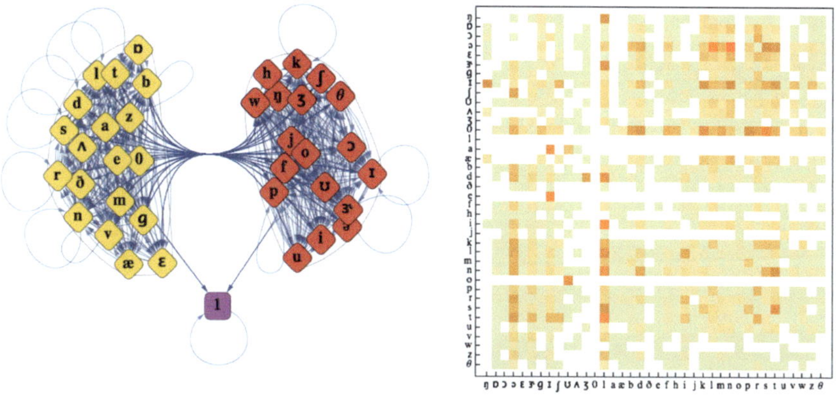

Fig. 19.1 *Left* Modularity based graph community plot on the Markov chain. *Right* Matrix plot of the transition matrix **W**. Colors range from the case $W_{ij} = 0$ (*white color*) to $W_{ij} = 1$ (*red color*)

discrete Markov process of order $m \in \mathbb{N}$ is a stochastic process where the transition probability to the next future state depends also on the m previous states. However, every Markov process of order m can be mapped into a first order Markov Chain, by expanding the state space [23]. Let $E_X = \{s_i\}_{i=1,\ldots,N}$ be the expanded state space obtained by considering all the dispositions of the states (i.e. phonemes) in the original Markov chain in groups of m. In our case, for a Markov process of order m, E_X has size $N = 38^m$ and it contains all the phonetic m-tuples $s_1, s_2, \ldots, s_N$. Using these m-tuples as new starting states allows for the construction of an expanded transition matrix $\mathbf{W}^X$, with rows being labelled by the e_i and columns being labelled by the s_i. Therefore, $\mathbf{W}^X$ is a rectangular matrix of dimension $38^m \times 38$.

Markov processes of a given order m partially reproduce the so-called phonotactic probabilities (i.e. the probability with which phonemes and phoneme $m-$tuples occur in words in a given language [17, 19]). Extensive research indicates that words having high phonotactic probability (i.e. more frequent) tuples are actually recognised more quickly [4, 16–18] in speech experiments. Contrary to the percolation experiments **UR** and **FR**, the pseudolexicons obtained via Markov processes (**MC**, **2MP** and **3MP**) do not incorporate any a-priori hard constraint on the word length distribution. Any artificial word of length more than 24 (the maximum in the empirical word length distribution) is discarded from the pseudolexica. A comparison of the word length distributions of **MC**, **2MP** and the empirical data is displayed in Fig. 19.2. Without hard constraints, it is evident that the Markov processes oversample longer words compared to the empirical frequency distribution. This finding points out the presence of constraints different from local phoneme correlations that act specifically on longer words.

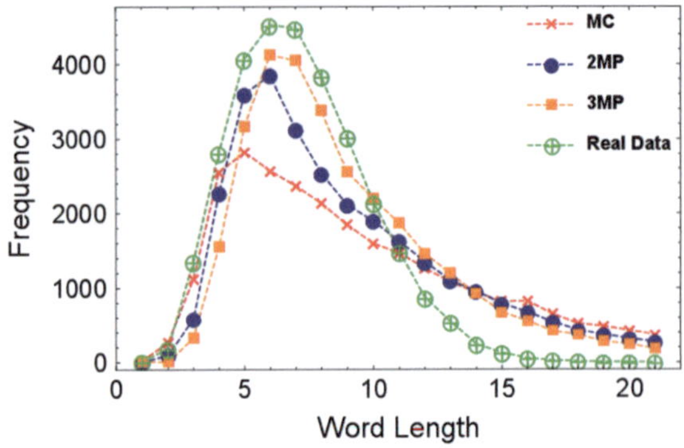

Fig. 19.2 Word length frequency distribution for the MC (×), 2MP (●), 3MP (■) and real data (⊕)

19.4 Network Analysis of Artificial PNs

There is a fundamental point that has to be underlined before proceeding with the comparison of our percolation and Markov null models with the English PN. The 36 English phonetic symbols constitute a finite alphabet $\mathcal{A} = \{s_i\}_{i=1}^{36}$ and, as in the Markov processes, sequences of these elements make up phonetic word-forms. Furthermore, all the sequences (i.e. words) of a given length constitute a labelling of a regular graph R_l that has $|V| = |\mathcal{A}|^l$ vertices, each one having $N_l = l(|\mathcal{A}| - 1)$ first neighbours. In [20], we introduced the concept of *phonological network layer* as the subgraph of R_l induced by the vertex labellings composed of all the words in the English PN of a given length. Therefore, a phonological network is a collection of interconnected subgraphs (i.e. layers) of regular graphs, where inter-layer and intra-layer connections represent phonological similarity. The numerical experiments presented in [20] strongly suggest that some network features such as the clustering and the assortativity coefficients of the real PN are inherited by the (percolating) layers of shorter word length: therefore they are not suitable indicators for testing the meaningfulness of null models in reproducing the topology of the empirical data, contrary to previous suggestions [22].

We adopted the phonological similarity metrics to produce artificial phonological networks (APN) from our artificial pseudolexica. A snapshot with some network statistics is reported in Table 19.1.

The performances of percolation null models are quite poor. The **UR** sampling retrieved networks with relatively small giant component sizes (620 ± 80 nodes). Inspired by the concept of percolating PN layers, we tested whether the size of the giant component (GC) was critically sensitive to the external constraints, namely the word length distribution. By using a maximum likelihood procedure, we fitted the empirical word length distribution to a generalised gamma distribution [23], obtaining parameters $(\alpha, \beta, \gamma, \mu) = (1.4, 5.1, 1.9, 1.5)$. These parameters correspond to an estimated mean word length $\langle \mathcal{L}_{gam} \rangle \simeq 7.1$ phonemes and to a standard deviation $\sigma_{gam} \simeq 2.5$ phonemes. Our choice of a gamma distribution was motivated by the need of having a function whose mean and variance could be easily tuned. The fitting

Table 19.1 Summary of word inventory, giant component (GC) size, total number of lexical hermits (i.e. unconnected nodes), total size of lexical islands, mean island size, mean node degree $\langle k \rangle$, mean node degree $\langle k \rangle$ in the giant component, mean geodesic distance $\langle \ell \rangle$ in the GC, global clustering coefficient CC, global clustering coefficient in the GC, assortativity coefficient AS, assortativity coefficient AS in the GC for the real PN and all our other artificial word inventories

	Empirical Data	UR	UR ($\mathcal{L}^*$)	FR	MC	2MP	3MP
Word Invent.	29048	29048	29048	29048	29050(50)	29050(50)	29050(50)
GC	9412 ~ 32 %	2 %	69 %	1 %	25 %	35 %	36 %
Lex. Her.	14459 ~ 50 %	97 %	19 %	96 %	73 %	62 %	57 %
Tot. Lex. Isl.	5177 ~ 18 %	1 %	13 %	3 %	2 %	3 %	7 %
Mean Isl. S.	2.5	2.2(1)	2.9(1)	3.0(1)	2.4(1)	2.6(1)	2.6(1)
$\langle k \rangle$	5.2	2.1(3)	2.6(3)	1.6(3)	10.2(4)	8.5(4)	6.3(4)
$\langle k \rangle$ in GC	7.5	2.1(2)	2.8(2)	2.6(2)	11.0(2)	9.1(2)	7.2(2)
$\langle \ell \rangle$ in GC	7.7	2.9(4)	10.4(4)	7.6(3)	5.7(3)	6.9(3)	8.5(3)
CC	0.21	0.01(1)	0.17(1)	0.09(1)	0.29(1)	0.29(1)	0.27(1)
CC in GC	0.28	0.40(1)	0.19(1)	0.23(1)	0.31(1)	0.31(1)	0.30(1)
AS	0.70	0.10(1)	0.36(1)	0.37(1)	0.50(1)	0.59(1)	0.63(1)
AS in GC	0.65	0.05(1)	0.31(1)	0.15(1)	0.49(1)	0.58(1)	0.61(1)

The error bars have been estimated on ensembles of 10 network realisations and they are reported in a compact way (i.e. 2.9(4) means 2.9 ± 0.4)

procedure was not statistically significant but the artificial networks generated with the resulting gamma distribution showed an average giant component size of 610±80, a value compatible with the percolation experiments with the empirical word length distribution. With the new parameters $(\alpha', \beta', \gamma', \mu') = (1.2, 2, 1.84, 2.65)$, we obtained another word length distribution, $\mathcal{L}^*$, having far smaller mean ($\langle \mathcal{L}^* \rangle \simeq 4.7$ phonemes) and variance ($\sigma^* \simeq 1$ phoneme) when compared to the empirical distribution. The artificial networks generated with $\mathcal{L}^*$showed a significantly larger *GC* average size of 19900 ± 200. This finding highlights a critical dependence of the *GC* size on the sampling of word lengths. In fact, shorter words live in combinatorics spaces of smaller size $|\mathcal{A}|^l$ with l being the word length. Sampling more words from these smaller spaces (this is what happened by using $\mathcal{L}^*$), without repetitions, means building layers that inherit some features of the original regular graph structure, i.e. clustering and assortative mixing by degree. In other words, varying the number of random words in layers is equivalent to a percolation experiment in which a giant component arises when the occupation density is large enough.

However, in spite of the appearance of a giant component, the networks generated from percolation (**UR**) have far less links than found in the English PN. The stark contrast in connectivity between artificial PNs and the real PN hints at peculiar properties of word organisation. Improving the null model by sampling phonemes at random according to their occurrence frequencies in the real data (experiment **FR**) only leads to minor changes in comparison to **UR**. In fact, including phoneme frequencies slightly boosts the likelihood for words to be phonologically similar and we can give an analytic explanation of this empirical finding.

Let us start from the remark that connections in the PN among words of the same length are formed between words differing for one phoneme only, independently on the phoneme itself. Let us compute the probability of finding words satisfying this instance. Let $p_i(l)$ be the probability for two words of length l, $w_l = (s_1, ..., s_l)$ and $w_l' = (s_1', ..., s_l')$ respectively, to have any given shared phoneme $s_i = s_i'$ at any position. On the one hand, because of the Bernoulli sampling, in **UR** phonemes are sampled independently on each other. Also, in the **UR** ensemble any phoneme s_i is sampled with uniform probability $f_i^{UR} = 1/|\mathcal{A}|$. As a consequence then $p_i^{UR}(l) = \sum_{\varphi=1}^{l} (f_i^{UR})^2 = l/|\mathcal{A}|^2$ and it is uniform over different phonemes, i.e. $p_i^{UR}(l) = p^{UR}(l)$. In order to quantify the probability of phonological similarity, let us denote with $P_{1,i}(l)$ the probability for two words of length l to differ on one given phoneme only. In both **UR** and **FR**, phonemes are sampled independently on each position, so that in both these ensembles phonemes have to be equal in $l-1$ positions and to differ on one position only. Therefore, in the **UR** ensemble $P_{1,i}(l)$ follows a Bernoulli distribution and summing over all the alphabet phonemes provides the probability $P_1(l)$ that two words of same length l are phonologically similar. In formulas:

$$P_{1,i}^{UR}(l) \propto (p^{UR})^{l-1}(1 - p^{UR}) = \frac{l^l}{|\mathcal{A}|^{2l}} \left(\frac{|\mathcal{A}|^2}{l} - 1 \right) \qquad (19.1)$$

$$P_1^{UR}(l) \propto |\mathcal{A}|\, P_{1,i}^{UR} = |\mathcal{A}|\, (p^{UR})^l \left(\frac{1}{p^{UR}} - 1\right).$$

On the other hand, in the **FR** ensemble $p_i(l)$ depends on the specific shared phoneme s_i. Hence, in **FR** the probability for two words to share a given phoneme at any position is $p_i^{FR}(l) = \sum_{\varphi=1}^{l}(f_i^{FR})^2$, where f_i^{FR} is the occurrence probability of phoneme $s_i \in \mathcal{A}$. From the empirical data, we approximate phoneme occurrences with a frequency-ranked power-law distribution, then $f_i^{FR} = Ar_i^{-a}$ for $i \in \{1, 2, ..., |\mathcal{A}|\}$, where r_i is the frequency rank of phoneme s_i and $a, A \in \mathbb{R}$. A power-law fitting of the empirical data corroborates this conjecture by retrieving a statistically significant power-law exponent $a = 1.10 \pm 0.08$ (Pearson χ^2 p-value $\simeq 0.021$). Therefore, we have that:

$$p_i^{(FR)}(l) = \sum_{\varphi=1}^{l}(f_i^{FR})^2 \simeq A^2 \sum_{\varphi=1}^{l} \frac{1}{r_i^{2\alpha}} = A^2 \frac{l}{r_i^{2\alpha}}, \tag{19.2}$$

Inserting $p_i^{(FR)}(l)$ in $P_{1,i}^{(FR)}(l)$, then the probability for two words of the same length in the **FR** ensemble to be phonologically similar is analytically given by:

$$P_{1,i}^{(FR)}(l) \propto \frac{(A^2 l)^{l-1}}{r_i^{2\alpha(l-1)}} - \frac{(A^2 l)^l}{r_i^{2\alpha l}} \tag{19.3}$$

$$P_1^{(FR)}(l) = \sum_{i=1}^{|\mathcal{A}|} P_{1,i}^{(FR)}(l) \propto (A^2 l)^{l-1} \left[H_{|\mathcal{A}|}^{[2\alpha(l-1)]} - A^2 l H_{|\mathcal{A}|}^{[2\alpha l)]}\right],$$

where $H_n^{[m]} - \sum_{k=1}^{n} k^{-m}$ is the generalised harmonic number [23]. Numerically, it can be checked that the estimated $P_1^{(FR)}$ is higher than $P_1^{(UR)}$ (their normalisation constant being the same) for words with length $l \subset [3, 7]$. This means that a power-law like behaviour in the individual phoneme sampling frequencies actually boosts phonological similarity in artificial repositories but mainly for words of intermediate lengths. This finding is compatible with the increased *GC* size in the **FR** ensemble when compared to the **UR** one.

Furthermore, in the **MC** ensemble constraining local correlations with Markov processes changes the situation dramatically. Using up to first-, second- and third-neighbour correlations, respectively, increments the sizes of the giant component and of the lexical islands, which gets closer to the empirical ones. Also the clustering and the assortativity coefficients are already retrieved by the first order Markov chain in **MC**. Increasing the order of the Markov process in **2MP** and **3MP** retrieves average node degrees, mean geodesic path length and link densities, that are similar to the empirical phonological network but not quite compatible. Also the empirical size of lexical hermits (i.e. unconnected words) is not matched by the Markov ensembles.

All these features constitute a qualitative indication that there are *long range correlations* in language, not fully approximated by the Markov processes, that influence the network structure of our given phonological similarities.

19.5 Conclusions and Future Directions

Phonological networks (PNs) can be used to analyse the phonological level of the human mental lexicon, which is an intricate repository of word-forms, relationships and correlations.

Within the idealisation of words acting as labels of subgraphs of a given collection of regular graphs, we discussed the concept of word layers, as in sets of words of a given length participating in pairwise interactions within and outside the layer itself. We proceeded with our analysis by proposing five different null models of phonological networks, using percolation-like experiments and Markov processes. Our artificial phonological networks indicate that the combinatorial structure of the PN layers can be retrieved also in the phonological networks, but always together with additional organisation patterns that represent the phonotactic constraints of human language. In fact, random percolation experiments and Markov chains are able to reproduce some network features such as the global clustering coefficient or the assortativity coefficient of the empirical data (as in [22]). However, percolation-like experiments are not able to reproduce other connectivity patterns such as the mean degree, the giant component size or the link density of the English PN. Even higher order Markov processes fail at this task, implying the importance of additional long range correlations in determining the topology of the English phonological network.

Interestingly, the analysis of the network structure at a microscopic level reveals that the majority of substitutions/additions/deletions of single phonemes happen in the first and in the last position of a given word, being the case for both intra- and inter-layer similarities. This feature is not reproduced by random null models but it is in good agreement with some linguistic conjectures according to which words sharing the same rhyming recognised preferentially as phonologically similar [16, 19].

Having successfully adopted the network paradigm for studying phonological similarities, there are still many open questions that are worth further investigation. In fact, it would be interesting to build and analyse phonological networks of different languages, in order to assess the universality of the network features or rather use network theoretic measures to quantify and distinguish between different languages. Also generalising the phonological similarity measure might lead to new insights about the outlay of real words in the underlying discrete word space.

Acknowledgments The authors acknowledge the Doctoral Training Centre in Complex Systems Simulation at the University of Southampton, in the completion of this work. MS was supported by an EPSRC grant (EP/G03690X/1).

References

1. Solé, R.V., Seoane, L.F.: Ambiguity in Language Networks. arXiv:1402.4802 (2014)
2. Kello, C.T., Beltz, B.C.: Scale-free networks in phonological and orthographic wordform lexicons. In: Approaches to Phonological Complexity (2009)
3. Kintsch, W.: The role of knowledge in discourse comprehension: a construction-integration model. Psychol. Rev. **95**(2), 163 (1988)
4. Aitchison, J.: Words in the Mind: An Introduction to the Mental Lexicon. Wiley (2012)
5. Ferrer-i-Cancho, R., Solé, R.V.: The small world of human language. Proc. R. Soc. Lond. Series B: Biol. Sci. **268**(1482), 2261–2265 (2001)
6. Baronchelli, A., Ferrer-i-Cancho, R., Pastor-Satorras, R., Chater, N., Christiansen, M.H.: Networks in cognitive science. Trends Cogn. Sci. **17**(7), 348–360 (2013)
7. Arbesman, S., Strogatz, S.H., Vitevitch, M.S.: The structure of phonological networks across multiple languages. Int. J. Bifurcat. Chaos **20**(03), 679–685 (2010)
8. Chan, K.Y., Vitevitch, M.S.: Network structure influences speech production. Cogn. Sci. **34**(4), 685–697 (2010)
9. De Deyne, S., Storms, G.: Word associations: network and semantic properties. Behav. Res. Meth. **40**(1), 213–231 (2008)
10. Griffiths, T.L., Steyvers, M., Firl, A.: Google and the mind predicting fluency with pagerank. Psychol. Sci. **18**(12), 1069–1076 (2007)
11. Vitevitch, M.S., Chan, K.Y., Goldstein, R.: Insights into failed lexical retrieval from network science. Cogn. Psychol. **68**, 1–32 (2014)
12. Vitevitch, M.S.: What can graph theory tell us about word learning and lexical retrieval? J. Speech, Lang. Hear. Res. **51**(2), 408–422 (2008)
13. Vitevitch, M.S., Chan, K.Y., Roodenrys, S.: Complex network structure influences processing in long-term and short-term memory. J. Mem. Lang. **67**(1), 30–44 (2012)
14. Elman, L.J.: An alternative view of the mental lexicon. Trends Cogn. Sci. **8**(7), 301–306 (2004)
15. Ke, J.: Complex Networks and Human Language. arXiv preprint cs/0701135 (2007)
16. Siew, C.S.: Community structure in the phonological network. Front. Psychol. **4** (2013)
17. Luce, P.A., Pisoni, D.B.: Recognizing spoken words: the neighborhood activation model. Ear Hear. **19**(1) (1998)
18. Sadat, J., Martin, C.D., Costa, A., Alario, F., et al.: Reconciling phonological neighborhood effects in speech production through single trial analysis. Cogn. Psychol. **68**, 33–58 (2014)
19. Vitevitch, M.S.: The neighborhood characteristics of malapropisms. Lang. Speech **40**(3), 211–228 (1997)
20. Stella, M., Brede, M.: Patterns in the English language: phonological networks, percolation and assembly models. J. Stat. Mech. P05006 (2015)
21. Newman, M.: Networks: An Introduction. Oxford University Press (2010)
22. Gruenenfelder, T.M., Pisoni, D.B.: The lexical restructuring hypothesis and graph theoretic analyses of networks based on random lexicons. J. Speech, Lang. Hear. Res. **52**(3), 596–609 (2009)
23. Grimmett, G., Stirzaker, D.: Probability and Random Processes. Oxford University Press (2001)

Chapter 20
An Agent-Based Model for Agricultural Supply Chains: The Case of Uganda

F. Caravelli and F. Medda

Abstract Uganda is a landlocked country in East Africa with a population estimated at 35 million. 85 % of the population still lives in rural areas and survives mainly on subsistence farming by growing crops such as matooke, beans, sweet potatoes, coffee (for export), cassava, maize, millet, groundnuts, sorghum, and sesame. There are many obstacles to moving towards sustainable, market oriented crop production. In this research study, we focus on the effect of logistics costs on crop prices from the farm gate through to markets.

20.1 Introduction

Against this background, in our study we develop an agent-based system [1–5] to simulate the supply chain of agricultural produce from farm gate to regional and export markets. A supply chain is a network of facilities that enable in our context the flow of goods from farmers to markets, and the flow of information from markets to farmers. It is a decentralized system with no designated command and control and the efficiency of the system depends on both the coordination between the actors/agents in the system and the robustness of the network. The supply chain can be described by identifying its actors, activities, interdependencies, and objectives. Agent based modelling provides a natural way to model such systems, and in particular allows for the clear identification of the effects due to different policies.

F. Caravelli (✉)
Invenia Technical Computing, 135 Innovation Dr., Winnipeg R3T 6A8, Canada
e-mail: francesco.caravelli@invenia.ca

F. Caravelli
Department of Computer Science, UCL, Gower Street, London WC1 E6BT, UK

F. Caravelli
London Institute of Mathematical Sciences, 35a South Street, London W1K 2XF, UK

F. Medda (✉)
QASER Lab, UCL, Gower Street, London WC1E 6BT, UK
e-mail: fmedda@ucl.ac.uk

S. Battiston et al. (eds.), *Proceedings of ECCS 2014*, Springer Proceedings in Complexity, DOI 10.1007/978-3-319-29228-1_20

We focused our study on 7 regions based on the proximity of the locations to Kampala and Entebbe, and the location of large farmers and exporting firms in the region [6]. The districts we considered were: Luwero, Mpigi, Masaka, Iganga, Mitiyana, Kamuli and Mukono. We additionally only considered the movement of 5 kinds of crops in the supply chain: hot pepper, matooke, okra, chillies and sweet potato.

The hypotheses we aim to test are the following. In the first instance, we test whether travel costs (comprised of transport costs and logistics/storage handling costs) are a major discriminant in the dynamics and ability of farmers to export, and if the quality of the infrastructures are, in the current conditions, a minor discriminant [7]. In addition to this, in our second hypothesis to test is that small scale farmers/outgrowers in the districts under consideration can considerably decrease bankruptcy by sharing risks, and by implementing coordination in the production and export processes. Our last hypothesis tests whether the implementation of GAP or standardization and control procedures leads to increased export and lower stress upon farmers. In this paper we describe the overall setting of the agent-based model and the main results obtained; the structure is organized as follows. In Sect. 20.2 we describe the data we had and how this was used in the model; in Sect. 20.3 we introduce the main agents being modelled; in Sect. 20.4 we study the relationship between the agents, and how this is encoded in a social network approach; in Sect. 20.5 we introduce our microscopic description of trading in terms of micromotives and risk adversity, meanwhile results and conclusion follow in Sects. 20.6 and 20.7 respectively.

20.2 Collected Data

Given the complexity of the model, one has to constrain the many parameters using realistic data. In order to anchor the model to reality, we have conducted a survey and collected data regarding various parameters which were considered in the model.

20.2.1 GPS and Road Data

In the first place, we have obtained GPS from the World Bank on the road infrastructure (paved vs. gravel), and used it to model a coarse grained version of the country roads (in between cities). Since we also modelled transportation means, through the survey we were able to obtain the data on the cost of transportation, measured in Ugsh/km for various transportation vectors used in the country Figs. 20.1 and 20.2. Among these, boda-boda, lorries, pickups and bikes.

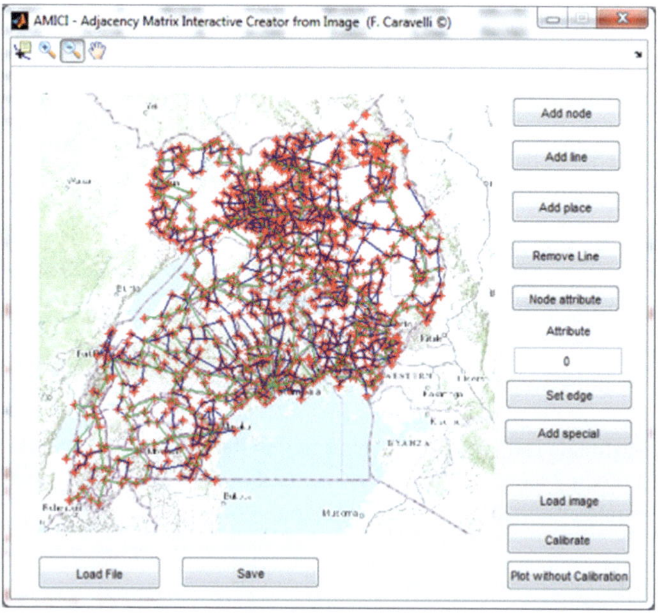

Fig. 20.1 Macroscopic model of the road infrastructure in Uganda. *Green roads* are paved, *blue* are gravel, meanwhile *red spots* are towns and villages

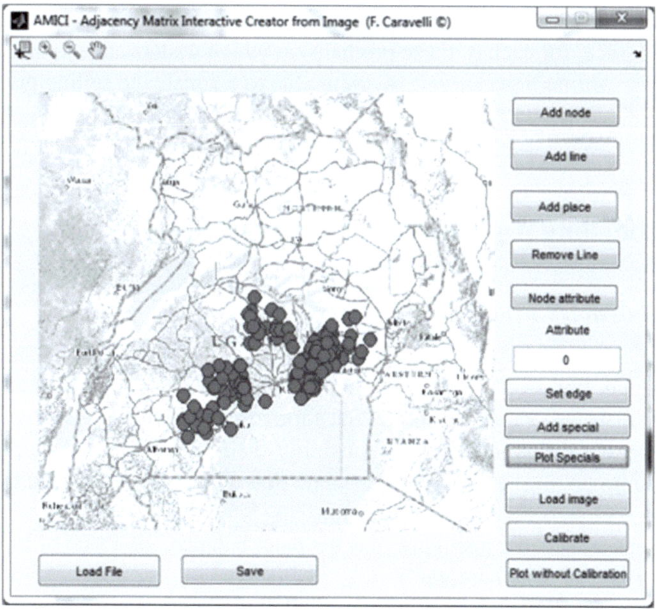

Fig. 20.2 Distribution of farmers considered in the model

20.2.2 Agents Behavior and Costs

We have conducted a survey among farmers and traders in cooperation with the World Bank. This survey was tailored to understand the behavior of the various agents in the model. We have obtained thus data on farmers' risk adversity, on the behavior of itinerant traders, exporters, and producers.

20.2.3 Products

Through the survey, we were able to identify the products which all farmers produce and sell. These are hot peppers, chillies, matooke, okra and sweet potatoes. We were then able to estimate few parameters connected to the spoilage and growing rates. In fact, to each product there is a quality parameter. For each product we have estimated a quality parameter which depends on time through:

$$Q(t) = q_0 e^{-d(T)t} \tag{20.1}$$

where T is the temperature and $d(T) = a_0 + a_1 T$, and a_i are estimated from tables of decay for different products. As an assumption, whenever $Q(t) < 0.2$, the product is considered spoiled and removed from the market [8]; $Q(t_0) = 1$, where t_0 is the time at which the product is harvested. From the survey, we were also able to estimate the average price, for each of these products, at which traders and exporters buy the product. Also, through the survey, we were able to estimate the selling price at local markets.

20.3 Modelling Agents and the Environment

20.3.1 Farmers and Production

A farmer agent has a certain amount of land and wealth that is assigned by using probability distributions. We assume that farmers are distributed at random in the ares of interest for this study. The probability distribution is parameterized using data about farmers in the region. We assume that all farmers adopt multi-cropping (Fig. 20.3).

The assets of a farmer agent are thus his bank balance, the value of his land and the value of his harvested crops.

Given the favouring weather of Uganda, cropping is ongoing at every time of the year; when a product is harvested, seeds are planted and harvesting reoccur after a certain number of months which is product dependent. They divide their land in a uniform manner to cultivate the five crops that were chosen for our study.

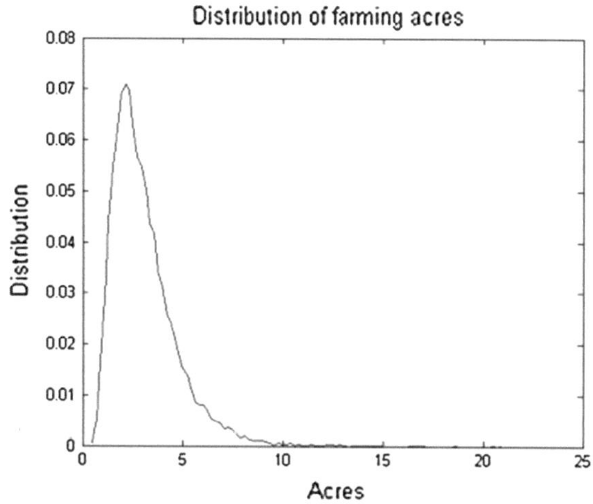

Fig. 20.3 Lognormal distribution of land in acres

While a farmer agent is busy in the cultivation of his crop, he is also engaged in other marketing activities, namely forming agreements. One can identify two typologies of farmers: large farmers (own more than a 5 acres of land) and the most frequently occurring small farmers, whom own on average about 4 acres of land. The activities of large farmers are production, solicit agreements with small farmers (in their social network) wherein they offer to buy produce at a fixed price, selling (using decision tree), and they use the market power and wealth to act as a trader and as a farmer. The activities of large farmers are production, solicit agreements with small farmers in their social network wherein they offer to buy products at a fixed price, selling and trade. The activities of small farmers are production, receive proposals for agreements from large farmers as well from traders, and selling. Some farmers are too poor to market their produce as they only produce enough for subsistence (here considered as bankruptcy). Once the production cycle is complete a farmer agent makes a decision on the sale of his produce using a decision tree process.

We build a social network of farmers that know each other and will go into the detail of its construction in Sect. 20.4. The utility of the social network is to get an idea of the average price, weighted by the quality, of the products to be sold on the market. The difference in price between the average price obtained through the social network and the one proposed by a trader is perceived as a cost.

The decision tree process involves evaluating multiple options based on logistic and processing costs, the expected market price (ascertained from his social network), and the risk-aversion utility function that is a function of the wealth of the farmer. The details of this process are described in Sect. 20.5.

20.3.2 Traders

At the beginning of the simulation, trader agents have a certain amount of wealth that is assigned by using probability distributions with gaussian distribution centered on the mean price obtained through the survey. Traders buy their goods from farmers and other traders. Each trader has a minimum volume to be traded, as well a minimum quality criterium. This ensures that produce moves upward in the supply chain, as once aggregated this can be sold only to a larger trader, eventually reaching the exporters.

Itinerant traders instead do not have any of the above selection criteria, and on top of this, they are travelling and buying at the farmer gate; this has the advantage for them to reduce the transaction cost to the farmers, by reducing the logistic costs and thus giving them an advantage. These are modelled as self-avoiding random walkers starting at a particular area (node), and where the assumption is that these will go around 10 villages in their trip (this fixes the loop length parameter of the self-avoiding random walk).

Trader agents are engaged in three kinds of activities: negotiating agreements (similarly to large farmers) buying and selling (using decision tree). Traders set their prices based on supply and demand as we will describe more in detail in Sect. 20.5.

An important aspect which has to be considered is the act of aggregation of produce which the traders perform. Whenever a trader agent (whether this is a large farmer or an actual trader) holds two similar products, we assume that the product are aggregated and sold together. This will allow the trader to move the product up in the supply chain. We assume the following rule for the aggregation of two products of the same time for volume and quality change:

$$V = V_1 + V_2 \tag{20.2}$$

$$Q = \frac{Q_1 V_1 + Q_2 V_2}{V_1 + V_2} \tag{20.3}$$

where we assume V and Q are the new quality of the product and V_i and Q_i are the volume and quality coefficient before aggregation.

Other important aspects of traders is that these can propose to buy a product and ship it to a warehouse.

20.3.3 Exporters

The role of exporters is to buy products which need to satisfy a minimum volume and a minimum quality to be exported from the country. We assume these are located in the nearby of Kampala and Entebbe, i.e. close to the airport. Their price is fixed and estimated from the survey, and it can be thought as a sink of the products, which once reached the exporters leave the market. This is the highest price in the model, and agents will have a price lower than this, as otherwise they would be making a loss on each transaction.

20.3.4 Logistic Service Provider

The logistics service provider (LSP) agent moves products between two locations. In the current situation of agricultural markets in Uganda, the logistic price is paid by the farmer if these do not possess transportation means. The transport costs incurred are proportional to distance, with a proportionality factor that was estimated from the survey. We assume the LSP has two types of service, one with refrigerators (for a higher price and a higher reliability, thereby implying a lower probability of failure) and one without. The difference between the two is in the spoilage rate of the produce which is temperature dependent, which is being simulated. In fact, we assume that agents can predict the spoilage rate of the transportation; the decrease in quality is considered as a cost and inserted in the utility function.

Also, we assume the path taken is always the shortest path between two points, and the simulation considers realistic movements of the transport mean on the road network, based on the speed of the carrier; the cost of the transport is carrier-dependent. The decision of choosing one carrier rather than another is based on the decision tree of Sect. 20.5.

The time taken to deliver the product depends on path taken, whether it is transported on gravel or paved roads. The quality of the transported goods depends on weather conditions (temperature), which is being simulated.

20.3.5 Local Markets

Domestic markets are located in each village. We assume that local markets are sinks for products (i.e. prices fixed and no demand restrictions). Farmers can, thus, sell at zero-kilometers; this price is however usually lower than the production costs, which gives an incentive to the farmers to sell elsewhere.

20.4 Social Network

One of the fundamental ingredients of this agent-based model, is the simulation of every detail regarding the interaction between farmers and traders. In the first place, we construct the network on which interaction will take place, as a Barabasi-Albert model [9] with an exponential smoothing dependent on the distance. We insert a farmer j at random at each step of the algorithm, and then the probability for two farmers i of being connected, is of the form:

$$P_{ji} = \frac{1}{Z} D_i e^{-\frac{d_{ij}}{K}} \tag{20.4}$$

where K is a constant set to 30 km, d_{ij} is the distance between the farmers and D_i is the degree of the farmer i; Z is a normalization constant which is updated at each step. A similar process is performed between traders and farmers, with traders being inserted after. A social network is thus a graph consisting of nodes, representing the agents, and links between nodes that represent the connections between the agents.

We now discuss the spreading of information. At each time step T, we assume that each agent has knowledge of the price and quality of all the products owned by all other agents in his social network. Therefore agent has an attribute that is a price vector indexed by all trading agent, all products owned by, and quality of the product. The graph then is dynamical. In fact, we assume that new connections are formed dynamically.

Information spreads in the following manner [10]. Let us assume that for instance a farmer i knows agent j and agent k, but agent j and k do know each other. With probability p a link between j and k is formed. The probability p is chosen such that this information is transmitted on average in 1 month. This process, due to competition is asymmetrical between farmers and traders, which is that traders do not connect farmer to other traders they know, but farmers can connect other farmers to traders.

The importance of the social network is twofold. In the first place, the assumption is that farmers have access to the information of other farmers on their selling price and quality. Once fixed the product to be sold, where P_i is the price of trader i and Q_i its quality, we use the following formula for the expected price:

$$EP_i = \frac{\sum_{j=farmer} A_{ij} \frac{P_j}{Q_j}}{\sum_{j=farmer} A_{ij}} Q_i \tag{20.5}$$

where A_{ij} is the adjacency matrix of the social network. Also, when one product is ready to be sold, farmers can choose between various offers and can propose to product to traders in their social network.

20.5 Trading and Decision Trees

At each time step in the trade process, price of crop that trader trades, is adjusted to reflect the imbalance of supply and demand. The price varies according to the Walras Law:

$$\frac{d}{dt} P_i^j = \xi(D - S) \tag{20.6}$$

where ξ is a constant to incorporate the action of the Walras Law and are the demand (D)and supply (S) of crop from traders. The demand is accounted from the volume of produce sold, meanwhile the supply is accounted from the amount of crop which is proposed and sold to the trader. This price will define the price at which the trader will buy products.

Decision trees form the core of the trading strategy. Whenever an agent has a product, he will try to sell it among the connections in the social network.

We now describe the decision making process in detail using Fig. 20.6. At the first stage, the trader must decide if he is going to sell his produce or store it to sell later. If he decides to store it, there are two options available to him: he can either store his produce for a week in a warehouse or on his farm. The former option incurs storage costs while the latter incurs costs in the loss of quality of the product, and hence a reduced expected price.

If he decides to trade, he checks his contract agreement, and either decides to honour it (the outcome is the fixed price agreed in the contract) or incur a fee to break the agreement.

If he breaks the contract, he then evaluates the other trading options available to him, based on logistics costs, the risks involved with trade, and his own propensity to risk his wealth. The decision at the end of each branch can be described using the following function:

$$D = (1 - U(C'))(R - C)F, \tag{20.7}$$

where F is the feasibility of the trade, C is the total cost of the transaction, R is the reward, U is the wealth utility function and C' is the implied cost. *Feasibility* In our simulation environment, any two traders can in theory trade with each other. A trader agent considers a trade with another trader agent provided he considers it a ?feasible? option. We incorporate this notion of feasibility using a function. *Cost.* This function can be broken into several pieces:

$$C = D * P_{log} + S + AG + Q \tag{20.8}$$

where D is the weighted Dijkstra distance on the infrastructure graph in kilometers, P_{log} is the transport cost in per unit of distance travelled for the mode of transport, S is the storage costs, which depends on whether the transaction involves sending the produce to a warehouse, and AG is the fee incurred for breaking the agreement with other traders. *Reward.* When a trader agent initiates a trade with trader agent, is the expected value of the product at the expected time of delivery. *Implied costs.* These are the costs which are effectively involved in the transaction plus those perceived by the agent. We can write the equation:

$$C' = C + Q + M \tag{20.9}$$

where C are the actual costs paid by the trader before receiving a payment. Q is the spoilage cost. One can evaluate how long the trip will take and the reduced quality Q'. The traders and exporters apply a reduction of the price due to the quality of the product at destination, and this cost can be perceived as δQPV, where $\delta Q = Q - Q'$ and V is the volume of the shipping. M is instead the market cost, i.e. the cost

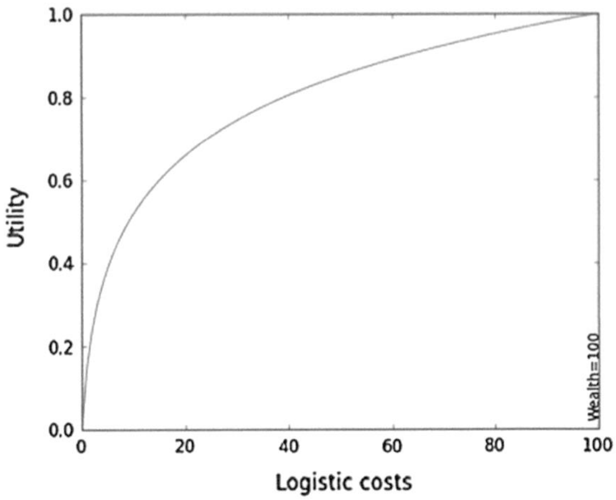

Fig. 20.4 Utility function based on wealth considered in the present model

perceived by the agent as the difference between the expected price EP, evaluated from the social network, and the selling price for this branch of the decision tree, $M = (EP - P) * V$. This term is an incentive for the agent to sell at market price.

Utility function. We associate a utility curve for each trading agent.

The utility curve represents the risk averseness of a trader to pay for logistic costs relative to his private wealth. He is less risk averse when the costs are low and is increasingly more risk averse as the logistic costs equal his total wealth.

We use a utility function of the following form:

$$U_B(x) = \frac{log(1 + sx)}{log(1 + sB)}, \quad 0 \leq x \leq B \tag{20.10}$$

where s is a parameter controlling the convexity of the function (risk-adversity) and B is the wealth of the agent, and is represented in Fig. 20.4.

20.6 Results

We now discuss the results obtained through this agent based model.

In the first place, all the results were obtained by averaging over various simulations, i.e. performing a Monte Carlo. Few parameters were not fixed through the survey, as for instance the wealth of the farmers and the reactivity constant in the Walras law. We averaged over various realizations of these by assuming these are distributed as gaussian around the mean that we were able to estimate by an independent reasoning. For the details on this construction we invite to read the original report [11]. We simulate the course of time at 6 hours interval, meaning that 4 time steps in the model correspond to one day.

As an order parameter to see the effects of the various policies, we use the number of bankrupt agents. These are defined as agents which, over the time of a simulated month, have a negative bank account. The result is that these agent then exit the market and are not considered anymore as active agents. We consider 400 hundred farming agents and 100 trading agents, of which 50 actual traders and 50 itinerant trades. Each curve is averaged over 100 simulations.

Our methodology is the following: we study the evolution of the bankrupt agents a function of time without any shocks and consider this as the closest to current situation in Uganda. We observed that the number of bankrupt agent in our simulation increases with time, which is a faithful representation of the current situation in the agricultural market. Thus, statements are made relatively to the current situation.

20.6.1 Transport

As a first shock, we consider a reduced transport cost of 10 %, and monitor the number of bankrupt agents as function of time. The time window is 1 year, and the results are shown in Fig. 20.5. The result is that in the long run, roughly 10 % less bankrupt agents are observed. This stresses the importance of current transport cost on the market.

As a second test, we gauge the importance of the road infrastructure through an improvement policy. The policies are shown in Fig. 20.6.

The infrastructure on the left is the one of the current road infrastructure, meanwhile the one at the center is our Policy 1 improvement, where we improve from gravel to paved roads (blue to green) only roads in the nearby of Kampala and Entebbe, meanwhile in the right figure we also includes the roads in the areas where

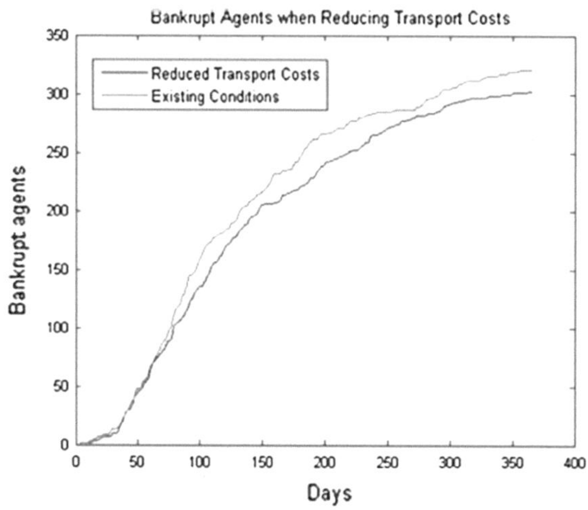

Fig. 20.5 Effect of the reduction of transport cost. The *green curve* represents the current situation, meanwhile the *blue curve* the effect of the reduced transport costs

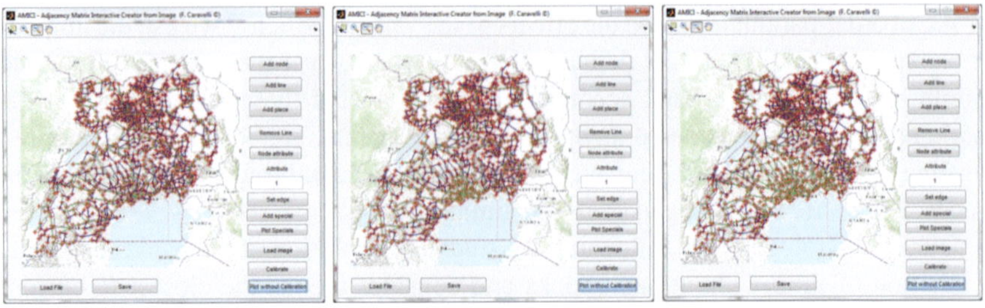

Fig. 20.6 *Left* current condition of road infrastructure in Uganda. *Blue roads* represent gravel roads meanwhile *green* are paved roads. *Center* A plausible infrastructure improvement policy for the road infrastructures around the cities of Kampala and Entebbe; each shipment has to pass through this area in order to reach exporters and the airport. *Right* A more substantial policy, where the improvement is extended also to the farming area where the agents of the model are located

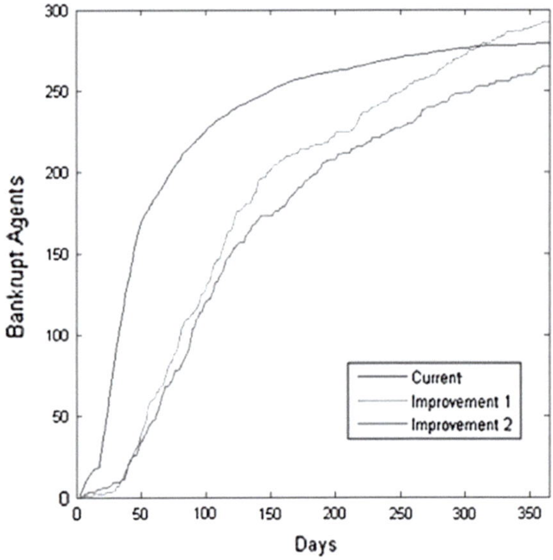

Fig. 20.7 Number of bankrupt agents for the infrastructure improvement shown in Fig. 20.6

farmers cultivate. The results are shown in Fig. 20.7. It is shown that the improvement of infrastructure does have a strong short term effect. Meanwhile the blue curve is the current situation, the green represents the case of Policy 1, meanwhile the red represents Policy 2. For both the cases of infrastructure Policy 1 and 2, in the short term the number of bankrupted agents is far lower than the current situation case, showing the importance of the infrastructure quality. In fact, for better infrastructure the speed of delivery is higher. This implies a better quality of the product, which thus can be exported. However, in the long run these policies does not change the number of bankrupt agents. In our opinion this shows that in order to improve the condition of Ugandan farmers one has to apply other policies rather than infrastructure interventions.

20.6.2 Cooperation

As a second set of policies, we consider the introduction of cooperation among farmers. In the current situation in the agricultural market there is little cooperation among farmers; these in fact produce and sell on their own to traders and itinerant traders. One of the main differences between itinerant and regular traders is in the logistic costs. Meanwhile regular traders have a higher price than itinerant traders, these require that the farmer ships at his own cost the produce. This cost affects dramatically the farmer, which then is inclined to get a lower pay, rather than anticipating the costs.

We consider, then, the case in which farmers cooperate in order to overcome the logistic costs. We consider the following policy: farmers within a radius of 30 kms aggregate the produce *in loco*, and share the cost logistic costs. Technically, this is achieved by aggregating the various farmer's production and wealth into a single, larger farmer.

The results for this policy are shown in Fig. 20.8. It is easy to see that the implementation of this policy result in a dramatic improvement in the short and log run of the number of bankrupt agents.

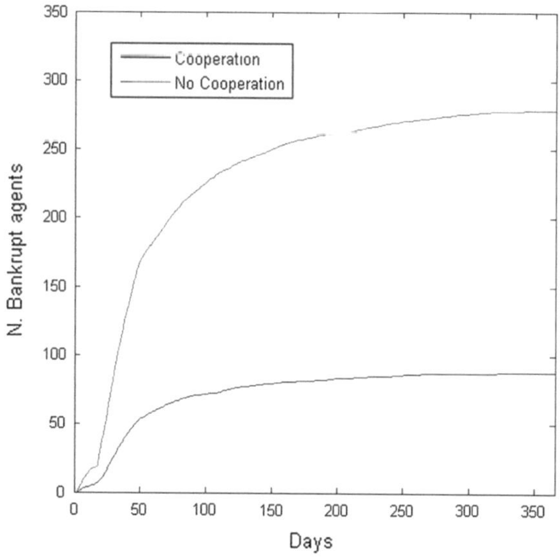

Fig. 20.8 Number of bankrupt agents the current situation (*green curve*) versus case of cooperation (*blue curve*)

Our interpretation of this result is that implementing policies in order to facilitate the cooperation among farmer agents could possibly result in less stress among Ugandan farmers.

20.6.3 GAP Policies

As a last policy test, we consider the case of GAP policies. GAP is a standard for farming in which, once exported, every single product can be tracked back to the farm gate. This is for instance a requirement in the European Union, and is a standard for exporters in african countries. At the moment, GAP policies in Uganda are implemented only by larger farmers, meanwhile itinerant traders do not use these policies and loose the trackability of the product by aggregating all the produce. We thus implement the policy of enforcing the GAP requirements at the level of the exporters. This implies that in the supply chain the itinerant traders have the disadvantage of not being able to sell to other traders, as the products do not fulfil the GAP standards. We show the result of this analysis in Fig. 20.9. Similarly to the case of cooperation, enforcing GAP standards does provide a possible solution to a long-term improvement of the Ugandan agricultural market, as itinerant traders are not favoured by these policy implementation. However, in the case of GAP itinerant traders could adapt and bypass this policy, unlike the case of cooperation.

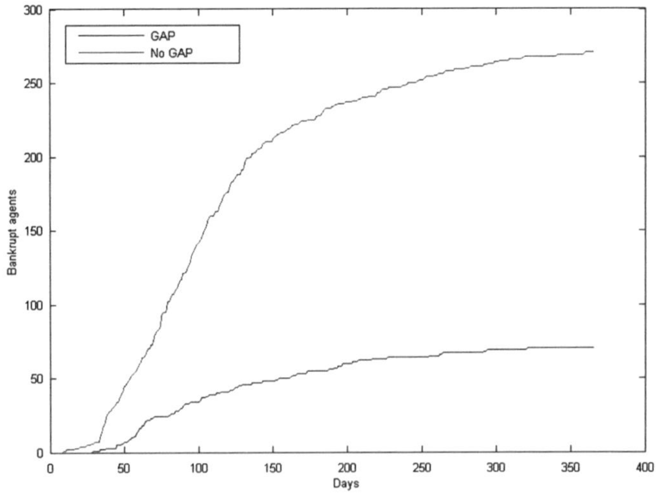

Fig. 20.9 Bankrupt agents in current situation (*green curve*) versus implementation of GAP policies (*blue curve*)

20.7 Conclusions

In this paper we have discussed how an agent based model can analyse policy alternatives in relation to the logistics of the agriculture trade supply chain, together with other agricultural and social policies. We have modelled at the microscopic level the interaction between farmers and traders, the production of fresh produce in the specific case of the Ugandan system, and we have tried to anchor the model as much as possible to the available data. In particular, we have modelled the road infrastructure system, at the macro level, keeping however the distinction between paved and gravel road, and tried to simulate logistic providers at the microlevel, by considering the product moving on the road infrastructure at realistic speed. In addition, we have considered the spoilage of the product due to weather conditions. Although we have not simulated rainfalls, we used a realistic model of temperatures based on the average seasonal minimum and maximum temperatures. This has allowed has to consider targeted policies aimed at improving the condition of farmers, which in the current situation are the weakest actors in the agricultural Ugandan market.

Specifically, we have considered three types of policy implementation. The first regards the reduction of the logistic costs, and second the improvement of the current road infrastructures. In the first case have observed that a reduction of logistic costs linearly affects the number of bankrupt agents in the long run. In addition, we have observed that the improvement of the infrastructures does provide a short term improvement of the number of bankrupt agents, but in the long run leads to a similar number of agents leaving the market.

We have however observed that cooperation among farmers and/or GAP policies can lead to a substantial improvement in the short and long-run. These results can be used as a test-base for policy recommendation to the Ugandan government.

Acknowledgments We thank the World Bank for the support in the course of this study.

References

1. Alfi, V., Cristelli, M., Pietronero, L., Zaccaria, A.: Mechanisms of self-organization and finite-size effects in a minimal agent based model. J. Stat. Mech. Theory Exp. (2009)
2. Axelrod, R.M.: The Complexity of Cooperation: Agent-Based Models of Competition and Collaboration. Princeton University Press (1997)
3. Epstein, J.M., Axtell, R.L.: Growing Artificial Societies: Social Science From the Bottom Up. Brookings Institution Press (1996)
4. Helbing, D.: Social Self-Organisation: Agent Based Simulations and Experiments to Study Emergent Social Behaviour. Springer (2012)
5. Schelling, T.C.: Micromotives and Macrobehaviour. WW Norton and Company (2006)
6. Africa Infrastructure Country Diagnostic (n.d.): Uganda: Interactive Infrastructure Atlas. Retrieved from Africa Infrastructure Knowledge Program
7. Blyde, J., Iberti, G.: A better pathway to export: How the quality of road infrastructure affects export performance. Int. Trade J. **28**(1), 3–22 (2014)

8. Hodges, R., Buzby, J.C., Bennet, B.: Postharvest losses and waste in developed and less developed countries:opportunities to improve resource use. J. Agric. Sci. **149**(S1), 37–45 (2011)
9. Albert, R., Barabasi, A.L.: Rev. Mod. Phys. **74** (2002)
10. Daley, D.J., Kendal, D.G.: Stochastic rumours. J. Inst. Math. Appl. **1** (1965)
11. Caravelli, F., Medda, F.: An Agent-Based Model for Agricultural Supply Chains in Uganda, to appear as a World Bank report

Chapter 21
Chimera States in Neuronal Systems of Excitability Type-I

Philipp Hövel, Andrea Vüllings, Iryna Omelchenko and Johanne Hizanidis

Abstract Chimera states is a fascinating phenomenon of coexisting synchronized and desynchronized behavior discovered in networks of nonlocally coupled identical phase oscillators. In this work, we consider a generic model for a saddle-node bifurcation on a limit cycle representative for neuron excitability type-I. It is given by N nonlocally coupled SNIPER oscillators in the oscillatory regime arranged on a ring. Depending on the system parameters we obtain chimera states with multiple coherent regions (clustered chimeras), coexisting traveling waves, and we observe a flip in the mean phase velocities of the coherent and incoherent regions.

21.1 Introduction

Among the many types of synchronization, chimera states have received much attention since their discovery a decade ago. These peculiar states, which are found for identical systems with strong symmetry in the coupling, exhibit a coexistence of spatially coherent (synchronized) and incoherent (desynchronized) domains. They were first reported by Kuramoto and Battogtokh in a model of densely and uniformly distributed oscillators, described by the complex Ginzburg-Landau equation,

P. Hövel (✉) · A. Vüllings · I. Omelchenko
Institut für Theoretische Physik, Technische Universität Berlin,
Hardenbergstraße 36, 10623 Berlin, Germany
e-mail: phoevel@physik.tu-berlin.de

P. Hövel
Bernstein Center for Computational Neuroscience Berlin, Humboldt-Universität zu Berlin, Philippstraße 13, 10115 Berlin, Germany

J. Hizanidis
National Center for Scientific Research "Demokritos", 15310 Athens, Greece

J. Hizanidis
Crete Center for Quantum Complexity and Nanotechnology,
Department of Physics, University of Crete, P.O.Box 2208, 71003 Heraklion, Greece

S. Battiston et al. (eds.), *Proceedings of ECCS 2014*, Springer Proceedings in Complexity, DOI 10.1007/978-3-319-29228-1_21

with nonlocal coupling of exponential form [1]. This seminal work was followed by the work of Abrams and Strogatz [2], who observed this phenomenon in a 1D ring continuum of phase oscillators assuming nonlocal coupling with a cosine kernel and coined the word "chimera" (after the creature of Greek mythology) for it. From the perspective of nonlinear dynamics, this surprising breaking of symmetry is observed through the coexistence of incongruent states of spatial coherence and disorder [3].

Experimental evidence of chimera states were first found in populations of coupled chemical oscillators and in optical coupled-map lattices realized by liquid-crystal spatial light modulators [4, 5]. Chimera states have also been reported in a mechanical experiment involving two subpopulations of identical metronomes coupled in a hierarchical network [6] and in experiments involving electrochemical oscillators [7] as well as electronic nonlinear delay oscillators [8]. Chimeras have also been observed in many other systems, including coupled chaotic logistic maps and Rössler models [9, 10], Van der Pol oscillators [11] or networks with time-varying topologies [12]. Together with numerical simulations, theoretical studies of chimera states have been recently provided, such as general bifurcation analysis for chimeras with one and multiple incoherent domains in the system of nonlocally coupled phase oscillators [13].

The importance of chimera states is also very relevant for brain dynamics, since it is believed that they could potentially explain the so-called "bumps" of neuronal activity (proposed in mechanisms of visual orientation tuning, the rat head direction system, and working memory [14]) as well as the phenomenon of *unihemispheric sleep* [15] observed in dolphins and other animals which sleep with one eye open, suggesting that one hemisphere of the brain is synchronous the other being asynchronous. For this reason, it is particularly interesting that such states were recently observed in leaky integrate-and-fire neurons with excitatory coupling [16], as well as in networks of FitzHugh-Nagumo [17, 18] and Hindmarsh-Rose [19] oscillators.

Excitability is an important feature of neuronal dynamics [20] as it determines the mechanism of the generation of action potentials (spikes) through which neurons communicate. There are two types of excitability: type I yields a response of finite amplitude and infinite period through a global bifurcation, and type II gives rise to zero-amplitude and finite period spikes via a Hopf bifurcation. Type-II excitability is often modeled by the FitzHugh-Nagumo system for which "multi-chimera" (or "clustered chimera" [21]) states, which consist of multiple coherent regions, were recently found slightly above the excitability threshold [17]. The Hindmarsh-Rose model, which is representative for both type-I and type-II excitability, exhibits very complex behavior including spiking, regular and chaotic bursting for which chimera states and other collective dynamics were identified [19].

In this work, we will focus on a generic model for type-I excitability and demonstrate the universal occurrence of chimera states to this class of models. The system under consideration is representative for a global bifurcation, namely a saddle-node bifurcation on a limit cycle also known as Saddle-Node Infinite PERiod (SNIPER) bifurcation, which is also known as Saddle-Node bifurcation on an Invariant Circle

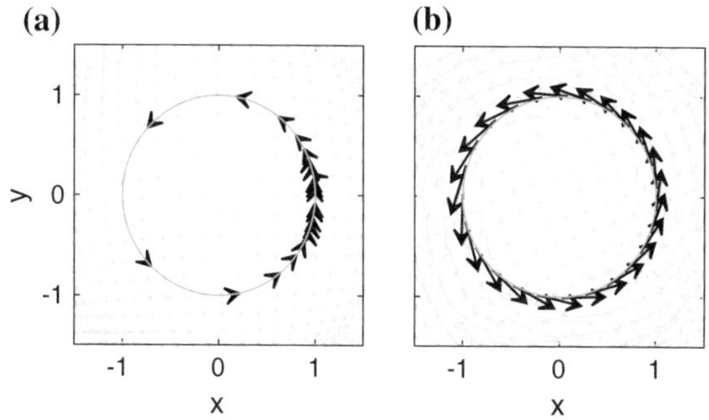

Fig. 21.1 SNIPER model in the oscillatory regime: numerical solution and vector field of (21.1) for two different values of the bifurcation parameter b: **a** $b = 1.05$, **b** $b = 9$

(SNIC). It studied neuroscience in various contexts [22–25] and is defined by the following equations:

$$\dot{x} = x(1 - x^2 - y^2) + y(x - b),$$
$$\dot{y} = y(1 - x^2 - y^2) - x(x - b), \quad (21.1)$$

with the state variables $x(t)$ and $y(t)$, and b is the bifurcation parameter. For $b < 1$, there are three fixed points: an unstable focus at the origin and a pair of a saddle-point and a stable node on the unit circle with coordinates $(b, +\sqrt{1 - b^2})$ and $(b, -\sqrt{1 - b^2})$, respectively. The latter two collide for $b_c = 1$ at $(x^*, y^*) = (1, 0)$ and a limit cycle with constant radius $\rho_c = \sqrt{x^2 + y^2} = 1$ is born. Above but close to the bifurcation, the frequency f of this limit cycle obeys a characteristic square-root scaling law $f \sim \sqrt{b^2 - 1}$.

In the following, we choose $b > b_c$ so that the system operates in the oscillatory regime. The system oscillates with constant amplitude $\rho_c = 1$ and the period T_0 is given by $2\pi/\sqrt{b^2 - 1}$. In Fig. 21.1 the numerical solution of x and y is shown for one period. For $b = 1.05$ Fig. 21.1a, the dense region (the so-called "ghost") where the system slows down marks the collision point of the saddle and the node, i.e. $(x^*, y^*) = (1, 0)$. For this parameter value, the system remembers the collision point because it is close to the critical value b_c. The phase velocity converges to a constant value as soon as b becomes large enough Fig. 21.1b.

The rest of this paper is organized as follows: In Sect. 21.2, we introduce the coupling topology and describe the main features of the observed dynamics. In Sect. 21.3, we scan the parameter plane spanned by the bifurcation parameter and coupling range. Section 21.4 focuses on coexistence of chimeras and other patterns and in Sect. 21.5, we address the role of the coupling strength. Finally, we conclude with a summary in Sect. 21.6.

21.2 The Model

We consider N nonlocally coupled SNIPER oscillators given by (21.1) arranged on a ring:

$$\dot{x}_k = x_k(1 - x_k^2 - y_k^2) + y_k(x_k - b) + \frac{\sigma}{2R}\sum_{j=k-R}^{k+R}\left[b_{xx}(x_j - x_k) + b_{xy}(y_j - y_k)\right],$$
$$\dot{y}_k = y_k(1 - x_k^2 - y_k^2) - x_k(x_k - b) + \frac{\sigma}{2R}\sum_{j=k-R}^{k+R}\left[b_{yx}(x_j - x_k) + b_{yy}(y_j - y_k)\right],$$
$$(21.2)$$

where $k = 1, 2, \ldots, N$, $\sigma > 0$ is the coupling strength, and $R \in [1, N/2]$ is the number of nearest neighbors of each oscillator on either side. The limit cases $R = 1$ and $R = N/2$ correspond to nearest-neighbor and all-to-all coupling, respectively. They can be also understood as local diffusion in one dimension and a fully connected network, respectively. It is convenient to scale this parameter by the system size, which defines a coupling radius $r = R/N \in [1/N, 0.5]$. The coefficients b_{lm}, where $l, m \in \{x, y\}$, are given by the elements of the rotational matrix:

$$\mathbf{B} = \begin{pmatrix} b_{xx} & b_{xy} \\ b_{yx} & b_{yy} \end{pmatrix} = \begin{pmatrix} \cos\phi & \sin\phi \\ -\sin\phi & \cos\phi \end{pmatrix}, \qquad (21.3)$$

where $\phi \in [-\pi, \pi]$. The matrix $\mathbf{B}$ allows for direct (xx)- and (yy)-coupling as well as cross coupling between x and y as in [17, 19].

Figure 21.2a shows a snapshot of the variables x_k at a fixed time, providing evidence of a classical chimera state: One group of neighboring oscillators on the ring is spatially coherent (shaded region) while the remaining elements form a second, spatially incoherent group (black dots). Note that the snapshot for y_k looks qualitatively the same. These two domains of coherent and incoherent oscillators can be distinguished from each other through the mean phase velocity of each oscillator $\omega_k = 2\pi M_k/\Delta T$, where M_k is the number of periods of the kth oscillator during a sufficiently long time interval ΔT [17].

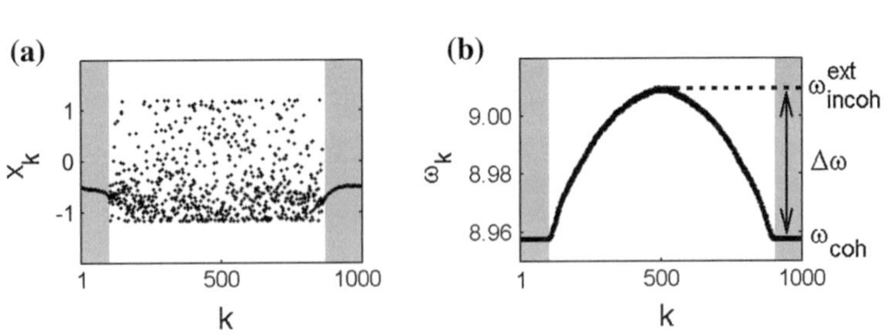

Fig. 21.2 Chimera state of nonlocally coupled SNIPER oscillators given by (21.2): **a** Snapshot of states x_k and **b** corresponding mean phase velocities ω_k (*shaded region* coherent, *white region* incoherent oscillators). Parameters: $b = 9$, $\sigma = 0.1$, $\phi = \pi/2 - 0.1$, $R = 350$, and $N = 1000$

Figure 21.2b shows the characteristic profile for the mean phase velocities ω_k corresponding to the chimera state of Fig. 21.2a. The oscillators in the coherent domain (shaded region) rotate along the unit circle at a constant speed ω_{coh}, whereas the incoherent oscillators (black) have different mean phase velocities ω_{incoh} with an extremum (in this case maximum) value denoted by $\omega_{\mathrm{incoh}}^{\mathrm{ext}}$. A non-zero difference

$$\Delta\omega = \omega_{\mathrm{incoh}}^{\mathrm{ext}} - \omega_{\mathrm{coh}} \tag{21.4}$$

is a good indicator for the existence of a chimera state. Note that, for the particular chimera state of Fig. 21.2a, it holds that $\omega_{\mathrm{incoh}}^{\mathrm{ext}} > \omega_{\mathrm{coh}}$.

In the following sections, we will systematically investigate the effect of the bifurcation parameter b as well as the coupling parameters R and σ on the chimera state. The initial conditions for x_k and y_k are randomly distributed on the unit circle and we discard transients of 1000 time units. For the mean phase velocities ω_k, we average over a time interval $\Delta T = 10.000$.

21.3 Impact of the Bifurcation Parameter and Coupling Range

A stability diagram for the chimera states is displayed in Fig. 21.3 where the dependence of the modulus of $\Delta\omega$ (21.4) is plotted with respect to the bifurcation parameter b and the coupling radius $r = R/N$.

Starting from the values $b = 9$, $r = 0.43$ and a certain set of initial conditions as described above, we perform a continuation on the direction of smaller r-values down

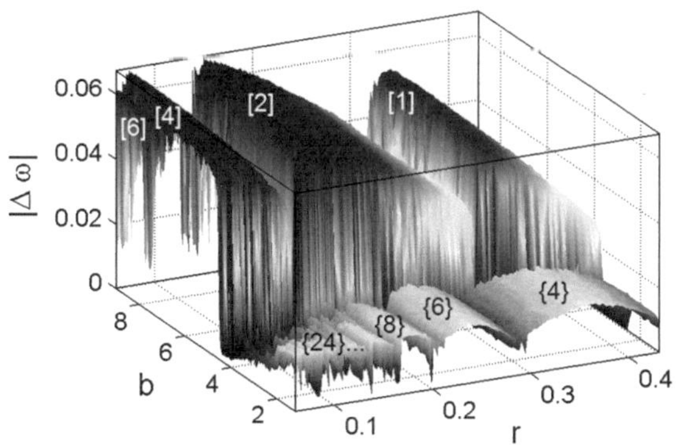

Fig. 21.3 Stability diagram in the (b, r)-plane: Modulus of the difference $|\Delta\omega|$ between the mean phase velocities of the coherent and incoherent oscillators (21.4) as a function of the bifurcation parameter b and the coupling radius r. The *numbers in the brackets* and *braces* denote the number of the (in)coherent domains of the corresponding chimera state. *Brackets* and *braces* refer to "normal" and "flipped" ω-profile, respectively. Parameters: $\sigma = 0.1$, $\phi = \pi/2 - 0.1$, and $N = 1000$

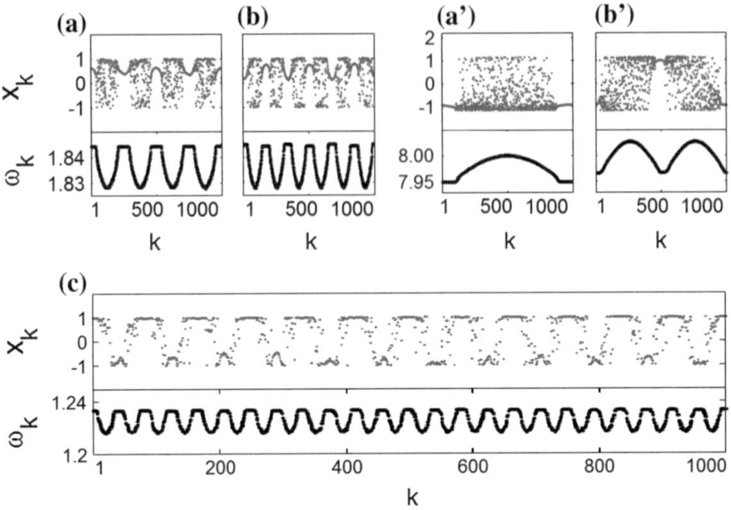

Fig. 21.4 Clustered chimera states of Fig. 21.3: snapshots of the states x_k at different points in the (b, r)-plane. Panels **a–c** and **a**$'$, **b**$'$ correspond to "flipped" and "normal" ω-profiles, respectively. **a** $r = 0.35$, $b = 2$, **b** $r = 0.24$, $b = 2$, **c** $r = 0.06$, $b = 2$, **a**$'$ $r = 0.35$, $b = 8$, **b**$'$ $r = 0.18$, $b = 8$. Other parameters: $\sigma = 0.1$, $\phi = \pi/2 - 0.1$, $N = 1000$

to $r = 0.06$ and calculate $\Delta\omega$ for each coupling radius. For this kind of continuation, we use the state (x_k, y_k) at $t = \Delta T$ obtained for one parameter set (b, r) as the initial condition for the next simulation with slightly changed parameters; here, a slightly reduced coupling radius r. Subsequently, for values of $r \in [0.04, 0.46]$ we perform a continuation in b-direction from $b = 9$ down to $b = 0.1$ starting again at $r = 0.43$. The coupling strength is fixed at a constant value $\sigma = 0.1$.

From Fig. 21.3 it is clear that $|\Delta\omega|$ has a non-monotonous behavior in the (b, r)-plane. Each "bump" in the 3D surface corresponds to a different type of chimera state associated to a different number of (in)coherent domains, marked in the square brackets/braces. Some of these states are explicitly shown below in Fig. 21.4 for certain combinations of b and r.

For large values of the bifurcation parameter (large "bumps" in Figs. 21.3 and 21.4a$'$) a classical chimera state with one group of (in)coherent oscillators exists. By decreasing r, which physically means removing more and more long-range connections, the number of clustered (in)coherent oscillators increases. In the large "bumps" of Fig. 21.3 these so-called "multi-chimera" states exhibit the characteristic feature that $\omega_{\text{incoh}}^{\text{ext}} > \omega_{\text{coh}}$ (i.e. $\Delta\omega > 0$), shown in the corresponding mean phase velocity profiles in Fig. 21.4b$'$. We denote these chimera states, for which $\Delta\omega > 0$, by the number of their (in)coherent domains in *square* brackets [1], [2], [4], and [6].

For lower values of b (small "bumps" in Figs. 21.3 and 21.4a–c), we exclusively find multi-chimera states. As in the case of larger b, the number of clustered (in)coherent oscillators increases with decreasing coupling radius r. However, there is a significant difference: The mean phase velocities of the incoherent oscillators is smaller than the velocity of the coherent ones, i.e. $\Delta\omega < 0$. Hence, there exists a critical value of the bifurcation parameter (found to be around $b = 4$), where $\Delta\omega$ changes its sign, resulting in a "flip" in the mean phase velocity profile. The chimera states with a "flipped" ω-profile are denoted by the number of (in)coherent domains in braces $\{4\}$, $\{6\}$, $\{8\}$, $\{10\}$, …, $\{24\}$.

The characteristic form of the average phase velocities profile has been used as a criterion to distinguish chimera states in the systems of coupled oscillators. In most systems, the coherent oscillators have smaller average phase frequencies than the incoherent ones. However, the opposite situation where the coherent oscillators are faster than the incoherent ones is also possible. In the system of nonlocally delay-coupled phase oscillators, two types of chimera states were distinguished depending on whether the effective frequencies of the incoherent oscillators are larger or smaller than the frequencies of the coherent ones [26, 27]. The regions of stability for these two types of chimera states depend on the time delay and strength of the coupling. Moreover, both types of chimera states can coexist.

The "flipped" phase velocities profile was also observed in systems, which do not consider time delay in the coupling, but has not been explained so far. The Kuramoto model with repulsive coupling allows for multi-chimera states for which the mean phase velocity profiles show larger average frequencies for oscillators that belong to coherent domain [28]. Similar behavior is also observed for chimera states with one incoherent domain in the complex Ginzburg-Landau equation with nonlocal coupling [29]. In that system, however, chimera states with multiple incoherent domains possess the usually observed mean phase velocity profiles. A flipped phase velocities profile was observed experimentally as well, in networks of electrochemical oscillators with nonlocal coupling, where the frequencies of the oscillators from the coherent domain are higher than the frequencies of the incoherent ones [7].

In our system, we observe direct dependence of the form of the mean phase velocity profile on the parameter b defining the frequency of the local uncoupled unit.

21.4 Multistability: Coexisting Chimeras and Traveling Waves

The coexistence of different multi-chimeras, traveling waves, and completely synchronized states in the phase space has been observed in many other systems of nonlocal coupled oscillators [17, 19, 30]. Depending on the initial conditions the stationary state can vary significantly. Such multistable solutions are also possible in system (21.2) as demonstrated in Fig. 21.5.

A schematic representation of the identified multi-chimeras in the (b, r)-plane is shown in Fig. 21.5a. Note that only the r-range for which chimera states are observed is shown. Each region has a different gray scale associated to a different chimera type

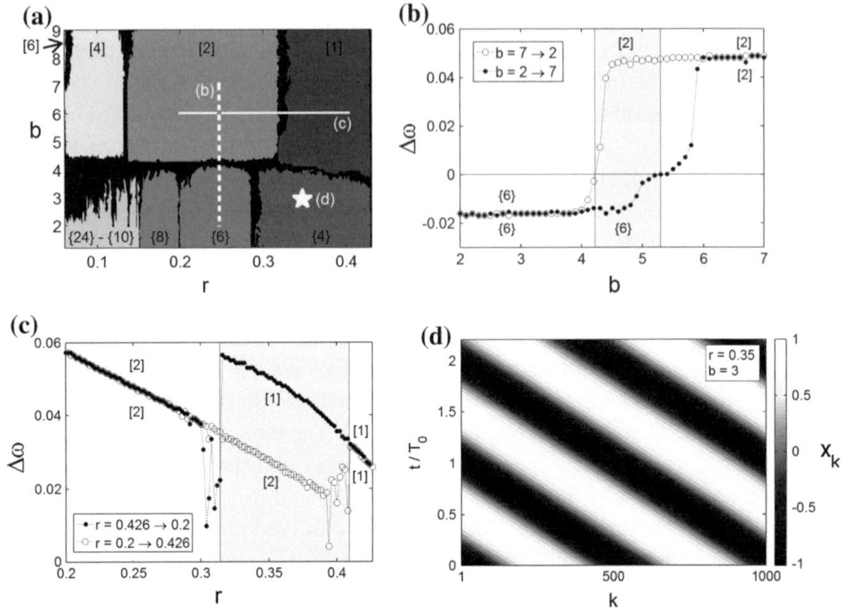

Fig. 21.5 Coexisting chimera states and traveling waves: **a** projection to the (b, r)-plane of Fig. 21.3. **b** Up and down sweep in b-direction as marked by the *dashed white line* in Fig. 21.5a for fixed $r = 0.25$. *[2]*- and *{6}*-chimera states coexist in the *shaded area*. **c** Up and down sweep in r-direction as marked by the *solid white line* in Fig. 21.5a for fixed $b = 6$. *[1]*- and *[2]*-chimera states coexist in the *shaded area*. **d** Traveling wave solution, which coexists with the *{4}*-chimera, for r and b marked by the *white star* in Fig. 21.5a. The time is scaled by the period T_0 of an uncoupled oscillator. Other Parameters: $\sigma = 0.1$, $\phi = \pi/2 - 0.1$, and $N = 1000$

as described in the previous section. The black regions correspond to intermittent states, which are mainly desynchronized. Along the white lines, Fig. 21.5b, c display the results of a continuation in b (dashed line) and r (solid line), respectively. The continuation is performed as down sweep in b (or r) and then repeated in the opposite direction. In both cases we find a region where different types of chimera states coexist.

In particular, for intermediate values of the bifurcation parameter b, there is coexistence of a [2]- and {6}-chimera state marked by the shaded area of Fig. 21.5b. This area of coexisting chimera states, moreover, marks the transition between "flipped" ($\Delta\omega < 0$) and "normal" ($\Delta\omega > 0$) mean phase velocity profile. This transition occurs at a different and, in particular, lower value of b when the continuation is performed in the direction of decreasing b (open dots) than when performed in the opposite direction (filled dots), i.e. our system exhibits, apart from multistability, hysteresis phenomena as well.

Coexisting chimera states may also be found by varying parameter r, as shown in Fig. 21.5c: Depending on the choice of initial conditions, one may observe either a [1]- or a [2]-chimera state (shaded area) both with $\Delta\omega > 0$. In both increasing (open dots) and decreasing r (filled dots) directions, there are deviations from the piecewise linear behavior of $\Delta\omega(r)$ which correspond to desynchronized states.

The observed multi-chimera states may also coexist with completely synchronized states and traveling waves. One example of such a point in parameter space is marked by the white star in Fig. 21.5a and the corresponding space-time pattern is shown in Fig. 21.5d. This is a traveling wave solution of wave number 2 coexisting with a {4}-chimera state. The time in the vertical axis is scaled by the period T_0 of the uncoupled oscillator. Multistability between traveling waves with a smooth profile and breathing states with a periodic motion of the coherent and incoherent domains have recently also been reported for chaotic systems with nonlocal coupling [30].

21.5 Role of the Coupling Strength

Again, we perform a parameter continuation and focus on the behavior of $\Delta\omega$ as the coupling strength σ increases for different multi-chimera states. Our findings show that even at large σ the corresponding multi-chimera state is preserved. However, we observe that, for certain values of the bifurcation parameter b and the coupling radius r, the coupling strength may induce a spatial motion of domains of the (in)coherent oscillators.

Figure 21.6a shows that the difference between the mean phase velocity of the coherent and incoherent oscillators $\Delta\omega$ linearly increases with the coupling strength, apart for a narrow range of $\sigma \approx 1$ where $\Delta\omega$ deviates. In this regime, the corresponding space-time plots of the [2]-chimera state reveal that the (in)coherent domains can start to move spatially with time (see Fig. 21.6b, c for $\sigma = 1$). Note that breathing chimera states, which are defined by a periodic motion of the coherent and incoherent domains, coexist with stationary ones due to multistability similar to the coexisting {4}-chimera state and traveling wave shown in Fig. 21.5. We observed stationary chimera states for $\sigma \approx 1$ (open dots in Fig. 21.6c). Beyond this regime of moving patterns, our continuation returns to the [2]-chimera state which is stationary.

In general, chimera states can be stationary or can perform two types of motion in space, in which the coherent and incoherent domains change their spatial position in time. The first one is a chaotic motion of the position of the chimera observed in nonlocally coupled phase oscillators. Such a motion shows a sensitive dependence on the initial conditions and is a finite-size effect that vanishes in the thermodynamic limit. It can be described as a Brownian motion and depends on the coupling radius, the phase lag parameter, and the shape of the coupling function [31]. The second type is the above described "breathing" chimera state exhibiting a periodic motion of the coherent and incoherent domains. Breathing chimeras were first observed in the system of two oscillator populations in which each oscillator is coupled equally to all the others in its group, and less strongly to those in the other group [32], and recently in the nonlocal complex Ginzburg-Landau equation in the limit of strong coupling [29].

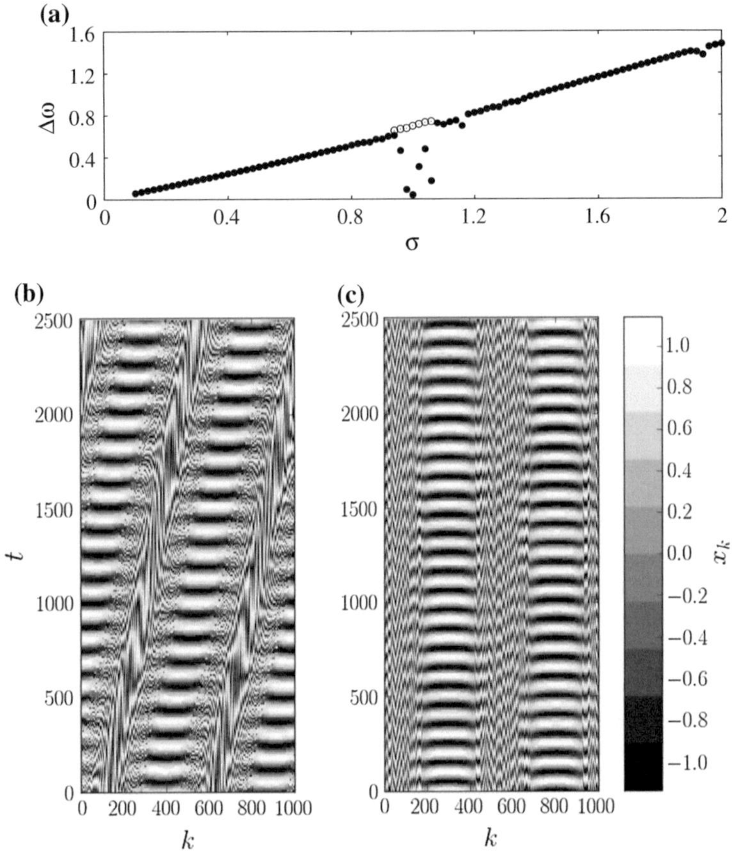

Fig. 21.6 Impact of the coupling strength on the [2]-chimera state: **a** $\Delta\omega$ as a function of the coupling strength (*filled dots* continuation, *open dots* different initial conditions). **b**, **c** Space-time plots of the x_k-variables for a fixed $\sigma = 1.0$ and two different initial conditions. Other parameters: $b = 6, \phi = \pi/2 - 0.1$, $R = 190$, and $N = 1000$

Based on our numerical simulations, we conclude that in principle, breathing chimera states exist for nonlocally coupled SNIPER models for some coupling parameters, but the exact parameter range needs to be determined in future studies.

21.6 Conclusions

In this work, we have verified the occurrence of clustered chimera states in a generic model for a saddle-node bifurcation on a limit cycle representative for neural excitability type-I. Together with recent reports on multi-chimera states in nonlocally coupled FitzHugh-Nagumo [17, 18] and Hindmarsh-Rose [19] oscillators, our findings provide strong, additional evidence that this kind of symmetry breaking is very relevant for applications in neuroscience.

In particular, we presented a detailed exploration of the parameter space where chimera states occur and investigated their dependence on the proximity to the excitability threshold and the range of the nonlocal coupling. We identified chimera states for which the mean phase velocity has a "flipped" profile. Findings of coexisting chimera states and traveling waves in the parameter space verify the existence of multistability in our model. Finally, it was shown that for increasing coupling strength the domains of coherent oscillators become bigger and at the same time spatial motion of the incoherent oscillators is observed.

Acknowledgments We thank A. Provata and E. Schöll for stimulating discussions. This work was supported by the German Academic Exchange Service *DAAD* and the Greek State Scholarship Foundation *IKY* within the PPP-IKYDA framework. IO and PH acknowledge support by BMBF (grant no. 01Q1001B) in the framework of BCCN Berlin (Project A13). PH, AV, and IO acknowledge support by DFG in the framework of the Collaborative Research Center 910. The research work was partially supported by the European Union's Seventh Framework Program (FP7-REGPOT-2012-2013-1) under grant agreement n316165.

References

1. Kuramoto, Y., Battogtokh, D.: Coexistence of coherence and Incoherence in nonlocally coupled phase oscillators. Nonlin. Phen. Complex Sys. **5**(4), 380–385 (2002)
2. Abrams, D.M., Strogatz, S.H.: Chimera states for coupled oscillators. Phys. Rev. Lett. **93**(17), 174102 (2004)
3. Panaggio, M.J., Abrams, D.M.: Chimera states: coexistence of coherence and incoherence in networks of coupled oscillators. Nonlinearity **28**, R67 (2015)
4. Hagerstrom, A.M., Murphy, T.E., Roy, R., Hövel, P., Omelchenko, I., Schöll, E.: Experimental observation of chimeras in coupled-map lattices. Nat. Phys. **8**, 658–661 (2012)
5. Tinsley, M.R., Nkomo, S., Showalter, K.: Chimera and phase cluster states in populations of coupled chemical oscillators. Nat. Phys. **8**, 662–665 (2012)
6. Martens, E.A., Thutupalli, S., Fourrière, A., Hallatschek, O.: Chimera states in mechanical oscillator networks. Proc. Natl. Acad. Sci. **110**, 10563 (2013)
7. Wickramasinghe, M., Kiss, I.Z.: Spatially organized dynamical states in chemical oscillator networks: synchronization, dynamical differentiation, and chimera patterns. PLoS ONE **8**(11), e80586 (2013)
8. Larger, L., Penkovsky, B., Maistrenko, Y.: Virtual chimera states for delayed-feedback systems. Phys. Rev. Lett. **111**, 054103 (2013)
9. Omelchenko, I., Maistrenko, Y., Hövel, P., Schöll, E.: Loss of coherence in dynamical networks: spatial chaos and chimera states. Phys. Rev. Lett. **106**, 234102 (2011)
10. Omelchenko, I., Riemenschneider, B., Hövel, P., Maistrenko, Y., Schöll, E.: Transition from spatial coherence to incoherence in coupled chaotic systems. Phys. Rev. E **85**, 026212 (2012)
11. Omelchenko, I., Zakharova, A., Hövel, P., Siebert, J., Schöll, E.: Nonlinearity of local dynamics promotes multi-chimeras. Chaos **25**, 083104 (2015)
12. Buscarino, A., Frasca, M., Gambuzza, L.V., Hövel, P.: Chimera states in time-varying complex networks. Phys. Rev. E **91**(2), 022817 (2015)
13. Omel'chenko, O.E.: Coherence-incoherence patterns in a ring of non-locally coupled phase oscillators. Nonlinearity **26**(9), 2469 (2013)

14. Laing, C.R., Chow, C.C.: Stationary bumps in networks of spiking neurons. Neural Comput. **13**(7), 1473–1494 (2001)
15. Rattenborg, N.C., Amlaner, C.J., Lima, S.L.: Behavioral, neurophysiological and evolutionary perspectives on unihemispheric sleep. Neurosci. Biobehav. Rev. **24**, 817–842 (2000)
16. Olmi, S., Politi, A., Torcini, A.: Collective chaos in pulse-coupled neural networks. Europhys. Lett. **92**, 60007 (2010)
17. Omelchenko, I., Omel'chenko, O.E., Hövel, P., Schöll, E.: When nonlocal coupling between oscillators becomes stronger: patched synchrony or multichimera states. Phys. Rev. Lett. **110**, 224101 (2013)
18. Omelchenko, I., Provata, A., Hizanidis, J., Schöll, E., Hövel, P.: Robustness of chimera states for coupled FitzHugh-Nagumo oscillators. Phys. Rev. E **91**, 022917 (2015)
19. Hizanidis, J., Kanas, V., Bezerianos, A., Bountis, T.: Chimera states in networks of nonlocally coupled hindmarsh-rose neuron models. Int. J. Bifur. Chaos **24**(03), 1450030 (2014)
20. Izhikevich, E.M.: Neural excitability, spiking and bursting. Int. J. Bifur. Chaos **10**(6), 1171–1266 (2000)
21. Sethia, G.C., Sen, A., Atay, F.M.: Clustered chimera states in delay-coupled oscillator systems. Phys. Rev. Lett. **100**(14), 144102 (2008)
22. Aust, R., Hövel, P., Hizanidis, J., Schöll, E.: Delay control of coherence resonance in type-I excitable dynamics. Eur. Phys. J. ST **187**, 77–85 (2010)
23. Ditzinger, T., Ning, C.Z., Hu, G.: Resonance like responses of autonomous nonlinear systems to white noise. Phys. Rev. E **50**, 3508 (1994)
24. Hizanidis, J., Aust, R., Schöll, E.: Delay-induced multistability near a global bifurcation. Int. J. Bifur. Chaos **18**(6), 1759–1765 (2008)
25. Hu, B.Y.K., Das Sarma, S.: Many-body exchange-correlation effects in the lowest subband of semiconductor quantum wires. Phys. Rev. B **48**, 5469 (1993)
26. Omel'chenko, O.E., Maistrenko, Y., Tass, P.A.: Chimera states: the natural link between coherence and incoherence. Phys. Rev. Lett. **100**(4), 044105 (2008)
27. Omel'chenko, O.E., Maistrenko, Y., Tass, P.A.: Chimera states induced by spatially modulated delayed feedback. Phys. Rev. E **82**, 066201 (2010)
28. Maistrenko, Y., Vasylenko, A., Sudakov, O., Levchenko, R., Maistrenko, V.L.: Cascades of multi-headed chimera states for coupled phase oscillators. Int. J. Bifur. Chaos **24**(8), 1440014 (2014)
29. Sethia, G.C., Sen, A., Johnston, G.L.: Amplitude-mediated chimera states. Phys. Rev. E **88**(4), 042917 (2013)
30. Dziubak, V., Maistrenko, Y., Schöll, E.: Coherent traveling waves in nonlocally coupled chaotic systems. Phys. Rev. E **87**(3), 032907 (2013)
31. Omel'chenko, O.E., Wolfrum, M., Maistrenko, Y.: Chimera states as chaotic spatiotemporal patterns. Phys. Rev. E **81**(6), 065201(R) (2010)
32. Abrams, D.M., Mirollo, R., Strogatz, S.H., Wiley, D.A.: Solvable model for chimera states of coupled oscillators. Phys. Rev. Lett. **101**(8), 084103 (2008)

Chapter 22
Multiobjective Optimization and Phase Transitions

Luís F. Seoane and Ricard Solé

Abstract Many complex systems obey to optimality conditions that are usually not simple. Conflicting traits often interact making a Multi Objective Optimization (MOO) approach necessary. Recent MOO research on complex systems report about the Pareto front (optimal designs implementing the best trade-off) in a qualitative manner. Meanwhile, research on traditional Simple Objective Optimization (SOO) often finds phase transitions and critical points. We summarize a robust framework that accounts for phase transitions located through SOO techniques and indicates what MOO features resolutely lead to phase transitions. These appear determined by the shape of the Pareto front, which at the same time is deeply related to the thermodynamic Gibbs surface. Indeed, thermodynamics can be written as an MOO from where its phase transitions can be parsimoniously derived; suggesting that the similarities between transitions in MOO-SOO and Statistical Mechanics go beyond mere coincidence.

22.1 Introduction

Optimization has always been a major topic in complex systems research. Optimality conditions are relevant for a wealth of biological [1–5] and other natural and synthetic systems [6–10]. Evolution through natural selection is a main driver of biological systems towards optimal designs [11, 12] and certain physical principles (e.g. maximum entropy or optimal diffusion structures) already introduce a bias

L.F. Seoane (✉) · R. Solé
ICREA-Complex Systems Lab, Universitat Pompeu Fabra-PRBB,
Dr. Aiguader 88, 08003 Barcelona, Spain
e-mail: luis.seoane@upf.edu

L.F. Seoane · R. Solé
Institut de Biologia Evolutiva, CSIC-UPF, Pg. Maritim de la Barceloneta 37,
08003 Barcelona, Spain

R. Solé
Santa Fe Institute, 1399 Hyde Park Road, Santa Fe, NM 87501, USA
e-mail: ricard.sole@upf.edu

S. Battiston et al. (eds.), *Proceedings of ECCS 2014*, Springer Proceedings
in Complexity, DOI 10.1007/978-3-319-29228-1_22

towards functional extrema. Human-made systems are equally constrained through cost-efficiency calculations—e.g. in transportation networks [9, 10].

Describing these situations requires optimal designs that often cope with interacting constraints. To give a good account of these selective forces, a *Pareto* or *Multi Objective Optimization* (MOO) approach can be useful. Let us introduce this theory through a recent relevant example [13]. (Technical definitions follow below.) Consider the set (Γ) of all connected networks with a fixed number of nodes ($\gamma \in \Gamma$, Fig. 22.1a). Among them we seek those minimizing the average path length $\langle l \rangle(\gamma)$ and the number of edges $\rho(\gamma)$. These are the *target functions* ($T_f(\gamma) \equiv \{t_1 = \langle l \rangle(\gamma), t_2 = \rho(\gamma)\}$) of our MOO problem. A fully connected clique minimizes the average path length, but we need to implement all possibles links, which is costly. The minimum spanning tree has the least number of edges possible but its average path length is quite large. Take networks γ_1 and γ_2 trading between these extremes and such that $\langle l \rangle(\gamma_1) < \langle l \rangle(\gamma_2)$ and $\rho(\gamma_1) < \rho(\gamma_1)$. This means that γ_1 implements a better tradeoff than γ_2 and we say then that γ_1 dominates γ_2 (Fig. 22.1c). A network $(\gamma_\pi \in \Pi) \subset \Gamma$ not dominated by any other $\gamma \in \Gamma$ is *Pareto optimal*. Often we cannot choose between a pair of networks because one is better than the other with respect to a target and worst with respect to the other—i.e. they are mutually not dominated. Because this situation is common, MOO solutions are often not a single global optimizer, but the collection of Pareto optimal (mutually non-dominated) networks that implement the most optimal tradeoff possible. We name this *Pareto optimal set* $\Pi \subset \Gamma$.

Target functions map each network $\gamma \in \Gamma$ into a point of $\mathbb{R}^2$: $(\langle l \rangle(\gamma), \rho(\gamma))$. This plane, with the relevant traits in its axes, constitutes a *morphospace* in which salient network topologies are located as a function of their morphology [14] (Fig. 22.1b). Morphospace of other systems visualize phenotypes or designs with respect to relevant properties. The Pareto optimal set is mapped onto $T_f(\Pi)$ and constitutes a boundary of the morphospace (Fig. 22.1b, c), also known as the Pareto front.

Some authors are beginning to explore the consequences of Pareto optimality in biological systems [3–5] or in relevant models such as networks [15, 16] or regulatory circuits [5, 17]. While they tackle relevant questions through MOO methods, the description of these optimal designs is often a qualitative account of the elements along the Pareto front (as in the study of a restricted morphospace—an interesting contribution nevertheless). The same qualitative bias appears in classic MOO literature. Is a more quantitative analysis possible? Are there *universal features* that reach through different MOO problems, thus uniting Pareto optimal systems despite their differences? Through our research [18] (sketched in Sect. 22.2) we have found a connection between MOO and statistical mechanics. Those *universal features* we were looking for are phase transitions and critical points, which leave clear imprints in the shape of the Pareto front. Some authors had explored MOO with Single Objective Optimization (SOO) methods—e.g. by integrating all targets linearly to define a *global energy function* $\Omega(\Lambda) = \sum_k \lambda_k t_k$, with $\Lambda = \{\lambda_1, \dots, \lambda_k\}$ arbitrary parameters that introduce a bias towards some of the targets. Such research often finds phase transitions and other phenomena [8, 10]. A parsimonious theory lacked as to why some systems would present such transitions and others would not.

To the best of our knowledge, authors researching MOO do not exploit this connection with thermodynamics which, we believe, much enriches the discussion of Pareto optimal designs. Two relevant examples from network theory: The efficiency of different topologies has been researched for the relay of information across a network using two distinct delivery heuristics [15]. It was made an exhaustive work in describing network topologies and locating them in a morphospace in which different network features are segregated. In that same morphospace, the Pareto fronts in [15] strongly indicate the presence of first and second order phase transitions. If the information theoretical aspect of the diffusion of messages across the network is considered, those transitions might become thermodynamically relevant. Similarly, the tradeoff of topological robustness when random or targeted nodes are taken away results in a Pareto front [16]. Under the light of our findings, second order transitions show up in that study. Also a first order transition exists that vanishes as the average degree of the network changes, suggesting a critical point. We further illustrate our findings with other two examples in Sect. 22.2.1.

A theory about phase transitions must fit within thermodynamics. For us, this is achieved due to the equivalence between the Pareto front and the Gibbs surface [18–20], an object known to embody phase transitions in its cavities and non-analyticities. We discuss thermodynamics in Sect. 22.3.1, not because our theory modifies previous knowledge about it, of course, but because in showing that phase transitions arise in thermodynamics *precisely in the same way* as in MOO, we place our findings for MOO on very solid ground.

22.2 Theoretical Framework

In this section we expand the loose introduction of MOO above. More details and methods can be found in the exhaustive literature [21–25]. We assume minimization unless indicated otherwise.

Consider a set X of possible designs $x \in X$. In the example above, $X = \Gamma$ is made of network designs. This will be used again later, along with another example in which $X = A$ stands for all possible languages $a \in A$ derived from a mathematical computational model of human communication [8]. Within X we seek those optimal designs $(x_\pi \in \Pi) \subset X$ that simultaneously minimize a series of *target functions* $(T_f \equiv \{t_1, \dots, t_K\})$. These $t_k \in T_f$ map each design $x \in X$ into *target space* $(T_f(x) = \{t_1(x), \dots, t_K(x)\} \in \mathbb{R}^K)$, a morphospace of the system under research.

Pareto dominance is defined in this target space. Take $x, z \in X$. x dominates z (noted $x \prec z$) if $t_k(x) \leq t_k(z)$ for all k and $t_{k'}(x) < t_{k'}(z)$ for at least one k'. This means that x is objectively better than z. If given two designs $(x, y \in X)$ none dominates the other $(x \nprec y \nprec x)$, we cannot chose one of them without introducing a bias towards some of the target functions. Pareto optimality is solved by putting choices between mutually non-dominated designs on hold.

The *Pareto optimal set* $\Pi \subset X$ is such that every element $z \in X$, $z \notin \Pi$ is dominated by some $x \in \Pi$ while any $x, y \in \Pi$ are mutually non-dominated. The projection $T_f(\Pi)$ conforms a $(D \leq K-1)$-dimensional surface in $\mathbb{R}^K$ that embodies the most optimal tradeoff possible between the targets. Moving along the front it is impossible to improve all targets at once: an increment in at least one t_k is necessary if we wish to decrease some other $t_{k'}$.

We sketch now the basic situations of our theoretical framework that connects the Pareto front and thermodynamics. We refer the reader to [18] for a more exhaustive discussion.

The simplest SOO problem that includes all MOO targets defines a linear *global energy function*:

$$\Omega(x, \Lambda) = \sum_k \lambda_k t_k(x), \tag{22.1}$$

where $\Lambda \equiv \{\lambda_k; k = 1, ..., K\}$ are parameters that bias the optimization towards some of the targets. We say that (22.1) has collapsed the MOO into an SOO. A set Λ with fixed values λ_k defines one single SOO, thus (22.1) (with free λ_k) produces indeed a family of SOOs whose members are parameterized through Λ. We will study: those SOOs, the constraints that the Pareto front imposes to their solutions, and the relationships between different SOOs of the same family. The validity of the results holds for any positive, real set Λ. For convenience, though: (i) We take $K = 2$, which simplifies the graphic representations and contains the most relevant cases. (ii) We require $\sum_k \lambda_k = 1$ without loss of generality. For $K = 2$ then $\lambda_1 = \lambda$, $\lambda_2 = 1 - \lambda$, and $\Omega = \lambda t_1 + (1 - \lambda)t_2$. (iii) We impose $\lambda_k \neq 0\ \forall k$, thus $\lambda \in (0, 1)$. Comments about fringe cases can be found in [18].

As said above, for given λ_k one fixed SOO problem is posed. Then, (22.1) with fixed Ω defines *equifitness surfaces* noted $\tau_\Lambda(\Omega)$. Each $\tau_\Lambda(\Omega)$ constitutes a $(K-1)$-dimensional hyperplane in target space. For $K = 2$ these surfaces become straight lines (Fig. 22.1b):

$$\tau_\lambda(\Omega) \equiv \left\{(t_1, t_2)|\ t_2 = \frac{\Omega}{1-\lambda} - \frac{\lambda}{1-\lambda} t_1\right\}. \tag{22.2}$$

The slope of $\tau_\lambda(\Omega)$ along each possible direction $\hat{t}_k$ in the target space only depends on λ (here, $dt_2/dt_1 = -\lambda/(1-\lambda)$). Different $\tau_\lambda(\Omega)$ for fixed λ are parallel to each other. The crossing of $\tau_\lambda(\Omega)$ with each axis is proportional to Ω (from (22.2), the crossings with the horizontal and vertical axes read: $\Omega/(1-\lambda)$ and Ω/λ). For a given SOO (constant λ), minimizing Ω means finding $\tau_\lambda(\tilde{\Omega})$ with $\tilde{\Omega}$ the lowest value possible such that $\tau_\lambda(\tilde{\Omega})$ still intersects the Pareto front (Fig. 22.1d). This is equivalent to *pushing* the equifitness surfaces against the Pareto front as much as possible thus lowering the crossings with the axes.

The SOO optimum $x_\lambda \in \Pi$ usually lays at the point $T_f(x_\lambda)$ at which $\tau_\lambda(\tilde{\Omega})$ is tangent to the front (Fig. 22.1d). The exceptions to this rule are the most interesting cases. The solutions to different SOOs (defined by different values of λ) are found in different points along the front. For $\lambda \in (0, 1)$, equifitness surfaces present a

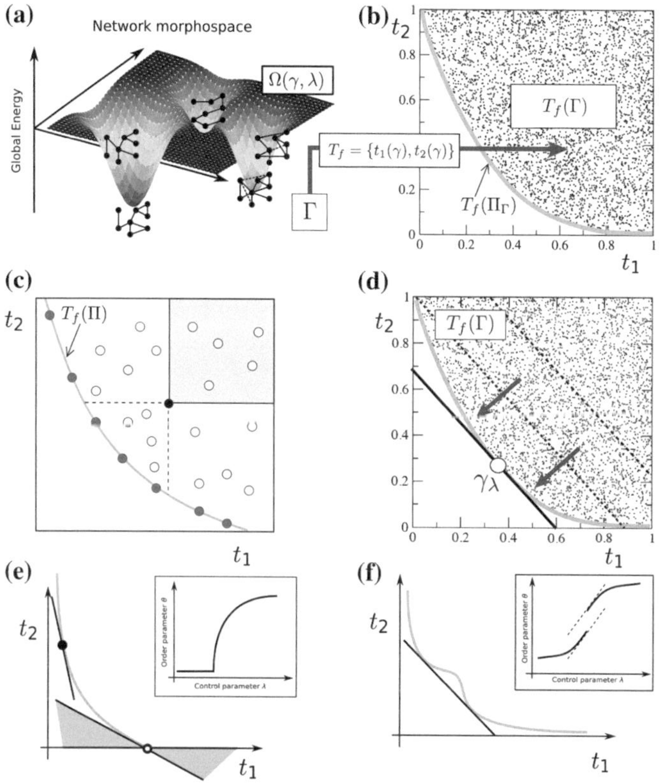

Fig. 22.1 Phase transitions in Pareto optimal systems. **a** In a design space of complex networks a set of weighted target functions defines a global energy (22.1) and renders a potential landscape (explored in [13, 18]). **b** Those same targets map the design space into the target space. The set of Pareto optimal designs is mapped onto a boundary of this morphoscpace: the Pareto front, which represents the most optimal tradeoff between the targets. **c** The concept of dominance is geometrically simple in target space. **d** Energy minimization for fixed λ returns a single point of the Pareto front. Changing λ we visit different solutions. Depending on the shape of the Pareto front, second (**e**) and first (**f**) order phase transitions arise as a function of λ

slope $-\lambda/(1-\lambda) = d \in (-\infty, 0)$ (d decreases as λ increases). Consider now differentially small modifications of λ. This allows us to drift infinitesimally slow through the SOO family. We could expect that solutions between different SOOs will change so gradually as well, but that is not always the case.

The front in Fig. 22.1d is convex (with respect to the optimization direction determined by $\lambda \in (0, 1)$). Its slope spans the whole range $d \in (-\infty, 0)$. This guarantees that, as we drift through λ, each different SOO problem has one characteristic solution laying exactly where the equifitness surface is tangent to the front. We can sample the front smoothly, thus anything that we measure on the SOO solutions (i.e. any order parameter) will be a smooth function of λ as well. Convex Pareto fronts whose slope span the whole range $d \in (-\infty, 0)$ do not present any *accident*.

Consider now the case in Fig. 22.1e. It represents a convex front whose slope spans $d \in (-\infty, d^* < 0)$. For $\lambda \in (\lambda^* \equiv -d^*/(1-d^*), 1)$, we can pose different SOOs whose solutions lay at different points of the convex part of the front. Varying λ within this interval renders a smooth sample of SOO solutions. However, for $\lambda \in (0, \lambda^*)$ we can pose different SOOs whose solution lays exactly at the same place, as indicated by the gray fan in Fig. 22.1c. If we measure anything about the SOO solutions, that quantity will be constant as a function of λ for $\lambda \in (0, \lambda^*)$ because we will persistently measure a property of the same design. That same property will vary smoothly over $\lambda \in (\lambda^*, 1)$. At λ^* this quantity will be continuous but its derivative will not (Fig. 22.1e, inset), as in second order phase transitions. In cases like this we say that the Pareto front ends abruptly at one of its extremes. Second order transitions also happen if the slope of the front spans $d \in (d^* > -\infty, 0)$ (i.e. if the opposite end of the front terminates abruptly) or if $d \in (-\infty, d^*_-) \cup (d^*_+, 0)$ (i.e. the front presents a sharp edge with an ill-defined derivative).

A cavity in the front leads to first order phase transitions. At either side of the cavity in Fig. 22.1f we find convex stretches whose points represent different solutions for different SOO problems posed by different $\lambda \in (0, \lambda^*)$ or $\lambda \in (\lambda^*, 1)$. But right at $\lambda = \lambda^*$ (represented by the thick straight line of Fig. 22.1d) two solutions are SOO optima at the same time. This is a phase coexistence phenomenon characteristic of first order transitions. Pareto optimal solutions laying inside the cavity are bypassed and never get to be SOO optima. If we measure an order parameter of the SOO solutions as a function of λ (Fig. 22.1f, inset), we find a gap resulting from the abrupt shift from one convex stretch of the front to the other at $\lambda = \lambda^*$.

22.2.1 Phase Transitions in Pareto Optimal Designs

As examples, we choose two problems that have recently been treated from an optimization perspective. Take Complex Networks first, which are good models of a series of natural systems such as vascular or nervous circuits [1, 2] that might be constrained by physical costs (available material) while seeking the efficient implementation of biological function (e.g. distribution of nutrients). Some human-made structures, such as transportation networks [9], would also benefit from optimal design.

In [13] we consider this problem to a greater extent. We take the cost $\rho(\gamma)$ of network γ as a function of its edges (number or length) and its efficiency is accounted for by the average path length $\langle l \rangle(\gamma)$, a naive proxy for how fast messages can be relayed across the network. These are the targets for minimization ($T_f = \{t_1 \equiv \rho(\gamma), t_2 \equiv \langle l \rangle(\gamma)\}$) that lead to a Pareto front and, depending on its shape, to phase transition and other interesting phenomena. In Fig. 22.2a we represent the front for such an MOO along with some Pareto optimal networks. In this example the nodes are spaced over a circle and the cost of each link is proportional to its Euclidean distance. This front ends abruptly (Fig. 22.2a, inset) and a cavity is present (Fig. 22.2b, see [13] for discussion). This implies, correspondingly, a second

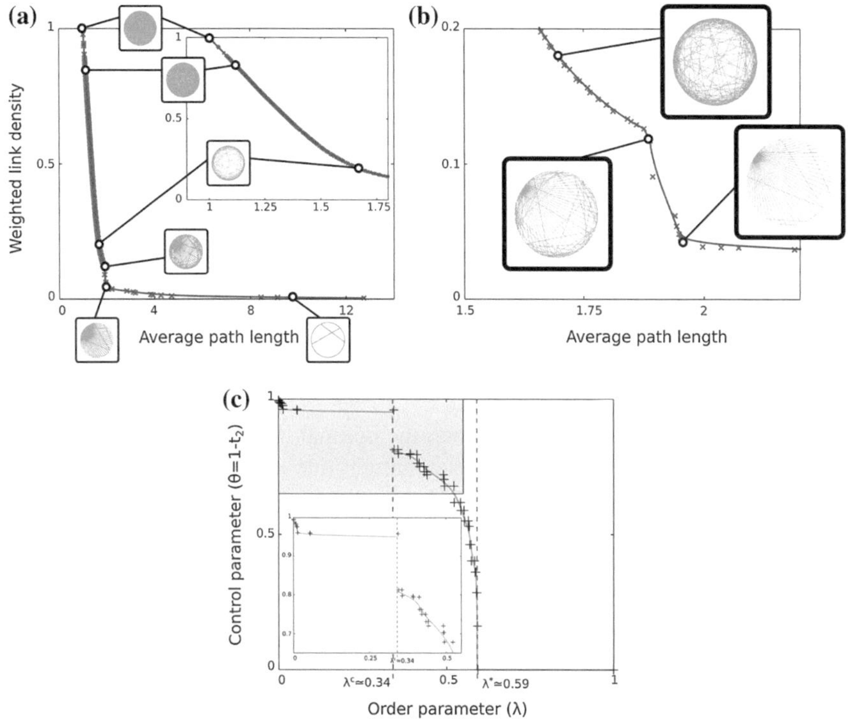

Fig. 22.2 Pareto optimal networks with nodes spaced over a circle. A genetic algorithm was used to approximate the Pareto front (*blue crosses* and *thick brown curve*) of networks that minimize the average path length and the cost of their links. **a** The front implements a tradeoff between the clique and the minimum spanning tree. (*Inset*) The clique extreme of the front ends abruptly (see [18]) indicating a second order phase transition. **b** A cavity is revealed at the center of the front, which implies a first order phase transition. **c** Both transitions are revealed in the plot of any order parameter (Second order at $\lambda_2^* \simeq 0.59$, first order at $\lambda_1^* \simeq 0.34$)

and a first order transitions at $\lambda_2^* \simeq 0.59$ and at $\lambda_1^* \simeq 0.34$. These transitions can be noted in the plot of any order parameter (Fig. 22.2c).

Our second example explores the evolutionary constraints of human language, an unsettled challenge for the scientific community. The optimization of linguistic structures brings together universal language properties (such as Zipf's law) and the presence of ambiguity, likely as a compromise between language economy and a large ability to talk about the outer world [26]. Such a tension was proposed by Zipf himself [6] and its mathematical formalization [8] leads to an MOO problem that was always treated as an SOO. Accordingly, phase transitions were readily identified but some debate lasted concerning its nature and meaning [27]. In [8], languages $a \in A$ are modeled through a set of signals S and objects R whose associations are encoded in a matrix $a = \{a_{ij}\}$ with $a_{ij} = 1$ if signal $s_i \in S$ names object $r_j \in R$ and 0 otherwise. This binary matrix presents many ones in a row if a signal is polysemous and many ones in a column if several words name the same object—i.e. if there are any synonymous. Every object is recalled equally often and, if an object has many names, a speaker chooses uniformly among them when necessary. Two quantities are relevant (see [8]): (i) one entropy $h_H(a)$ associated to the uncertainty of a message when a hearer has to decode it—i.e. what object it is meant after the speaker has uttered a signal; and (ii) another entropy $h_S(a)$ associated to the speaker choosing the right word to name an object among those available. A speaker might be allowed to be vague (as in "it" referring to any object) or she might be requested to be specific (perhaps finding the precise technicism in a scientific context).

These two entropies represent the effort made by hearers or speakers when using a language. They act as minimization targets ($T_f = \{t_1 \equiv h_H(a), t_2 \equiv h_S(a)\}$), so that languages $a \in A$ are subjected to a MOO. A subset ($a_\pi \in \Pi) \subset A$ of object-signal associations implements the Pareto front, the optimal tradeoff between the efforts of a hearer and a speaker. This front is a straight line in target space (Fig. 22.3a). Attending to the theory sketched above this is akin to a first order phase transition (see [18]) with one end of the front being the global optimum for $\lambda < \lambda^c$ and the other extreme of the front being optimum for $\lambda > \lambda^c$. Right at $\lambda = \lambda^c$, a sudden jump happens between these two, very distinct phases. This can be appreciated in any order parameter as a sharp discontinuity (Fig. 22.3b, c). The extremes correspond to (i) a language that minimizes the speaker's efforts for $\lambda < \lambda^c$ (one single signal names every object, as in the "it" example before, so that the speaker does not need to

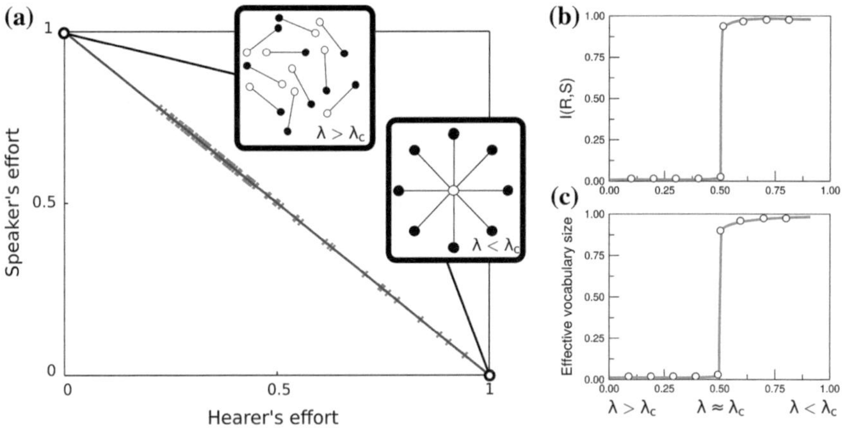

Fig. 22.3 Least effort languages. **a** Arbitrary Pareto optimal languages (*blue crosses*) lay on the straight line $t_2 = 1 - t_1$. A straight front is a sign of criticality along a first order phase transition scenario. Either phase represents respectively the best scenario for the speaker ($\lambda < \lambda^c$, where communication is impossible unless through the context) and for the hearer ($\lambda > \lambda^c$, with high memory demands). Only at the critical point is a wide complexity available. Any order parameter (**b** mutual information between the signals and the external world; **c** effective vocabulary size) reflects the phase transition

think the right association every time) and (ii) a language that minimizes the hearer's effort for $\lambda > \lambda^c$, with perfect pairings between signals and objects so that there is not any ambiguity when decoding the messages.

Communication is difficult in both extremes, either because the signals convey little information about the objects ($\lambda < \lambda^c$, Fig. 22.3b), or because of the memory needs to browse a vast vocabulary ($\lambda > \lambda^c$, Fig. 22.3c). Besides, we know that more complicated structures exist in real languages. These structures can be found right at $\lambda = \lambda^c$. A straight Pareto front is an indication of criticality [18, 28]. In such cases, exactly at the critical value λ^c, the whole front is also SOO optimal. Note that in usual first order transitions those solutions laying at the cavity are skipped altogether, while here a plethora of them becomes available. In [27] the authors proved that the global SOO minimizers at $\lambda = \lambda^c$ consist of all possible languages without synonyms, hence these must constitute the Pareto front. More importantly, among these possible languages it exists one such that the frequency of the signals obeys Zipf's law, as in natural human languages.

22.3 Discussion

22.3.1 Thermodynamics as an MOO-SOO Problem

Thermodynamics is one of the best established branches of physics and dates back to more than two centuries ago. In its modern form—as statistical mechanics—it allows us to make precise predictions about diverse macroscopic physical phenomena. Its applications extend beyond physics, as complex systems are increasingly being investigated through maximum entropy models [29, 30]. In [18] we rewrite thermodynamics as an MOO-SOO problem, not to suggest that our theoretical framework modifies it in any way. Rather the opposite: By checking that our framework reproduces a robust physical theory, we strand our findings in a more solid ground.

Phase transitions in complex systems often raise heated debate: being strict, phase transitions are defined for thermodynamics alone, through partition functions, and involve fluctuations that compel us to take a thermodynamic limit. Little can be done against such epistemological stand. This is yet another reason why we undertake the task of writing thermodynamics as an MOO-SOO. Such a formalization of statistical mechanics reproduces all the results concerning phase transitions *in the exact same way* that transitions arise in other MOO-SOO scenarios. We suggest and support that the phase transition phenomenology arising in other MOO-SOO systems is more than a qualitative similarity.

The argument is not repeated here because of space constraints, but the idea is to show that the *independent, simultaneous* minimization of internal energy and maximization of entropy leads to a Pareto front subjected to the phenomenology found in Sect. 22.2. In thermodynamics we deal with given physical systems that cannot be modified. We test probabilistic descriptions that tell us how likely it is to find the system in each part of its phase space. We wonder which of these descriptions present a lower internal energy and larger entropy. Thus our design space X is the set of all possible probabilistic descriptions of the system under research. From that optimization we obtain a Pareto front whose shape (through cavities) and differential geometry (through sharp edges) imply phase transitions if the targets ($t_1 \equiv U$, $t_2 \equiv S$) were collapsed into an SOO problem. But that is precisely what happens in equilibrium thermodynamics through the minimization of the free energy $F = U - TS$ [20]. We identify $\Omega \equiv F$, $\lambda_1 \equiv 1$, and $\lambda_2 \equiv -T$, and the theory exposed above applies with transitions at singular temperature values.

We insist that the optimization operates *upon* probabilistic descriptions of the thermodynamic species—while the shape of the front is determined by the properties of the physical system. It might be interesting to segregate what thermodynamic phenomenology happens because thermodynamic systems *are* probabilistic ensembles (in this regard they are unlike Pareto optimal networks or least effort languages, as much as networks and languages are unlike each other) and what phenomenology arises because of the shape of a Pareto front (that would yield the same phenomenology irrespective of the kind of designs considered—were they networks or languages—as long as the front had the same shape).

This interpretation of statistical mechanics systems is illustrated with two very simple examples with first and second order transitions and one critical point in [18]. As stated above, this is not to prove new thermodynamic results, but to provide more solid basis for this theory regarding MOO-SOO situations. Indeed, the role of cavities in first order phase transitions dates back to Gibbs [19, 20], whose *Gibbs surface* represented the states of a thermodynamic species. That surface is associated to the microcanonical ensemble [18] and may be concave or convex. Its convex hull is associated to the canonical ensemble (hence to equilibrium at given temperature through free energy minimization), which is always convex. At cavities in the Gibbs surface, the description of both ensembles must differ (as noted by the theory of ensemble inequivalence [31]) and first order transitions occur.

22.3.2 Closing Remarks

With our recent findings [18] we close a gap between the MOO literature, research on SOO tradeoffs, and statistical mechanics. On the one hand, standard MOO analysis does not take into account phenomena like phase transitions or criticality which, we believe, add up to our knowledge and enrich the description of Pareto optimal designs. On the other hand, analyses of the Pareto front are often qualitative or based on subjective appreciations of its shape. The formalism developed in Sect. 22.2 allows us to locate quantitatively very relevant details of the systems under research. These features shall persist under transformations of the targets and, if not, the qualitative description would tell us *how* do these phenomena disappear. Furthermore, a solid connection to thermodynamics has been established. We are pretty confident of the immutable, lasting nature of thermodynamics; thus we can guess that, through the

Pareto formalism, we have located broad features that unite the description of diverse MOO problems.

A prominent field for MOO application is biology [3–5]. Thermodynamic-like phenomenology is not discussed in these references, but the stage looks great: Is there a place for true MOO in biology? Against this, natural selection concerns itself with fitness maximization alone. This feels like an exciting MOO-SOO picture, but we cannot guarantee linear global functions as in (22.1). Beyond linearity, new phenomenology might be uncovered.

Finally, an important, though conceptually difficult issue was left aside in [18] and only incidentally dealt with here. How do critical systems look like under an MOO perspective? Can we recover the astounding phenomena of criticality? This is studied in [28]. Other theoretical aspects of MOO remain open to research.

Acknowledgments This work has been supported by an ERC Advanced Grant, the Botín Foundation, by Banco Santander through its Santander Universities Global Division and by the Santa Fe Institute. We thank CSL members for insightful discussion.

References

1. West, G.B., Brown, J.H., Enquist, B.J.: A general model for the structure and allometry of plant vascular systems. Nature **400**, 664–667 (1999)
2. Pérez-Escudero, A., de Polavieja, G.G.: Optimally wired subnetwork determines neuroanatomy of Caenorhabditis elegans. Proc. Natl. Acad. Sci. USA **104**(43), 17180–17185 (2007)
3. Shoval, O., Sheftel, H., Shinar, G., Hart, Y., Ramote, O., Mayo, A., Dekel, E., Kavanagh, K., Alon, U.: Evolutionary tradeoffs, Pareto optimality, and the geometry of phenotype space. Science **336**, 1157–1160 (2012)
4. Schuetz, R., Zamboni, N., Zampieri, M., Heinemann, M., Sauer, U.: Multidimensional optimality of microbial metabolism. Science **336**, 601–604 (2012)
5. Szekely, P., Sheftel, H., Mayo, A., Alon, U.: Evolutionary tradeoffs between economy and effectiveness in biological homeostasis systems. PLoS Comput. Biol. **9**(8), e1003163 (2013)
6. Zipf, G.K.: Human Behavior and the Principle of Least Effort (1949)
7. Maritan, A., Rinaldo, A., Rigon, R., Giacometti, A., Rodrígued-Iturbe, I.: Scaling laws for river networks. Phys. Rev. E **53**(2), 1510 (1996)
8. Ferrer i Cancho, R., Solé, R.V.: Least effort and the origins of scaling in human language. Proc. Natl. Acad. Sci. **100**(3), 788–791 (2003)
9. Barthelemy, M.: Spatial networks. Phys. Rep. **499**, 1–101 (2011)
10. Louf, R., Jensen, P., Barthelemy, M.: Emergence of hierarchy in cost-driven growth of spatial networks. Proc. Natl. Acad. Sci. **110**(22), 8824–8829 (2013)
11. Dawkins, R.: The Blind Watchmaker: Why the Evidence of Evolution Reveals a Universe Without Design. WW Norton & Company (1986)
12. Dennett, D.C.: Darwin's Dangerous Idea. The Sciences (1995)
13. Seoane, L.F., Solé, R.: Phase transitions in Pareto optimal complex networks. Phys. Rev. E **92**, 032807 (2015)
14. Avena-Koenigsberger, A., Goñi, J., Solé, R., Sporns, O.: Network morphospace. J. Royal Soc. Interface **12**(103), 20140881 (2015)
15. Goñi, J., Avena-Koenigsberger, A., de Menizabal, N.V., van den Heuvel, M., Betzel, R., Sporns, O.: Exploring the morphospace of communication efficiency in complex networks. PLoS ONE **8**, e58070 (2013)

16. Priester, C., Schmitt, S., Peixoto, T.P.: Limits and trade-offs of topological network robustness. PLoS ONE **9**(9), e108215 (2014)
17. Otero-Muras, I., Banga, J.R.: Multicriteria global optimization for biocircuit design. BMC Syst. Biol. **8**, 113 (2014)
18. Seoane, L.F., Solé, R.: A multiobjective optimization approach to statistical mechanics. http://arxiv.org/abs/1310.6372 (2013)
19. Gibbs, J.W.: A method of geometrical representation of the thermodynamic properties of substances by means of surfaces. Trans. Conn. Acad. **2**, 382–404 (1873)
20. Maxwell, J.C.: Theory of Heat. Longmans, Green, and Co., pp. 195–208 (1904)
21. Fonseca, C.M., Fleming, P.J.: An overview of evolutionary algorithms in multiobjective optimization. Evol. Comput. **3**, 1–16 (1995)
22. Dittes, F.M.: Optimization on rugged landscapes: a new general purpose Monte Carlo approach. Phys. Rev. Lett. **76**(25), 4651–4655 (1996)
23. Zitzler, E.: Evolutionary algorithms for multiobjective optimization: methods and applications. A dissertation submitted to the Swiss Federal Institute of Technology Zurich for the degree of Doctor of Technical Sciences (1999)
24. Coello, C.A.: Evolutionary multi-objective optimization: a historical view of the field. IEEE Comput. Intell. M. **1**(1), 28–36 (2006)
25. Konak, A., Coit, D.W., Smith, A.E.: Multi-objective optimization using genetic algorithms: a tutorial. Reliab. Eng. Syst. Safe. **91**(9), 992–1007 (2006)
26. Solé, R., Seoane, L.F.: Ambiguity in language networks. Linguist. Rev. **32**(1), 5–35 (2014)
27. Prokopenko, M., Ay, N., Obst, O., Polani, D.: Phase transitions in least-effort communications. J. Stat. Mech. **2010**(11), P11025 (2010)
28. Seoane, L.F., Solé, R.: Systems poised to criticality through Pareto selective forces. http://arxiv.org/abs/1510.08697 (2015)
29. Harte, J.: Maximum Entropy and Ecology: A Theory of Abundance, Distribution, and Energetics. Oxford University Press (2011)
30. Mora, T., Bialek, W.: Are biological systems poised at criticality? J. Stat. Phys. **144**(2), 268–302 (2011)
31. Touchette, H., Ellis, R.S., Turkington, B.: An introduction to the thermodynamic and macrostate levels of nonequivalent ensembles. Phys A. **2004**(340), 138–146 (2004)

Chapter 23
Power-Laws as Statistical Mixtures

M. Patriarca, E. Heinsalu, L. Marzola, A. Chakraborti and K. Kaski

Abstract Many complex systems are characterized by power-law distributions. In this article, we show that for various examples of power-law distributions, including the two probably most popular ones, the Pareto law for the wealth distribution and Zipf's law for the occurrence frequency of words in a written text, the power-law tails of the probability distributions can be decomposed into a statistical mixture of canonical equilibrium probability densities of the subsystems. While the interacting units or subsystems have canonical distributions at equilibrium, as predicted by canonical statistical mechanics, the heterogeneity of the shapes of their distributions leads to the appearance of a power-law.

M. Patriarca (✉) · E. Heinsalu · L. Marzola
NICPB–National Institute of Chemical Physics and Biophysics,
Rävala 10, 10143 Tallinn, Estonia
e-mail: marco.patriarca@kbfi.ee

E. Heinsalu
e-mail: els.heinsalu@kbfi.ee

L. Marzola
Laboratory of Theoretical Physics, Institute of Physics,
University of Tartu, Ravila 14c, 50411 Tartu, Estonia
e-mail: luca.marzola@ut.ee

A. Chakraborti
SCIS-School of Computational & Integrative Sciences,
Jawaharlal Nehru University, New Delhi 110067, India
e-mail: anirban@jnu.ac.in

K. Kaski
Department of Computer Science, Aalto University School of Science,
P.O. Box 15500, FI-00076 Aalto, Finland
e-mail: kimmo.kaski@aalto.fi

S. Battiston et al. (eds.), *Proceedings of ECCS 2014*, Springer Proceedings in Complexity, DOI 10.1007/978-3-319-29228-1_23

23.1 Introduction

Power-laws are characteristic to many complex systems. Due to their scaling properties, they are signatures of some underlying processes of self-organization. The first examples of power-law distributions were encountered in fields far from the targets of traditional physics research, however, nowadays considered standard topics in complex systems theory, e.g., the Pareto law of wealth distribution [9] and Zipf's law for the occurrence frequency of words in a written text [15]. The origin of power-law tails and their microscopic interpretation is still an open question and various mechanisms responsible for their appearance have been proposed. Some of them, such as the extensive generalizations of the Boltzmann entropy [24] or alternative forms of the Gibbs distribution [23], suggest that the basic assumptions of statistical mechanics should be reformulated.

The goal of the present contribution is to illustrate how the appearance of power-law distributions in some specific complex systems can be explained through the diversity of the components of the system under study within the framework of canonical statistical mechanics. In such systems, the power-law tail in the equilibrium distribution $f(x)$ of the relevant variable x is the outcome of the superposition of different bell-shaped equilibrium distributions of the subsystems. Far from being a strictly technical mechanism or a phenomenon arising in some rare circumstance, it is a general effect taking place in many *heterogeneous* complex systems, recently identified. Such systems can be characterized by a true equilibrium state which is usually described by a canonical Boltzmann-Gibbs-type distribution.

The perspective suggested in the present paper simplifies and unifies the understanding of the power-law tails appearing in the distributions of many systems as a diversity-induced phenomenon.

23.2 Statistical Formulation

The aim of this section is to provide a general theoretical framework for describing the heterogeneity-induced appearance of a power-law tail in the probability distribution. This is done in a simple and general way relying on well-known concepts of statistics and statistical mechanics.

In the general problem considered here, one is interested in the probability distribution function $f(x)$ of a random variable x of a system $\mathcal{S}$ composed of N different types of units that can be correspondingly be grouped into N disjoint subsystems $\mathcal{S}_i$, $i = 1, \dots, N$, so that $\mathcal{S} = \cup_{i=1}^{N} \mathcal{S}_i$.

From an operative point of view, the probability density $f(x)$ can be constructed by measuring the frequency of occurrence of the value x (independently of the type i of the subsystem observed) in the limit of a large number of measurements. The partial probability densities $f_i(x)$ can be constructed in a similar way recording in each measurement both the occurrence frequency x and the subsystem type i. The distribution $f(x)$ can be expressed as the weighted sum of the partial distributions $f_i(x)$, i.e., $f(x) = \sum_i f_i(x)\, p_i$, where each partial probability density $f_i(x)$ is assumed to be normalized, $\int dx\, f(x) = 1$, and the statistical weight p_i represents the fraction of units of type i present in the system, i.e., the probability to pick up in a measurement a unit of the i-th type.

From the point of view of probability theory, the global probability density $f(x)$ is interpreted in terms of conditional probabilities, through *the law of total probability* [11], by rewriting it as

$$f(x) = \sum_i f_i(x)\, p_i \equiv \sum_i P(x|i) P(i)\,. \tag{23.1}$$

Here $P(i) \equiv p_i$ is the probability to extract a unit belonging to the subpopulation $\mathcal{S}_i$ and $P(x|i) \equiv f_i(x)$ is the conditional probability that, if the unit belongs to subpopulation $\mathcal{S}_i$, the value x is found for the variable x. For the sake of clarity, in the examples considered here, the partition of the system $\mathcal{S}$ is fixed, i.e., the populations N_i of each subset $\mathcal{S}_i$ are constant in time. Therefore, the set of fixed parameters $\{p_i\}$ defines the heterogeneity of the total system $\mathcal{S}$, i.e., the level and type of quenched disorder.

Equation (23.1) is the reference formula in the following, since it links directly the total probability distribution $f(x)$ to the *diversity* of the system defined by the $\{p_i\}$. Notice that in the most general case one cannot make any prediction on the total distribution $f(x)$ if no further information is available, apart from the relevant fact that the total equilibrium distribution will have a shape different from the canonical one, even if the partial distributions of the subsystems $f_i(x)$ do. Whether the statistical mixture in (23.1) will produce a distribution with a power-law or not depends on the details of the system considered, i.e., on (a) the form of the equilibrium partial distributions $f_i(x)$ and (b) the heterogeneity of the system as defined by the p_i's. Section 23.3 on the Pareto law and (23.5) on the heterogeneous gas present two solvable models in which the partial distributions $f_i(x)$ are known and the weight distribution $\{p_i\}$ are input parameters—in this situation the necessary conditions for the appearance of a power-law tail can be formulated exactly. Instead, in Sect. 23.4 about the Zipf law, a similar statistical decomposition of a power-law is presented in a phenomenological way based on data.

The mechanism outlined above, describing how the appearance of a power-law can be interpreted as a heterogeneity-induced effect, has similarities with the so-called super-statistics introduced in 2003 by Beck and Cohen in the framework of non-equilibrium statistical mechanics [1, 2]. In super-statistics, a power-law can arise from a compound distribution, expressed in a way similar to (23.1) where, however, the sum represents a *randomization procedure* [11] of the probability distribution function of a single system over the values of some system parameter(s) varying slowly and randomly on a long time scale. In the present paper, we study what can be considered to be the complementary (noise-less) case in which the appearance of a power-law is instead due to quenched disorder. This mechanism was proposed in 2005 [9] by various groups [3, 7, 19] in the framework of kinetic wealth-exchange models, as a possible explanation of the Pareto power-law in economics. The formal similarity with super-statistics suggests, as a natural generalization, a mechanism of power-law formation involving both stochastic fluctuations of some external parameters as well as some degree of internal heterogeneity. However, this topic will be considered elsewhere.

23.3 Kinetic Exchange Models

The Pareto law of wealth distribution is probably the first example of power-law distributions reported. An experimentally-based interpretation of it as a mixture of distributions associated to, e.g., categories of companies with different features is still missing. However, such an interpretation is suggested by some theoretical models such as the *kinetic wealth exchange models* (KWEM) [8, 20, 22], which mimic in a very simplified way the exchanges of wealth in an economy system.

Let us consider a heterogeneous KWEM, in which M agents $i = 1, \dots, M$ are assigned initially a certain wealth x_i and then begin to exchange wealth through pair-wise interactions. At each time step two agents i, j are extracted randomly and their wealths x_i, x_j updated depending on the saving parameters λ_i, λ_j, representing the minimum fractions of wealth saved (see [21, 22] for details),

$$\begin{aligned} x_i' &= \lambda_i x_i + \epsilon[(1-\lambda_i)x_i + (1-\lambda_j)x_j]\,, \\ x_j' &= \lambda_j x_j - (1-\epsilon)[(1-\lambda_i)x_i + (1-\lambda_j)x_j]\,. \end{aligned} \tag{23.2}$$

Here x_i', x_j' are the wealths after a trade and ϵ is a uniform random number in (0, 1). The total wealth is conserved during each transaction, $x_i + x_j = x_i' + x_j'$, just as energy in molecular collisions. This analogy, that was already noticed by Mandelbrot [14], is in fact quite close. In the homogeneous version of KWEMs, i.e. when $\lambda_k \equiv \lambda$ [4], the dynamics is equivalent to that of a perfect gas in an effective number N of dimensions: the equilibrium wealth distribution $f(x)$ is the energy distribution of a perfect gas in a number of dimensions $N(\lambda)/2 = 1 + 3\lambda/(1-\lambda)$, i.e., a Γ-distribution $\gamma_{N/2}(x)$ of order $N(\lambda)/2$ [18]. This results has been demonstrated to be exact for some particular KWEMs by Katriel, see [13] for details. In fact, by inverting $N(\lambda)$, one obtains an average fraction of wealth exchanged during one trade $1-\lambda \propto 1/N$ ($N \gg 1$), similarly to the energy exchanges during molecular collisions of a N-dimensional gas [5, 21]. Then a heterogeneous system composed of agents with different λ_i is analogous to a dimensionally heterogeneous system. The relevance of the heterogeneous KWEM with λ_i distributed in the interval (0, 1) is in the fact that they relax toward realistic wealth distributions $f(x)$ with a Pareto tail, as it was shown numerically and analytically in [8]. In the case of a uniform distribution for the saving parameters, $\phi(\lambda) = 1$ if $\lambda \in (0, 1)$ and $\phi(\lambda) = 0$ otherwise, setting $n = N/2$, the dimension density has a power-law $\sim 1/n^2$, $P(n) = \phi(\lambda)d\lambda/dn = 3/(n+2)^2$ $(n \geq 1)$.

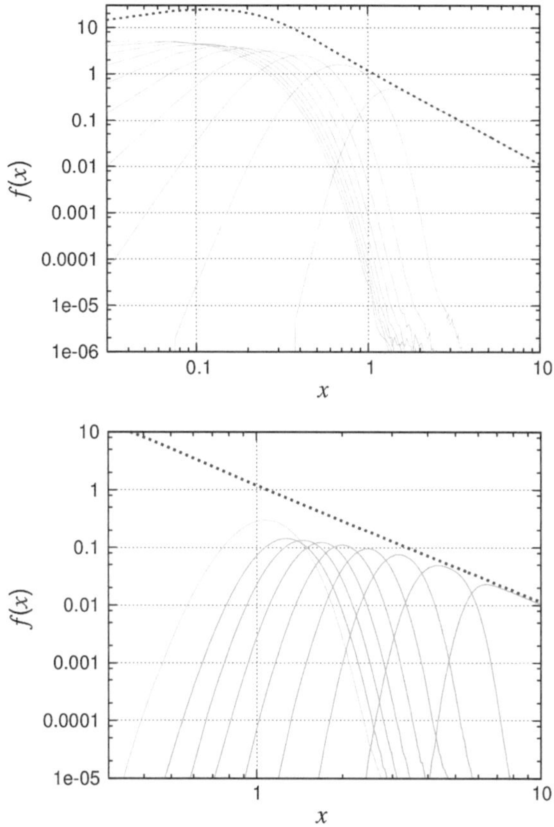

Fig. 23.1 The power-law of the wealth distribution $f(x)$ for agents with uniformly distributed λ_i in the interval (0,1) is decomposed into partial distributions $f_i(x)$. *Above* Decomposition of $f(x)$ into ten partial distributions $f_i(x)$ of agents with $\lambda \in (0, 0.1), (0.1, 0.2) \ldots (0.9, 1)$. *Bottom* The rightmost distribution with $\lambda \in (0.9, 1)$ on the upper plot, still exhibiting a power-law, is here further decomposed into ten partial distributions of agents with $\lambda \in (0.9, 0.91), (0.91, 0.92) \ldots (0.99, 1)$; again the power-law only remains in the rightmost distribution which could in turn be further decomposed

The way in which the single Γ-distributions of the subsystems comply to generate a power-law distribution is illustrated in Fig. 23.1, taken from [16, 19].

23.4 Zipf's Law

In this section, we consider the other most famous power-law distribution, which is Zipf's law of occurrence frequency of the words in a written text. Zipf originally demonstrated the presence of a power-law in the rank plot of the words appearing in a text. Since the rank plot is closely related to the cumulative distribution function, a power-law in the rank plot implies a power-law (with a different exponent) in the cumulative as well as in the probability distribution function. In the present study, the probability function or histograms like the ones shown in Fig. 23.2 are preferable (to rank plots) since they are more suited for a decomposition of the

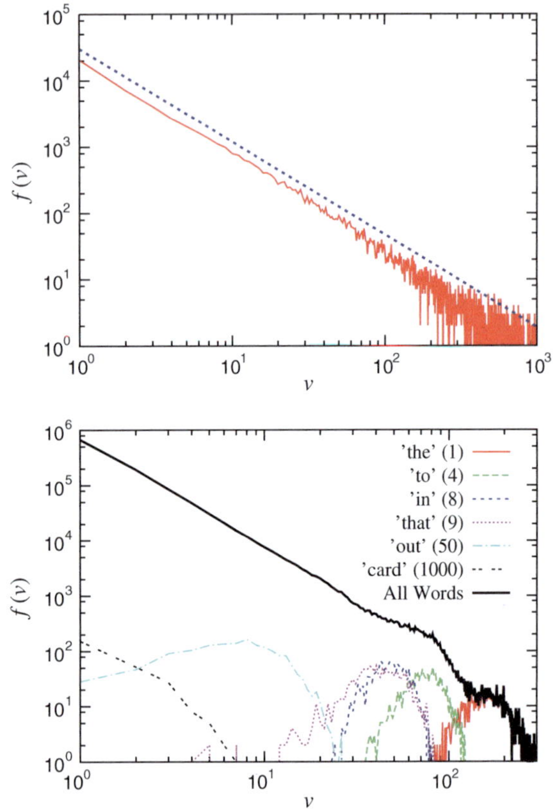

Fig. 23.2 *Top* Histogram of word occurrence frequencies (*red continuous line*) of the text considered (see Appendix for details) and comparison with the power-law $f \propto x^{-1.4}$ (*dotted blue line*). *Bottom* Comparison of the total probability distribution function with a few partial distributions corresponding to some single words (rank shown in parentheses, see text for details)

power-law distribution into a statistical mixture of localized probability distributions or histograms, as done in the example discussed in the previous section.

The text considered here was obtained by merging a collection of novels (see Appendix for details). In Fig. 23.2-top, the histogram of the occurrence frequency of words through the whole text is depicted. The histogram shown, that is proportional to the corresponding probability distribution, exhibits a power-law $f \propto x^{-1.4}$.

For the sake of simplicity, our starting point is the partition of the text into words, i.e., words are considered as the basic units of the text, as in Zipf's approach. We investigate the behavior of *single words* and inquiry about the type of their frequency distribution in the text. That is, we ask what is the probability distribution $f_i(x)$ that the *single* i-th word has an occurrence frequency x in a text with some given properties. When trying to do this, however, one immediately realizes that while the rank plot or the occurrence-frequency distribution of words can also be defined for a single text, for the definition of the occurrence frequency probability distribution $f_i(x)$ of the single i-th word the counting of the occurrence frequency through many different texts is needed—for each word a rank plot only provides a single number equal to the occurrence frequency in that text. In order to make a sensible comparison between the single-word distributions $f_i(x)$, we have to consider some other text properties.

First, the word occurrence frequency presents large *topical* variations when going from one text to another one [10, 12]. In order to minimize possible issues related to this variability we analyzed a single text obtained by merging novels of the same genre (as detailed in the Appendix).

Furthermore, the number of different words N_W in a given text depends on (and rescales with) the text length measured by the total number of words N, i.e., Heap's law holds, stating that $N_W \propto N^\alpha$, with typical values of α in the range $\alpha \in (0.5, 1)$ [12]. For this reason, in order to construct the occurrence frequency distributions $f_i(x)$, we used texts with the same (fixed) length N constructed by dividing the total text into parts containing the same number N_W of words. Therefore, the following results do not refer to the original text but to an effective (shorter) sample text long N_W words. We have chosen $N_W = 3000$ which is a good compromise between the length of a single part and the total number of parts. Measuring the occurrence frequencies of each word i, the corresponding occurrence frequency distribution $f_i(x)$ was computed, providing the (relative) probability to find the i-th word in one of the text parts.

Results are presented in Fig. 23.2-bottom. While the histograms of the occurrence frequency of *any* word (black continuous line) presents a power-law tail, *all* the distribution of single words present a localized shape. Many words, especially those with similar rank, have histograms similar to each other and for reason of clarity only the distributions of a few words have been plotted (the rank is shown in parentheses). This shows that a power-law tail in the occurrence-frequency distribution emerges only as a collective phenomenon due to the interplay of the various (different) words composing a text, when the histograms of single words are summed up building the total distribution.

Finally, it is to be noticed that we have not processed the text before the analysis in order to remove e.g. prepositions, article, and other non-topical words, keeping all the words for the analysis. In fact, we are interested in the Zipf's law encountered in a general text—the aim of the example above was to carry out a statistical analysis of the original text to show that the properties discussed are general and valid also for standard written text. The detailed comparison with different types of processed text will be done elsewhere.

23.5 A Mechanical Model: An Assembly of Heterogeneous Polymers

As another prototypical example of a system presenting a diversity-induced power-law distribution, we consider a theoretical model system made up of an assembly of harmonic polymers. Such a model is simple and exactly solvable, yet it is general in the sense that the inter-particle harmonic potentials can be thought to describe the

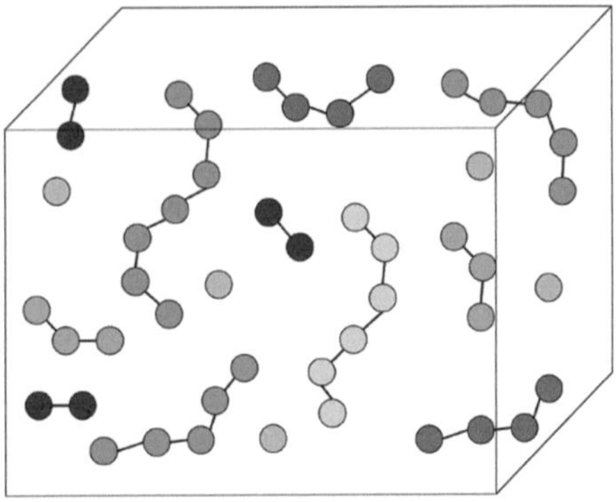

Fig. 23.3 A prototypical model of system presenting a diversity-induced power-law tail in the (kinetic as well as potential) energy distribution is an assembly of harmonic polymers with different numbers of monomers, see text for details

small displacements of the normal modes with respect to the equilibrium configuration of a more general nonlinear system. We assume that polymers consist of different numbers of monomers, i.e., they have different numbers of degrees of freedom, see Fig. 23.3, and study the potential energy distribution (similar considerations hold also for the distribution of kinetic energy or velocity). Notice that such a model can also be used to study a general system composed of subsystems with different dimensions or numbers of degrees of freedom. The hypothesis of non-interacting polymers is made, in the same spirit of the statistical mechanical treatment of a perfect gas, even if a weak interaction is understood to be present in order to bring the system toward thermal equilibrium, implying that each polymer undergoes independent statistical fluctuations.

It is convenient to start from the homogeneous system, composed of identical subsystems with N harmonic degrees of freedom. Using suitably rescaled coordinates $\mathbf{q} = \{q_i\} = \{q_1, q_2, \dots, q_N\}$, the energy function can be written in the form $x(\mathbf{q}) = (q_1^2 + \cdots + q_N^2)/2$. The equilibrium energy distribution coincides with the standard Gibbs energy distribution of an N-dimensional harmonic oscillator. After integrating out the angular variables in the space $\mathbf{q}$, it reduces to a Γ-function of order $n = N/2$ [18],

$$f_n(x) = \beta\,\gamma_n(\beta x) \equiv \frac{\beta}{\Gamma(n)}(\beta x)^{n-1}\exp(-\beta x)\,,\quad n = N/2\,; \tag{23.3}$$

β is the inverse temperature. The same result is obtained through a variational principle from the Boltzmann entropy that after integration of the $(N-1)$ angular variables becomes

$$S_n[f_n] = \int_0^{+\infty} dx\; f_n(x)\left\{\ln\left[\frac{f(x)}{\sigma_{2n}\,x^{n-1}}\right] + \mu + \beta x\right\}. \tag{23.4}$$

Where σ_N is the hyper-surface of a unitary N-dimensional sphere; μ, β are Lagrange multipliers determined by the constraints on the conservation of the total number of units and energy, respectively; the power x^{n-1} comes from the angular integration. The result of the variation is the Γ-distribution in (23.3), if one defines $\beta^{-1} = 2\langle x\rangle/N = \langle x\rangle/n$ as the temperature, in agreement with the equipartition theorem [17].

In the heterogeneous case, the statistical independence of the polymers allows one to write the total entropy of the heterogeneous system as the sum of the entropies of the different units,

$$S[\{f_n\}]=\int dn\ P(n)\int_0^{+\infty} dx\ f_n(x)\left\{\ln\left[\frac{f_n(x)}{\sigma_{2n}\,x^{n-1}}\right]+\mu_n+\beta x\right\}, \tag{23.5}$$

where the fractions $P(n)$ of units with dimension $N = 2n$ have been introduced, with $\sum_n P(n) = 1$. Notice that different Lagrange multipliers μ_n have been used for each n since the fractions $P(n)$ of polymers with $2n$ degrees of freedom are conserved separately, but a single temperature parameter β is present as the total energy is conserved. The equilibrium probability density obtained is again the Γ-distribution $f_n(x)$ in (23.3), as in the homogeneous case, but the corresponding β is now given by a generalization of the equipartition theorem to a dimensionally heterogeneous system,

$$\langle x\rangle = \int dn \int_0^{\infty} dx\ f_n(x)\,x = \frac{\langle N\rangle}{2\beta}\,. \tag{23.6}$$

Here $\langle N\rangle = 2\langle n\rangle = 2\int dn\,P(n)n$ is the average dimension (a finite value of $\langle N\rangle$ and of the average energy $\langle x\rangle$ are obtained when the dimension density $P(n)$ has a finite cutoff or decreases faster than $1/n^2$ for $n \gg 1$). The probability of measuring a value x of energy *independently of the unit type* is given by the statistical mixture [11]

$$f(x) = \int dn\ P(n)\,f_n(x) = \int dn\ \frac{P(n)\beta}{\Gamma(n)}(\beta x)^{n-1}\exp(-\beta x). \tag{23.7}$$

While the distributions $f_n(x)$ have exponential tails, the asymptotic shape of the function $f(x)$ can be in general very different. By making a saddle-point approximation it is possible to show that $f(x)$ coincides with the dimension density $P(n)$ apart from a scaling factor (needed also for dimensional reasons), i.e.,

$$f(x \gg \beta^{-1}) \approx \beta P(\beta x)\,, \tag{23.8}$$

if $P(n)$ decreases fast enough with increasing n. Some examples are shown in Fig. 23.4, taken from [6]. For a detailed demonstration of (23.8), see [6]. The particular and relevant result here is that if $P(n)$ has a power-law tail (as a function of n) then also $f(x)$ have a power-law tail in x (with the *same* exponent).

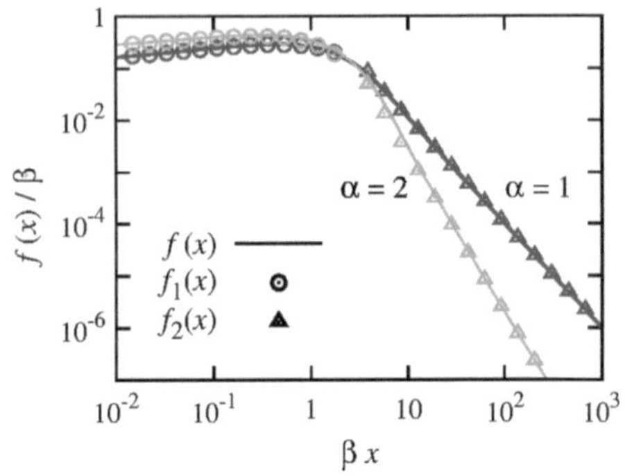

Fig. 23.4 Distribution $f(x)$ in (23.7) with $P(n) = \alpha/n^{1+\alpha}$ $(n \geq 1)$, $P(n) = 0$ otherwise, for $\alpha = 1$ (red), $\alpha = 2$ (*green*). *Continuous lines* Numerical integration of (23.7). *Triangles* Saddle point approximation. *Circles* Small-x limit. See text for details

23.6 Conclusions

We have discussed and tried to justify the hypothesis that the origin of power-law distributions in many complex systems is in the heterogeneity of their constituent units, i.e., power-law tails emerge as a collective diversity-induced effect. For constraints of space we limited ourselves to two famous examples of power-law tailed distributions, namely the Pareto law of wealth distributions and Zipf's law for the occurrence frequency of words in a written text. We provided theoretical reasons in the first case and phenomenological ones in the second case that the respective power-law distribution can be decomposed into a statistical mixture of localized distributions. As a further theoretical model which may hopefully find new applications, we have also illustrated an elementary and exactly solvable model of mechanical system for which canonical statistical mechanics based on the Gibbs distribution or the Boltzmann entropy predicts a power-law tail in the energy distribution.

While much remains to be done for confirming and understanding the basis of the results obtained and their consequences, this exploratory work suggests that diversity may represent the right track toward a deeper understanding of many instances of power-law distributions.

The mechanism discussed in this contribution is based on the presence of a quenched heterogeneity in the system. We noticed how such a mechanism is complementary with respect to the super-statistics introduced by Beck and Cohen [2] that is mostly based on the random, slow fluctuations in time of some parameters of a single unit. The relation and possible merging of these different mechanisms will be discussed in detail elsewhere.

Acknowledgments M.P., E.H., and A.C. acknowledge support from the Estonian Science Foundation Grant no. 9462 and the Institutional Research Funding IUT (IUT39-1) of the Estonian Ministry of Education and Research. A.C. acknowledges financial support from grant number BT/BI/03/004/2003(C) of Government of India, Ministry of Science and Technology, Department of Biotechnology, Bioinformatics Division. L.M. acknowledges the Estonian Research Council for supporting his work with the grant PUTJD110.

Appendix: Text analyzed

We constructed the text to be analyzed by merging in a single text file all the crime stories available on-line from the Gutenberg Project (www.gutenberg.org), mostly novels, by e.g. Christie, Collins, Davies, etc. The final file was a plain text file with a size of about 27 MB, containing about 50 millions words of which about 57,000 words were different from each other.

In order to extract the probability distribution of the occurrence frequency, we constructed a set of texts of equal size by dividing the original file into parts, each one containing a number $N_W = 3{,}000$ of words. The last part containing less than N_W words was neglected.

References

1. Beck, C.: Stretched exponentials from superstatistics. Physica (Amsterdam) **365A**, 96 (2006)
2. Beck, C., Cohen, E.: Superstatistics. Physica A **322**, 267–275 (2003)
3. Bhattacharya, K., Mukherjee, G., Manna, S.S.: Detailed simulation results for some wealth distribution models in econophysics. In: Chatterjee, A., S.Yarlagadda, Chakrabarti, B.K. (eds.) Econophysics of Wealth Distributions, p. 111. Springer (2005)
4. Chakraborti, A., Chakrabarti, B.K.: Statistical mechanics of money: how saving propensity affects its distribution. Eur. Phys. J. B **17**, 167–170 (2000)
5. Chakraborti, A., Patriarca, M.: Gamma-distribution and wealth inequality. Pramana J. Phys. **71**, 233 (2008)
6. Chakraborti, A., Patriarca, M.: Variational principle for the Pareto power law. Phys. Rev. Lett. **103**, 228701 (2009)
7. Chatterjee, A., Chakrabarti, B.: Ideal-gas like markets: effect of savings. In: Chatterjee, A., S.Yarlagadda, Chakrabarti, B.K. (eds.) Econophysics of Wealth Distributions, pp. 79–92. Springer (2005)
8. Chatterjee, A., Chakrabarti, B.: Kinetic exchange models for income and wealth distributions. Eur. Phys. J. B **60**, 135 (2007)
9. Chatterjee, A., Yarlagadda, S., Chakrabarti, B.K. (eds.): Econophysics of Wealth Distributions—Econophys-Kolkata I. Springer (2005)
10. Church, K.W., Gale, W.A.: Poisson mixtures. J. Natur. Lang. Eng. **103**(2), 163 (1995)
11. Feller, W.: An Introduction to Probability Theory and its Applications, vol. 1 and 2. John Wiley & Sons, 2nd edn. (1966)
12. Gerlach, M., Altmann, E.G.: Scaling laws and fluctuations in the statistics of word frequencies. New J. Phys. **16**, 113010 (2014)
13. Katriel, G.: Directed random market: the equilibrium distribution. Eur. Phys. J. B **88**, 19 (2015)
14. Mandelbrot, B.: The Pareto-Levy law and the distribution of income. Int. Econ. Rev. **1**, 79 (1960)
15. Newman, M.E.J.: Power laws, Pareto distributions and Zipf's law. Contemp. Phys. **46**, 323 (2005)
16. Patriarca, M., Chakraborti, A., Germano, G.: Influence of saving propensity on the power law tail of wealth distribution. Physica A **369**, 723 (2006)
17. Patriarca, M., Chakraborti, A., Kaski, K.: Gibbs versus non-Gibbs distributions in money dynamics. Physica A **340**, 334 (2004)
18. Patriarca, M., Chakraborti, A., Kaski, K.: Statistical model with a standard gamma distribution. Phys. Rev. E **70**, 016104 (2004)

19. Patriarca, M., Chakraborti, A., Kaski, K., Germano, G.: Kinetic theory models for the distribution of wealth: Power law from overlap of exponentials. In: Chatterjee, A., S.Yarlagadda, Chakrabarti, B.K. (eds.) Econophysics of Wealth Distributions. p. 93. Springer (2005)
20. Patriarca, M., Chakraborti, A.: Kinetic exchange models: From molecular physics to social science. Am. J. Phys. **81**(8), 618–623 (2013)
21. Patriarca, M., Chakraborti, A., Heinsalu, E., Germano, G.: Relaxation in statistical many-agent economy models. Eur. J. Phys. B **57**, 219 (2007)
22. Patriarca , M., Heinsalu, E., Chakraborti, A.: Basic kinetic wealth-exchange models: common features and open problems. Eur. Phys. J. B **73**, 145–153 (2010)
23. Treumann, R.A., Jaroschek, C.H.: Gibbsian theory of power-law distributions. Phys. Rev. Lett. **100**, 155005 (2008)
24. Tsallis, C.: Possible generalization of Boltzmann-Gibbs statistics. J. Stat. Phys. **52**, 479 (1988)

Chapter 24
A Network-Based Analysis of the European Emission Market

Andreas Karpf, Antoine Mandel and Stefano Battiston

Abstract This paper analyses the European Emission Trading System (ETS) from a network perspective. It is shown that the network exhibits a strong core-periphery structure also reflected in the network formation process. Due to a lack of centralized market places, operators of installations which fall under the EU ETS regulations have to resort to local networks or financial intermediaries if they want to participate in the market. This undermines the central idea of the ETS to exploit marginal abatement costs.

24.1 The European Emission Trading System

24.1.1 Background

The Kyoto Protocol (KP) [1] from 1998 extended the United Nations Framework Convention on Climate Change (UNFCCC) [2] negotiated in 1992 during the UN Conference on Environment and Development by defining targets for the reduction of green house gas (GHG) emissions into the atmosphere. These targets follow the principle of "*common but differentiated responsibilities*" as outlined in Article 3 of the KP [1]. Accommodating the responsibility of industrialized countries for the contemporary levels of GHG emissions, these targets were determined to be binding for the group of developed signatory states referred to as the Annex 1 parties. The

A. Karpf (✉)
Université Paris 1 Panthéon-Sorbonne, Centre d'Économie de la Sorbonne/Paris School of Economics, Paris, France
e-mail: andreas.karpf@univ-paris1.fr

A. Mandel (✉)
Université Paris 1 Panthéon-Sorbonne, Centre d'Économie de la Sorbonne, Paris, France
e-mail: antoine.mandel@univ-paris1.fr

S. Battiston (✉)
Department of Banking and Finance, University of Zurich, Zurich, Switzerland
e-mail: stefano.battiston@uzh.ch

S. Battiston et al. (eds.), *Proceedings of ECCS 2014*, Springer Proceedings in Complexity, DOI 10.1007/978-3-319-29228-1_24

protocol was signed and ratified by 191 parties of which one was the European Union.[1,2]

The Annex 1 parties comprise 37 industrialized countries of which 28 are now members of the European Union. The legally binding commitment of the signatory countries concerns the most relevant greenhouse gases and gas groups.[3] The targets themselves are however quantified in CO_2 equivalents with regard to global warming potential and as percentages of the emissions in a base year, which, for the majority of the Annex 1 parties, is 1990. The European Union as a whole committed itself to collectively reduce CO_2 emissions by 8 % until 2012 and 20 % until 2020 with respect to base year emissions. Under the premise of "*common but shared responsibilities*" member state specific reduction goals were defined, which take into account the different levels of economic development within the union, the respective structures of the national economies as well as early measures to reduce GHG emissions.

To keep the costs of limiting CO_2 emissions as small as possible for the signatory countries the KP allows for so called "*flexible mechanisms*" which serve as an alternative to traditional approaches like carbon taxes or compensating measures as reforestation (Art. 3.3) [1]. These mechanisms comprise *International Emission Trading* (IET), *Clean Development Mechanisms* (CDM) [1, Art.12] and *Joint Implementation* (JI) (Art. 6) [1].

Emission Trading The concept of IET plays the central role of flexible emission reduction instruments. It exploits differing marginal abatement costs (MAC)[4] between countries, firms, industries or even between different branches within a company [4]. The system bases on a "*cap-and-trade*" principle in which permitted emission units, so called allowance units,[5] are allocated to emitters of green house gases. These assigned allowance units (AAU) normally depend on historical yearly green house gas emission data and are capped with regard to committed emission reduction targets. Thereby allowance units become a scarce good which participants can exchange in a market context. Periodically market participants which are legally committed to reduce their emissions have to surrender the amount of allowance units in their possession. These are subsequently compared with the realized emissions which are permanently recorded at the respective installations to check if the emission reduction targets were met. Installations can be factories, power plants or even aircrafts. If the available allowance units fall short of the realized emissions the obliged market participants have to pay a fine proportional to the allowance units by which the emission reduction obligations were missed.

Clean Development Mechanism and Joint Implementation The system of emission trading is complimented by the CDM and the JI mechanism. In contrast to emission

[1]Council Decision of 15 December 1993 [3].

[2]Noteworthy exceptions are the United States which signed but never ratified the KP and finally withdrew in 2001 and Canada which quit the treaty in 2011.

[3]Carbon dioxide, methane, nitrous oxide, sulphur hexafluoride, hydrofluorocarbons and perfluorocarbons.

[4]This is the marginal cost of reducing green house gas emissions by one unit.

[5]One emission allowance unit typically corresponds to one metric ton of CO_2-equivalent.

trading these mechanisms are project based. Predicated on the assumption that actions which lead to the reduction of GHG eventually have positive effects in slowing down global warming no matter where on the planet they are conducted, Annex 1 countries can engage in GHG emission reducing projects abroad in order to earn emission reduction units (ERU) which in turn can be traded on the emission market or used when surrendering one's allowances at the end of a compliance period. While the JI mechanism is supposed to foster cooperation between Annex 1 countries[6] to meet their GHG reduction targets, the CDM aims to stimulate GHG reducing investments and projects in non-Annex 1 countries (mainly developing countries) to promote sustainable development (Art. 12) [1] and to help Annex 1 countries to meet their emission reduction commitments with the lowest possible costs.

24.1.2 *The Adoption of European Emission Trading System and Its Functioning*

Since Japan rejected all attempts to give the UN the legal instruments to enforce the emission reduction commitments in the KP and the United States withdrew from the protocol in 2001 it became soon clear that the EU had to find an internal solution if it wanted to stick to the GHG reduction targets to which it committed itself in the KP [6]. After an understanding was found between member states to differentiate the GHG reduction targets with regard to the level of economic development in the form of the "*burden sharing agreement*" (BSA) [7], the initial resistance with regard to the implementation of a European emission trading scheme (ETS) began to crumble. The European Union emission trading scheme was finally legally implemented by directive 2003/87/EC [8][7] and subsequently adopted into national laws.

The ETS covers factories, power stations, and other installations with a net heat excess of 20 MW in emission intensive industries responsible for roughly 50 % of the GHG emissions in the concerned 31 countries (EU plus Switzerland, Norway and Liechtenstein). With directive 2008/101/EC the aviation industry was also included into the ETS. The emission allowance units (EUA) are allocated to each of the approximately 11,000 installations in February each year on a national level in line with the respective BSA and KP reduction targets and have to be surrendered by the operator holding accounts (OPA) at latest end of April in the subsequent year. The fine for each EUA after surrendering that falls short of the verified emissions amounts to EUR 100. Operators are allowed to bank and respectively borrow allowances within a trading period. It was however not permitted to carry allowances from Pilot Phase I (2005–2007) to Phase II (2008–2012), and from there to Phase III (2013–2020) [9].

[6]The majority of currently ongoing Joint Implementation projects are situated in transition economies with Annex 1 obligations like the Russian Federation and Ukraine [5].

[7]The directive was later amended by Directive 2004/101/EC, Directive 2008/101/EC, Regulation (EC) No 219/2009 and Directive 2009/29/EC.

The ETS is not only open to OHAs, but also private entities which don't fall under the ETS regulation are allowed against a fee to trade on the emission market. These entities are referred to as private holding accounts (PHA). EUAs can be traded bilaterally, over the counter (OTC) via a broker or on one of Europe's climate exchange markets like the European Climate Exchange (ECX), the European Energy Exchange AG (EEX) etc. For the time for which the transaction data set is available the most common form of transactions was OTC.

As prescribed by 2003/87/EC and 2009/29/EC every transaction in the ETS has to be recorded in some sort of accounting system (registries) and is accessible to the public with an embargo of three years. At the beginning these registries were organized on a national level. Since 2008 this function is resumed by a central Community Independent Transaction Log (CITL) accessible online under http://ec.europa.eu/environment/ets/. The transaction data from the CITL form the base of our network-based analysis of the EU ETS.

24.2 The Data Set

The transaction data set containing the exact time stamp of the transaction and its volume as well as information about the accounts active in the ETS and data with regard to the allowance allocation, the surrendering of the allowances as well as the verified emissions were scraped from the CITL. The raw data set contains approximately 520,000 transactions to which we added spot price information downloaded from Bloomberg as well as data about the ownership structure and the type of companies in the ETS from the "Ownership Links and Enhanced EUTL Dataset" [10]. In our analysis we concentrate only on the market movements which are relevant for the price formation of the EUA certificates (transaction types 3-0, 3-21 and 10-0). Transactions connected to the administration of the ETS as for the allowance issuance, retirement, cancellation, surrender, allocation, and correction were discarded. The remaining 364,810 transactions are analyzed in what follows.

24.3 Methodology and Research Questions

A network based analysis of the European Emission market is performed. A network based on the transaction data set is therefore constructed. Thereby agents active in the emission market are regarded as vertices. These vertices are connected by directed edges in the form of transactions from the seller (the source vertex) to the buyer (the target vertex). The edges are weighted by the volume of EUAs transferred in the respective transaction. Figure 24.1 shows a plot of the resulting network graph.

The aim is to investigate the connection between the network structure and the functioning of the market. In this context the following research questions are to be addressed: (1) Is the organization of the market reflected in the structure of the

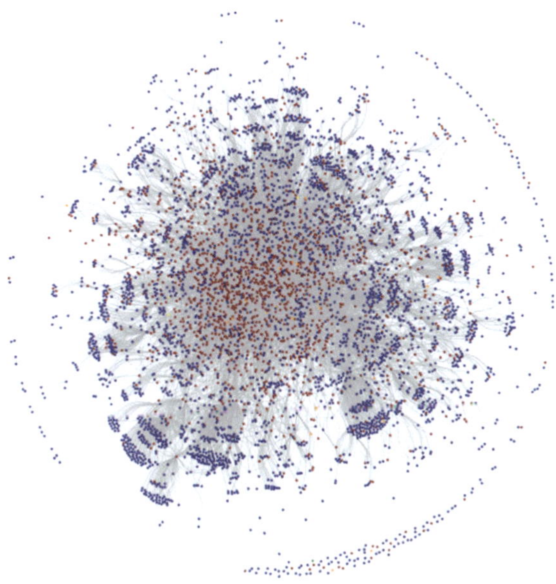

Fig. 24.1 The CO_2 trading network (CDM (*green*), finance (*red*), foundation (*yellow*), government (*orange*), industry (*blue*))

network? (2) Which factors are relevant for the matching process on the EU ETS? (3) Is the network structure supporting the idea of emission markets to exploit differences in marginal abatement costs? (4) Does the position of an agent within the network have an implication for its ability to create revenues out of a trade?

24.4 The Network Structure of the European Emission Markets

Following Li et al. [11] some tests with regard to the market structure were conducted. Figure 24.2 plots the in- and out-degrees[8] versus the cliquishness[9] of agents. The downward sloping cloud implies a hierarchy in the market with a strong core of highly connected nodes and clusters of nodes on the periphery. This phenomenon can also be observed in the plot of the emission market network in Fig. 24.1.

The core periphery structure of the trading network is also observable when looking at the degree distribution directly (see Fig. 24.3): The distributions of the in-, out- and total-degrees follow a power law i.e. there are agents whose in-, out- or total

[8]The in- and out-degrees of each agent: this means the active and passive connectedness of agents.

[9]Be the k-core of graph a maximal subgraph in which each vertex has at least degree k. The cliquishness or coreness of a vertex is then k if it belongs to the k-core but not to the (k+1)-core. [12].

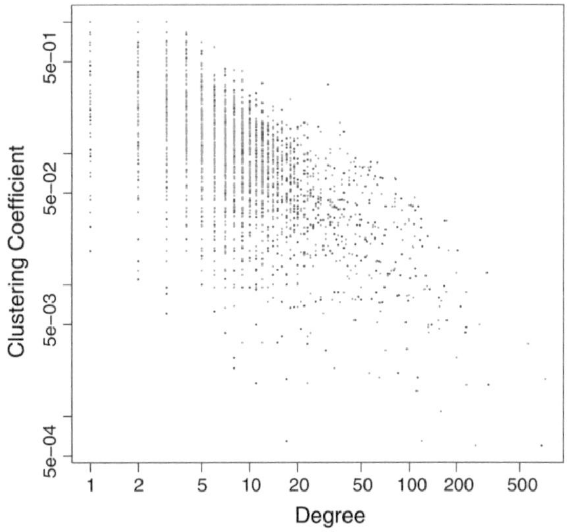

Fig. 24.2 In-/out-degree versus cliquishness (in-degree: *blue*; out-degree: *red*)

degrees strongly exceed the average. The exponents of the power-law distributions fitted to the in-, out- and total-degree-distribution are 2.25, 2.29 and 2.21 respectively. A conducted Kolmogorov-Smirnov test on the degree distributions resulted in p-values of 0.76, 0.97 and 0.96 respectively, indicating that the hypothesis that the original data could have been drawn from the fitted power-law distributions cannot be rejected in several cases. The observed network thus falls into the category of scale free networks.

We further computed the density distribution of multiple network statistics (in-, out-degree as well as eigenvector centrality[10]) as well as for profits of companies in the European emission market combined with informations about their type.

In this case the core-periphery structure is observable in the wave-like forms of the density plots displayed in Fig. 24.4. This structure is also reflected in the network plot in Fig. 24.1 which has a highly connected center which is dominated by nodes from the finance sector (red) surrounded by concentric circles of nodes from the industries (OHA). Going from the inside to the outside the nodes are lesser connected and thus exhibit a lower degree of centrality. Looking at the plot in the lower right of Fig. 24.4 this, at least for the group of agents which can be attributed to the sector government, seems to have an influence on the profits these respective agents are able to derive from trading on the emission market.

[10]Eigenvector centrality: the first eigenvector of the adjacency matrix giving the centralities for each vertex. It can be understood as a reciprocal process in which the centrality of a vertex depends proportionally on the centralities of other vertices to which it is connected.

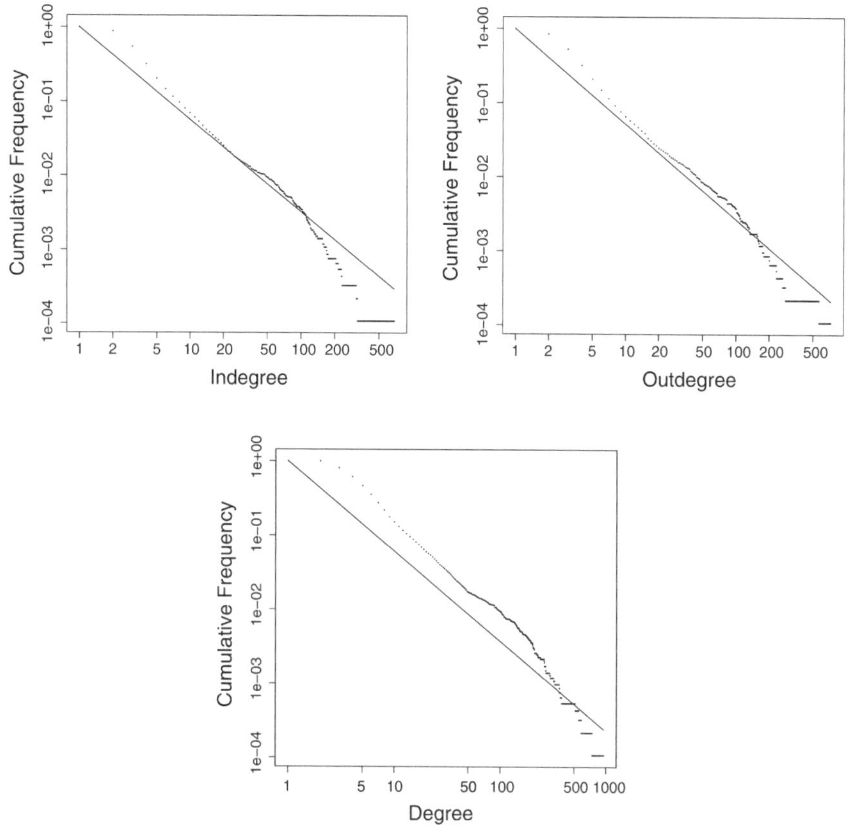

Fig. 24.3 In-/out-/total-degree distribution with fitted power law

24.5 Network Position, Trading Volume and Profits

To further investigate the connection between the position of an individual company within the network and its market participation (trade volume) on the one hand and its ability to derive profits from the market on the other hand we follow Boyd et al. [13] in computing more sophisticated individual coreness values than the k-core (cliquishness) measure which was used above. Boyd et al. [13] show that a singular value decomposition (SVD) of the adjacency matrix combined with a prior imputation of missing values on the diagonal represents a fast and reliable method to compute the out- (u) and in-coreness (v) of individual agents within a large graph. The coreness of an agent is high, if an agent is well connected with other well connected agents. The SVD is methodologically and in terms of interpretation similar to the eigenvector centrality discussed above. The so computed coreness values as well as information about agents profits and volumes traded in the market were then used

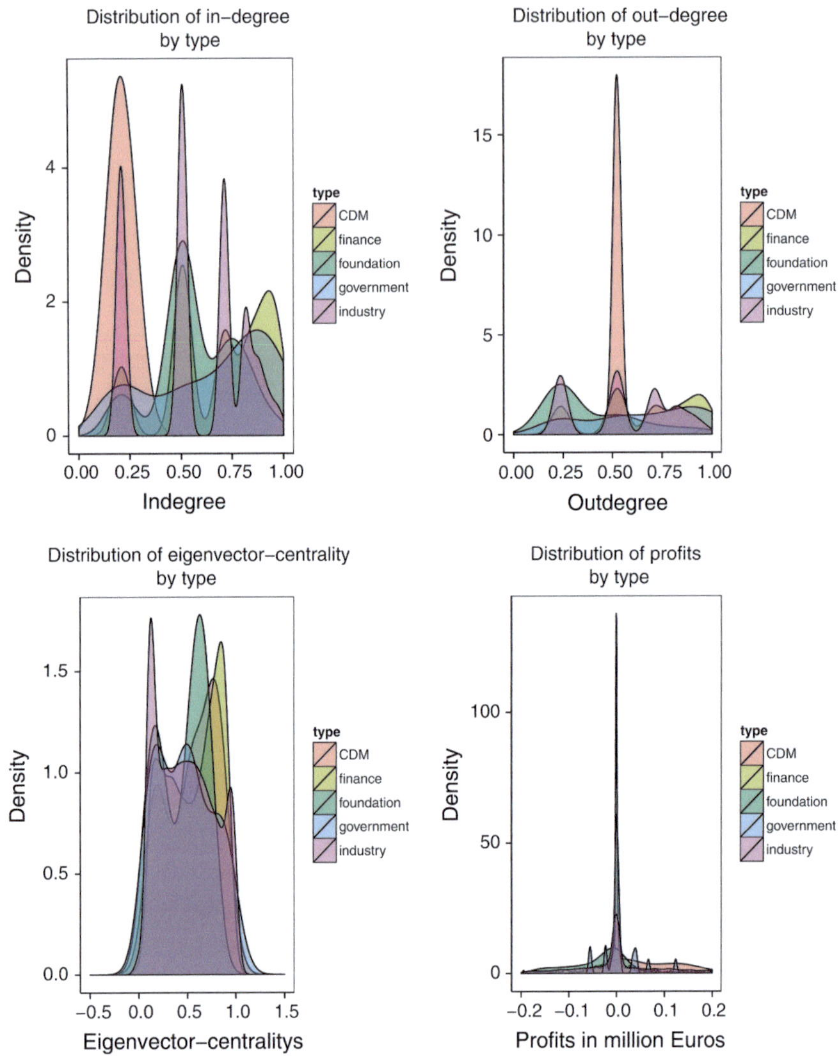

Fig. 24.4 Densities plots for various trade network statistics and company types in the ETS

in combination with interpolation by means of local polynomial regression fitting to create the elevation plots displayed in Fig. 24.5.

Looking at the plot on the left-hand side in Fig. 24.5 it appears as the ability to generate profits on the ETS positively depends on the out-coreness of an individual. The market participation on the other hand depends positively on both the in- and out-coreness of an agent. As far as the in-coreness is concerned this effect seems to be slightly weaker.

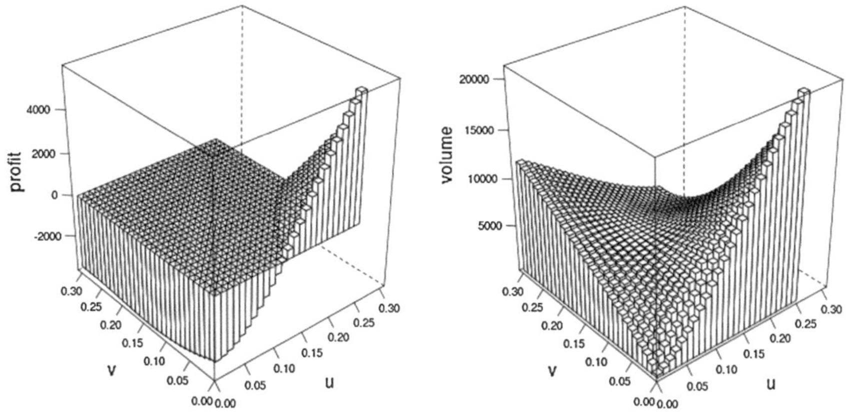

Fig. 24.5 Elevation plots of out- (u) and in- (v) coreness values with respect to the generated profits (*left*) and market participation (total volume traded) (*right*) of individual agents

24.6 Network Formation

The results from the last two sections can be interpreted as some kind of informational asymmetry in the market. OHAs which are legally forced to participate in the emission market and are seeking to buy and sell emission certificates resort to local networks (firms from the same parent company, country, industry etc.) or to huge financial players which form the center of the trading network. This undermines the central idea of the emission market to take advantage of differentials in abatement costs. This interpretation is further supported when we take a closer look at the network formation process.

A basic method within this class of approaches to investigate the formation process of a network is the Maslov-Sneppen [14] algorithm: comparing the empirical network with a quantity of random networks with the identical degree sequence and distribution allows us to generate degree-degree correlation profiles which permit to identify connectivity patterns between nodes of different degrees. The so called null-model is generated by systematically rewiring the original network. Two pairs of connected nodes $A->B$ and $C->D$ are randomly selected from a network and rewired in the fashion $A->D$ and $C->B$. If the thereby generated new connections already exist the procedure is aborted and two new pairs of connected nodes are selected and the rewiring attempt is repeated. Doing this sufficiently often, a rule of thumb suggests a number as high as ten times the number of edges, one obtains a random model with the same degree sequence and distribution as the original graph. This procedure is repeated multiple times. Then the generated null-models are compared with the original network. More precisely, we compare the number of edges between two nodes with degrees K_1 and K_2 in the empirical network $N(K_1, K_2)$ and the mean in the generated random networks $\bar{N}_r(K_1, K_2)$:$R(K_1, K_2) = N(K_1, K_2)/\bar{N}_r(K_1, K_2)$. Whether the deviance of

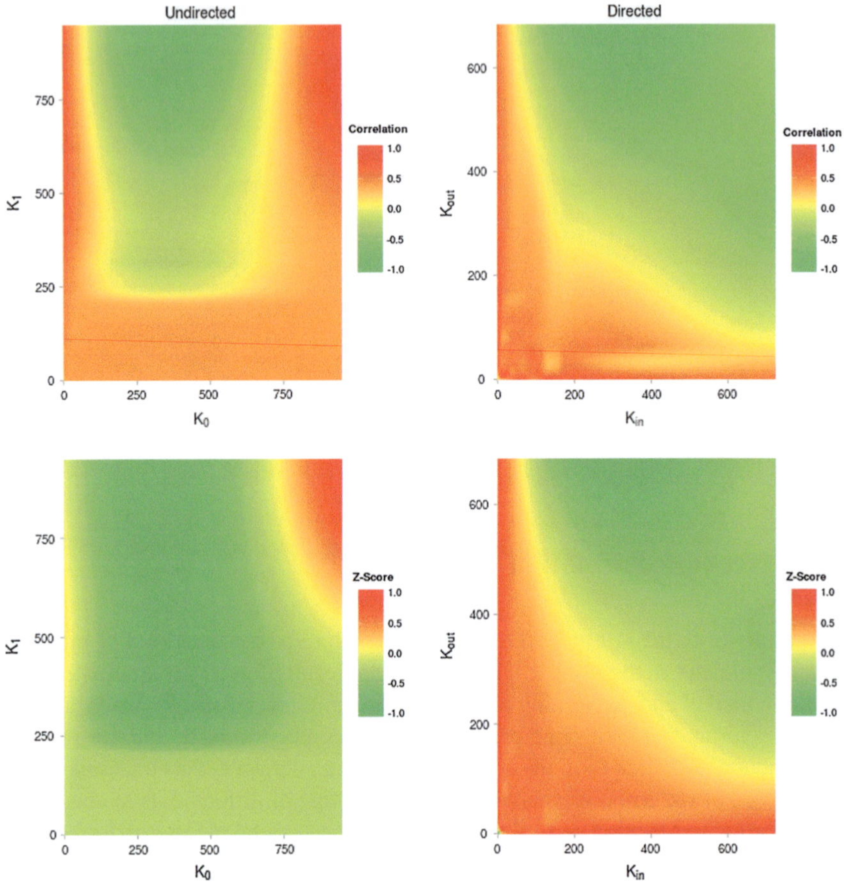

Fig. 24.6 Degree-degree correlation profiles generated by the Maslov-Sneppen algorithm

the empirical network from the null-model is significant can be assessed by computing the Z-scores: $Z(K_1, K_2) = (N(K_1, K_2) - \bar{N_r}(K_1, K_2))/sigma(K_1, K_2)$, where $sigma(K_1, K_2)$ is the standard deviation of $\bar{N_r}(K_1, K_2)$. This method works for directed and undirected networks. The results of the Maslov-Sneppen approach for the emission trading network are presented in Fig. 24.6.

The interpretation of the degree-degree correlation profiles is twofold: (1) When interpreting the emission trading network as an undirected graph one recognizes a compared to the null model significantly increased connectedness between highly connected nodes (the red area in the upper right corner of the LHS plot). (2) In both the undirected and the directed case (RHS) we note a significantly increased degree of asymmetric connectedness, i.e. between low- and high-degree nodes (the orange to red area along the axes). This is in line with the results of a strong core-periphery structure presented earlier in the paper.

A bit more involved but based on a similar idea is the class of Exponential Random Graph models (ERGM). A random graph Y is made up by a set of n nodes and e edges $\{Y_{ij} : i = 1, \ldots, n;\ j = 1, \ldots, n\}$ where, similar to a binary choice model, $Y_{ij} = 1$ if the nodes (i, j) are connected and $Y_{ij} = 0$ if this is not the case. One can thus model the given network by

$$P(Y = y|\theta) = \frac{\exp(\theta^T g(y))}{c(\theta)}$$

where θ and $g(y)$ are vectors of parameters and network statistics respectively and $c(\theta) = \sum \exp\{\theta^t g(y)\}$ is a normalizing constant corresponding to all possible networks. Evaluating above expression (as the number of possible outcomes vastly exceeds the number of constraining parameters this is usually done by Gibbs sampling) allows us to make assertions whether and how certain nodal attributes influence the network formation process. These nodal attributes can be endogenous to the network (like the in- and out-degrees of a node) or exogenous as in the context of the trading network for example the country in which a specific company is registered [15].

We ran a basic ERGM model over the emission trading network. The results are presented in Table 24.1. The most important features of the results are as follows: We observe positive log-odds for the closing of triangles (clusters), homophily for country and general ultimate owner (GUO) respectively. We however remark negative log-odds for the formation of ties between agents of the same type (i.e. OHA vs. PHA). We thus see what we already observed graphically earlier in the paper:

Table 24.1 A simple ERGM model applied to the ETS network

	Dependent variable
	Carbon network
Edges	−4.785***
	(0.172)
Triangle	0.321***
	(0.026)
Asymmetric	−3.711***
	(0.166)
Nodematch.type	−0.099***
	(0.019)
Nodematch.country	0.392***
	(0.031)
Nodematch.guo	0.757***
	(0.285)
Akaike Inf. Crit.	527,052.300
Bayesian Inf. Crit.	527,150.300

Note $^{*}p<0.1$; $^{**}p<0.05$; $^{***}p<0.01$

OHAs who seek to sell or buy EUAs have to address themselves to local networks (homophily as far as origin [country] and ownership [GUO] is concerned) or to financial institutions or brokers (heterophily with regard to type).

24.7 Conclusion

The EU ETS network is characterized by a significant core-periphery structure which is also reflected in an asymmetry within the degree-degree correlation profiles computed by the Maslov-Sneppen algorithm. This has effects on the profits agents are able to derive from the market and their market participation in general. An ERGM analysis shows that OHAs have to resort to local networks or financial intermediaries when they want to participate in the market. This might be due to the fact that the EU ETS is not organized in a central market place but based to a large extent (for the time the data was available) on OTC transactions. This in our opinion violates the central idea of exploiting differences in marginal abatement costs, imposes unnecessary additional costs on the OHAs who often don't possess the resources to collect informations about the market and thus undermines the goal of the EU ETS.

Acknowledgments The authors acknowledge the support of the EU FP7 FET project SIMPOL.

References

1. United Nations: Kyoto Protocol to the United Nations Framework Convention on Climate Change. http://unfccc.int/resource/docs/convkp/kpeng.pdf (1998)
2. United Nations: United Nations Framework on Climate Change. https://unfccc.int/files/essential_background/background_publications_htmlpdf/application/pdf/conveng.pdf (1992)
3. European Council: 94/69/EC: Council Decision of 15 December 1993 concerning the conclusion of the United Nations Framework Convention on Climate Change. http://eur-lex.europa.eu/legal-content/EN/TXT/?uri=CELEX:31994D0069 (1993)
4. Ellerman, D., Decaux, A.: Analysis of post-Kyoto CO2 emissions trading using marginal abatement curves. MIT Joint Program on the Science and Policy of Global Change (1998)
5. UNEP Risoe Centre: JI & CDM projects. http://www.cdmpipeline.org/ (1998)
6. Ellerman, A.: Pricing carbon: the European Union emissions trading scheme. Cambridge University Press (2010)
7. European Council: 2106th Council meeting ENVIRONMENT Luxembourg, 16–17 June 1998. http://europa.eu/rapid/press-release_PRES-98-205_en.pdf (1998)
8. European Parliament: Directive 2003/87/EC of the European Parliament and the Council of 13 October 2003 establishing a scheme for greenhouse gas emission allowance trading within the Community and amending Council Directive 96/61/EC. http://eur-lex.europa.eu/legal-content/EN/TXT/PDF/?uri=CELEX:02003L0087-20090625&from=EN (2003)
9. Ellerman, A., Joskow, P.: The European Union's emissions trading system in perspective. Pew Center on Global Climate Change Arlington (2008)
10. Jaraite, J., Jong, T., Kazukauskas, A., Zaklan, A., Zeitlberger, A.: Ownership Links and Enhanced EUTL Dataset. European University Institute, Florence. http://fsr.eui.eu/CPRU/EUTLTransactionData.aspx (2013)

11. Li, D., Schürhoff, N.: Dealer networks. SSRN paper 2023201 (2014)
12. Seidman, S.: Network structure and minimum degree. Soc. Netw. **5**, 269–287 (1983)
13. Boyd, P., et al.: Computing continuous core/periphery structures for social relations data with MINRES/SVD. Soc. Netw. **32**, 125–137 (2010)
14. Maslov, S., Sneppen, K.: Specificity and stability in topology of protein networks. Science **296**, 910–913 (2002)
15. Butts, C. et al.: Introduction to Exponential-family Random Graph (ERG or p*) modeling with ergm. http://cran.r-project.org/web/packages/ergm/vignettes/ergm.pdf (2014)
16. Ellerman, D., Buchner, B.: The European Union emissions trading scheme: origins, allocation, and early results. Rev. Environ. Econ. Policy **1**(1), 66–87 (2007)
17. Schreurs, M., Tiberghien, Y.: Multi-level reinforcement: explaining European Union leadership in climate change mitigation. Glob. Environ. Politics **7**(4), 19–46 (2007)
18. Freeman, L.C.: Centrality in social networks I: conceptual clarification. Soc. Netw. **1**, 215–239 (1979)

Chapter 25
Dynamics of Commodity Price Fluctuations in Japan

Yoshi Fujiwara, Hideaki Aoyama, Hiroshi Iyetomi and Hiroshi Yoshikawa

Abstract Deflation is a most important economic problem having been faced by Japan, and also developed countries in Europe, under zero interest rate. How individual prices influence each other, namely the dynamics of a large number of commodity prices and fluctuations plays a crucial role there. By using hundreds of individual commodities and their prices that comprise the import price, corporate goods price and consumer price indices, we show that price fluctuations have frequencies and synchronizations specific to space and time. Space means industrial sectors for the commodities and how they are located in the supplier-customer network. Temporal structure includes background movement due to deflation, inflation and exogenous shocks such as VAT (consumption tax) rate changes, the Lehman shock and so forth, but also endogenous shocks or mutual influences among the commodities.

25.1 Introduction

Japan has suffered deflation more than a decade from the late 1990s under the zero interest rate. European Union faces low inflation rates recently. See Fig. 25.1 for the annual inflation of consumer prices in Japan, European countries and USA for the past 15 years. How prices behave is of primary importance in economics, particularly because real interest rate is nominal interest rate minus inflation rate. Under zero interest rate, deflation can be a threat to the macro-economy. Many central banks are

Y. Fujiwara (✉)
Graduate School of Simulation Studies, University of Hyogo, Kobe 650-0047, Japan
e-mail: yoshi.fujiwara@gmail.com

H. Aoyama
Graduate School of Sciences, Kyoto University, Kyoto 606-8502, Japan

H. Iyetomi
Department of Mathematics, Niigata University, Niigata 950-2181, Japan

H. Yoshikawa
Graduate School of Economics, The University of Tokyo, Tokyo 113-0033, Japan

S. Battiston et al. (eds.), *Proceedings of ECCS 2014*, Springer Proceedings in Complexity, DOI 10.1007/978-3-319-29228-1_25

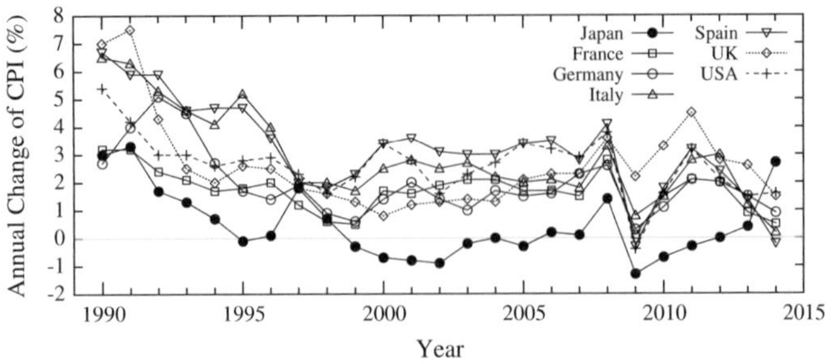

Fig. 25.1 Consumer prices in Japan, European countries and USA, annual inflation from 1990 to 2014. *Source* OECD Statistics

resorting to quantitative easing and related monetary policy including inflation targets, but its efficacy under zero interest rate has been much argued among researchers.

Deflation and inflation are changes in the aggregate price index over time, which is a weighted average of micro prices for individual commodities in consumption. It is therefore important to understand how individual prices change in an interacting way. Recent empirical works on micro price dynamics (see [7] for a survey) have uncovered hitherto little known dynamics of micro prices. See, however, the literature of dynamic factor models including [1, 5, 9, 10].

In this paper, studying 830 prices of goods and services in Japan for more than 30 years, we demonstrate that the frequency of individual price changes and synchronization are, in fact, not constant but time-varying. The existing literature routinely assumes that distribution of micro price changes is constant. However, this assumption is not borne out by data. See [7] and references for the literature. Frequency, synchronization, and size of price changes are all time-varying. Moreover, they change in clusters, *not* simultaneously in the economy as a whole. In this respect, there is a significant gap between observed facts and theory because in standard theory, changes in money, supposedly the most important macro disturbance, affect more or less uniformly all the prices.

In a separate paper [11], we presented a new method of extracting information on the systemic changes of aggregate price based on micro price data as a true "core" price index, and found that they are not significantly correlated with money supply. The present paper is a study on the spatio-temporal structure of the micro price dynamics. The absence of significant correlation with money supply is our new finding in that paper, crucially different from the literature.

In Sect. 25.2, we describe our data of individual prices which constitute the import price index. corporate goods price index, consumer price index. These prices of individual commodities range from the upstream to downstream of production. In Sect. 25.3, we show the spatio-temporal pattern in the fluctuations of these prices. By "spatio" we mean industrial sectors for the commodities, each of which is located in the production network. "Temporal" pattern shows that the fluctuations contain "background" movement due to deflation, inflation, exogenous shocks such as VAT (consumption tax) rate changes and so forth, but also endogenous shocks or mutual influences among commodities, typically from the the upstream to the downstream of production. We discuss in Sect. 25.5, and summarize in Sect. 25.6.

25.2 Individual Prices and Data

We employ monthly data of the following three categories of individual prices in Japan for the period from January 1980 to June 2013 (402 months).

IPI Import Price Index, compiled by the Bank of Japan (BoJ) consists of prices of imports at the stage of entry into Japan. It covers 75 goods [2].

DCGPI Domestic Corporate Goods Price Index, compiled by the BoJ, surveys on the prices of goods traded among companies, specifically domestically produced goods for domestic markets, mainly at the stage of shipment from producers and partly from wholesalers. It covers 420 goods [2].

CPI Consumer Price Index, compiled by the Statistics Bureau of the Ministry of Internal Affairs and Communications covers 335 consumption goods and services [8].

We denote the individual price by $p_\alpha(t)$, where $\alpha = 1, 2, \ldots, 830$ ($:= N$) is the kind of goods and services, and $t = 1, 2, \ldots, 402$ is the time of month.

Figure 25.2a shows the time-series of monthly price indices from 1980 up to present (all indices; 2010 base). Note that the IPI is shown in a different scale (right axis) because it has much greater volatility than those of DCGPI and CPI. While a considerable part of the import goods of raw materials is highly volatile, it is not immediately reflected to the downstream commodities corresponding to DCGPI and CPI. For reference three epochs in which VAT was raised, in Aprils of 1989, 1997 and 2014 (VAT 3 %, 5 %, 8 % respectively) and the epoch of the Lehman shock in September 2008 are marked by vertical dashed lines in Fig. 25.2.

Figure 25.2b is the plot of annual (year-to-year) changes of the monthly aggregate price indices. The IPI has a different scale as depicted by the right axis for the reason explained above. One can observe that Japan suffered from deflation for a more than a decade from 1999 to 2013 in terms of domestic price indices, namely DCGPI and CPI.

Now we study the behavior of individual prices, namely monthly changes of them as defined by

$$r_\alpha(t) := \log_{10}\left[\frac{p_\alpha(t+1)}{p_\alpha(t)}\right]. \tag{25.1}$$

Heterogeneity of micro prices found in the existing literature can be easily confirmed for the Japanese data we analyze.

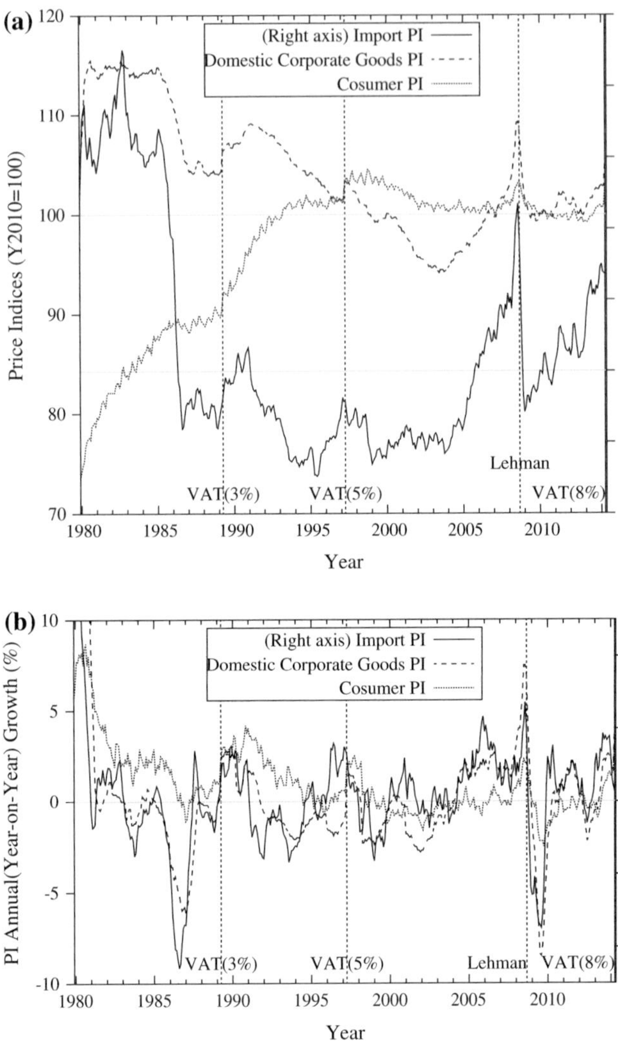

Fig. 25.2 Time-series of monthly price indices (*PI*) for Import PI (*solid line* and *right-axis*), Domestic Corporate Goods PI (*dashed line*), and Consumer PI (*dotted line*) from 1980 up to present (all indices; 2010 base). *Dashed vertical lines* correspond to the 3 months in which VAT was raised, namely April of 1989, 1997 and 2014 (VAT 3, 5, and 8 % respectively), and September 2008 of the Lehman Brothers failure

Table 25.1 shows classification of sectors, and the mean duration (in months) of the period during which individual prices remains unchanged, averaged over each sector.

Denote by λ the monthly frequency or probability that the price changes in a month. λ is not directly observed, but can be estimated by the mean duration as follows. Assume that the price changes according to a homogeneous Poisson process

Table 25.1 List of IDs, classification of sectors, the numbers of goods, the durations and frequencies of price changes for the commodities of IPI, DCGPI and CPI

ID	Classification of sector	#Goods	Months	Freq
IPI—Import PI				
01	Foodstuffs and feedstuffs	17	1.04	61.80
02	Textiles	6	1.26	55.25
03	Metals and related products	19	1.06	61.11
04	Wood, lumber and related products	3	1.02	62.66
05	Petroleum, coal and natural gas	8	1.04	61.94
06	Chemicals and related products	9	1.50	53.20
07	General purpose, production and business oriented machinery	2	1.14	58.47
08	Electric and electronic products	2	1.13	58.84
09	Other primary products and manufactured goods	9	1.09	60.15
–	All	75	1.13	59.75
DCGPI—Domestic Corporate Good PI				
01	Food, beverages, tobacco and feedstuffs	78	3.29	32.92
02	Textile products	19	8.04	21.35
03	Lumber and wood products	8	3.16	33.45
04	Pulp, paper and related products	20	3.22	30.96
05	Chemicals and related products	55	6.32	24.65
06	Petroleum and coal products	11	2.01	42.88
07	Plastic products	8	3.75	26.94
08	Ceramic, stone and clay products	25	5.12	24.62
09	Iron and steel	26	4.49	27.51
10	Nonferrous metals	19	1.54	51.38
11	Metal products	26	5.21	22.78
12	General purpose machinery	20	6.13	19.24
13	Production machinery	16	4.77	26.90
14	Business oriented machinery	6	10.41	12.01
15	Electronic components and devices	5	2.22	37.59
16	Electrical machinery and equipment	20	4.65	22.79
17	Information and communications equipment	4	2.95	33.57
18	Transportation equipment	11	10.26	10.78
19	Other manufacturing industry products	15	8.87	16.91
20	Agriculture, forestry and fishery products	17	5.18	40.18
21	Minerals	3	10.18	13.67
22	Electric power, gas and water	3	8.31	16.07
23	Scrap and waste	5	1.09	60.27
–	All	420	4.95	28.37

(continued)

Table 25.1 (continued)

ID	Classification of sector	#Goods	Months	Freq
CPI—Consumer PI				
01	Goods related to Food	132	1.24	57.67
02	Goods of house materials, household utensils (incl. electronics)	35	1.16	57.88
03	Goods of clothes and footwear	22	1.28	55.06
04	Goods of medical care	11	1.55	48.09
05	Goods of automobiles, car equipments, misc.	6	3.19	36.81
06	Goods related to education, culture, recreation and misc.	44	6.61	39.55
07	Services in CPI	85	6.85	31.40
–	All	335	3.41	47.79

Months is the mean duration of the period during which individual prices remains unchanged, averaged over each sector. Freq is the constant monthly frequency of price changes or probability (in percent) that the price changes in a month, calculated under the assumption of homogeneous Poisson process

with parameter θ, namely a constant probability of change at any instance of time. For a realization of n changes of the price at times $0 \equiv t_0 < t_1 < t_2 < \cdots < t_n \equiv T$, the likelihood function is given by $L(\theta) = \theta^n \exp(-\theta\, T)$, because the inter-occurrence times $T_k = t_k - t_{k-1}$ $(k = 1, 2, \ldots, n)$ are independent and identically distributed by an exponential distribution with parameter θ. The maximum likelihood estimate is then obtained by $\theta = n/T = 1/d$, where d is the average of inter-occurrence times. On the other hand, the probability that the price changes in a month, λ, is related to the parameter θ by $\lambda = 1 - e^{-\theta}$ as easily shown. It therefore follows that $d = -1/\ln(1 - \lambda)$. See [3, Chap. 6.2] for example.

The mean duration varies from 10 months for business machinery and transportation equipment to 1 month for food, cloths and most imported goods and materials. In between is 6 months for chemicals in DCGPI and services in CPI. They are broadly consistent with the results obtained in previous works.

25.3 Spatio-temporal Dynamics of Price Fluctuations

Micro prices of individual goods and services have different volatilities. They must reflect differences in industrial organization and the nature of goods and services. To take into account these differences in volatility, in what follows, we need to consider the normalized price change. Denoting by $\langle r_\alpha \rangle_t$ and σ_α the sample average and standard deviation of the time-series $r_\alpha(t)$ in (25.1), respectively, we define normalized time-series by

$$w_\alpha(t) := \frac{r_\alpha(t) - \langle r_\alpha \rangle_t}{\sigma_\alpha} \tag{25.2}$$

We examine "spatio-temporal patterns" of the individual price changes. Figure 25.3 shows the normalized changes $w_\alpha(t)$ defined by (25.2). Figure 25.3 focuses on "significant" changes of prices in the sense that the data for $|w_\alpha(t)| < w_*$, namely changes smaller than a threshold, are shown as blank space where $w_* = 2$. Blue and red colors of each point indicate significant positive and negative changes, $w_\alpha(t) > w_*$ and $w_\alpha(t) < -w_*$, respectively.

Figure 25.3 demonstrates that the simultaneous changes of individual prices or the synchronization occasionally occur *without any clear periodicity*. The April 1989 and the April 1997 are two examples of significant synchronization as indicated by the arrows in Fig. 25.3. In Japan, the 3 % value added tax (VAT) called the consumption tax was introduced in April 1989, and the tax rate was raised from three to 5 % in April 1997. Almost all the prices were raised then. Note, however, that individual prices were not mechanically raised by three and 2 %, respectively. Evidently, many firms found good opportunities to adjust their prices when the VAT rate was changed.

One can quantify the degree of synchronization of micro price changes by examining the numbers of positive, negative and zero price changes for each month. Let us denote such numbers by $n_+(t), n_-(t), n_0(t)$, and the sum of them is the total number of goods and services, N. Figure 25.4a–c show the fractions $n_+(t)/N, n_-(t)/N, n_0(t)/N$, for IPI, DCGPI and CPI (from top to bottom), respectively. Not to mention volatile IPI, one can observe that DCGPI and CPI prices are also raised or lowered in *time-varying* way. The number of prices that are raised is larger than those that are lowered under (even mild) inflation, while the converse is true under deflation. For example, in the plot for CPI, the fraction $n_-(t)/N$ exceeds $n_+(t)/N$ persistently from 1999 up to 2007 when deflation continued.

25.4 Case Study: The Lehman Shock in 2008

As a case study, let us examine the pre- and post-Lehman Brothers failure during the period from 2007 to 2009. In the first half of the year 2008, import prices and prices of intermediate products in DCGPI significantly went up. In CPI, food prices also rose. Deflation appeared to change into mild inflation during this period.

The bankruptcy of the Lehman Brothers in September 2009 turned the tide. The fraction of price decline suddenly went up. This sudden change is clearly seen in Fig. 25.5a, b which enlarges Fig. 25.3 for the period during 2007–09. The figures show how mild inflation up to the first half of the 2008 abruptly changed to deflation in the course of the Great Recession triggered by the bankruptcy of the Lehman Brothers in September 2008. Evidently, changes in price during this period have little to do with money, because the growth rate of money during May 2008 to March 2009 had hardly changed within narrow limits of 1.9 and 2.4 %. They basically reflect a fall of real economic activity, namely the Great Recession.

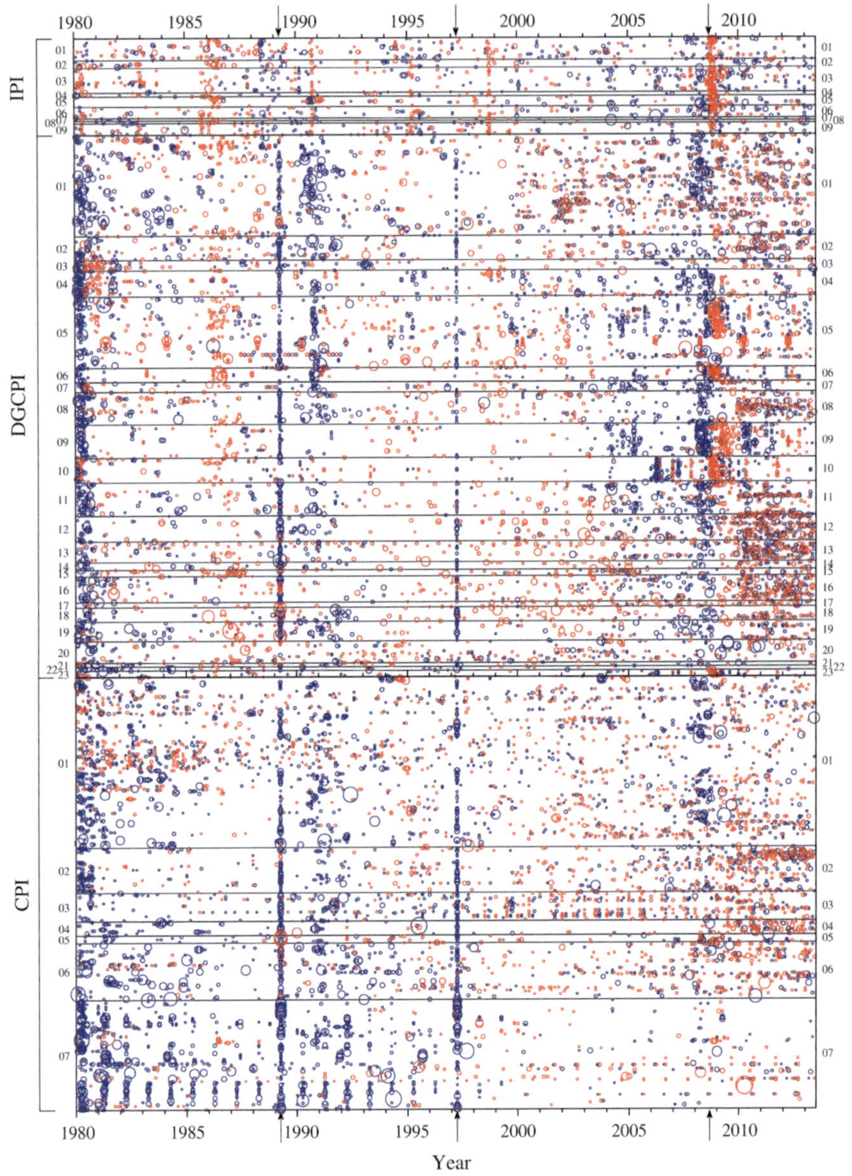

Fig. 25.3 (Color on-line) Normalized price changes $w_\alpha(t)$ calculated from the monthly changes $r_\alpha(t)$ by (25.2). Segments denoted by labels starting from "01" in each PI are classification of sectors. *Blue* and *red colors* correspond to $w_\alpha(t) > w_*$ and $w_\alpha(t) < -w_*$, respectively (*ups* and *downs*), where the threshold is $w_* = 2.0$. Each circle has a radius proportional to the absolute magnitude of change. *Blank areas* correspond to no "significant" change in the sense that $|w_\alpha(t)| < w_* = 2.0$. *Three arrows* are drawn at the epochs of VAT 3 % (Apr 1989), VAT 5 % (Apr 1997), Lehman shock (Sep. 2008)

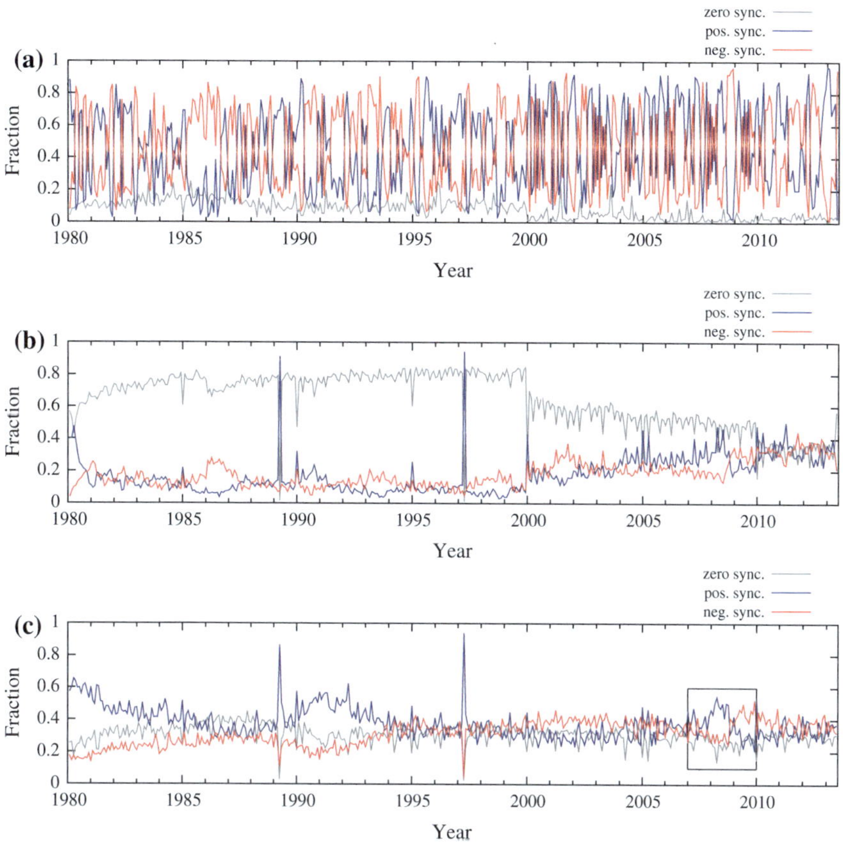

Fig. 25.4 (Color on-line) The fraction of the numbers of goods and services for which we observe positive (*blue*), negative (*red*) and zero (*gray*) price changes, respectively. From *top* to *bottom*: **a** IPI. **b** DCGPI. **c** CPI

25.5 Discussion

First, the fact that the average frequency of price change of individual goods and service gives only a limited information on the behavior of aggregate price because price change is not time homogeneous, as shown for the year 2008, rejects the popular assumption that the firm's price changes strategy is time-invariant. For example, [4] assumes that for each firm, and opportunity to change its price arrives at random with a given probability, while others in the more recent literature assume that the hazard rate of price is time-invariant. See [7] and references therein. More generally, the existing literature focuses on *cross-sectional* distribution of micro prices, and assumes that it is given and time-invariant. We note that micro optimization exercise results in a particular pattern of price setting which is time-invariant. This assumption

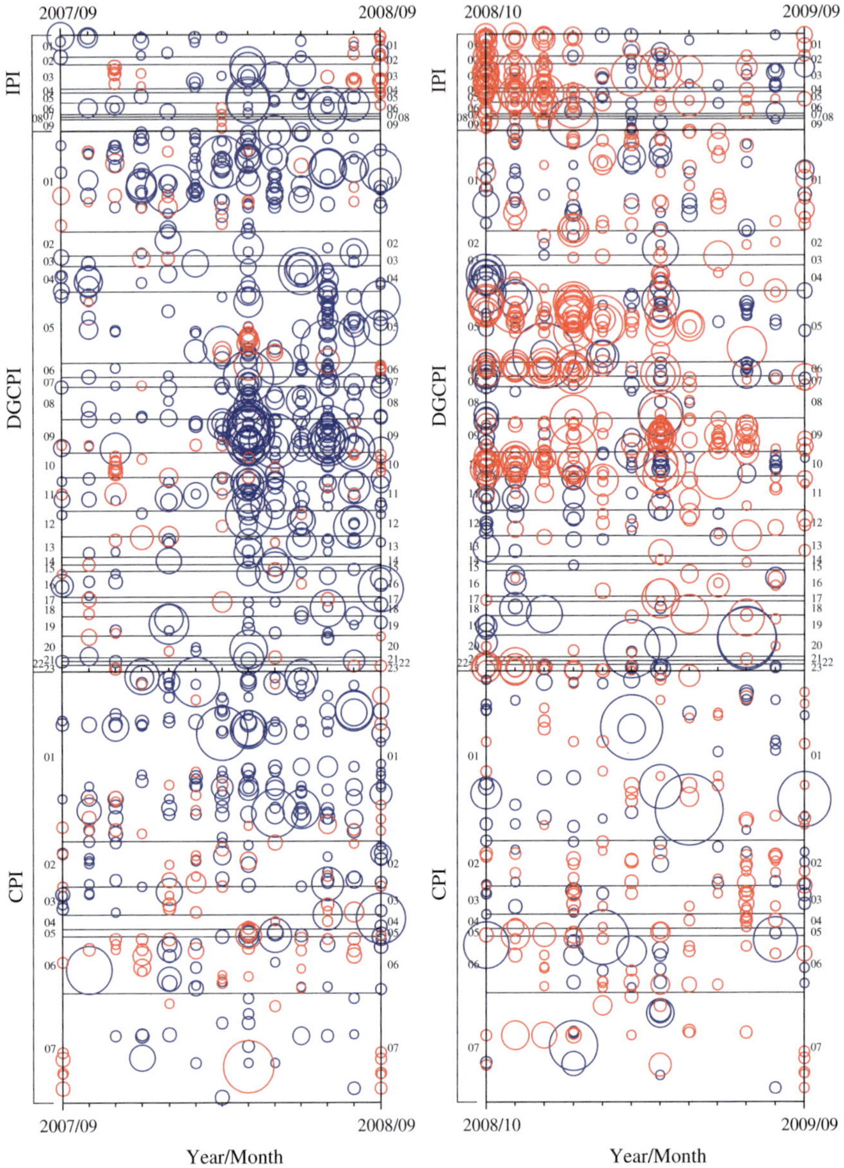

Fig. 25.5 (Color on-line) Enlarged view of Fig. 25.3 for the period **a** 1 year before the Lehman shock in September 2008 and **b** 1 year after the shock respectively. The threshold $w_* = 2.0$, and the *colors* and radius of *circles* are the same as given in the caption of Fig. 25.3

of time-invariance of price setting is not borne out by the data. Instead, it is important to explore what macro variables drive individual prices to synchronized actions. See the separate paper [11] for our recent finding on this point.

Second, to understand the behavior of aggregate price, we must explicitly consider subsets or clusters of prices, not just a single macro-group of prices. For example, as shown in Fig. 25.3 for the period during 1995–2000 vertically. Except for April 1997 when VAT was raised, prices of some goods went up or down in clusters while others remained unchanged.

This point casts doubt on the existing theories of price setting such as menu cost and contract models. In most theoretical models, and individual firm is assumed to strategically set or reset its price considering the behaviors of *all the other firms*. It is commonly assumed that firm j is interested in P_j/P where P_j is the firm j's price and P is the aggregate price index. In other words, it is routinely assumed that the universe in which each firm optimizes is the economy as a whole. However, the behavior of individual prices shown in Fig. 25.3 does not support this presumption; it shows that there is a significant tendency that a cluster of price changes change together while at the same time prices which belong to other clusters do not. The standard theoretical model takes the macro-economy as if it were a single industry or a group of retailers in a region. Such a model may server for the purpose of industrial organization, but does not fir the purpose of macroeconomics and monetary policy.

Generally, we can consider how N commodities' prices are determined by J firms. A firm changes the prices of goods and services which it produces in response to the changes in other prices. However, firm is not interested in all prices, but only in a subset of prices. Obvious examples are prices of intermediate goods and services used in production, and also prices of close substitutes produced by rival firms. The response is not usually taken based on the single information of aggregate price P in a synchronized way, but on a partial information among the N prices that are relevant to the firm. See [6, pp. 32–33] for the same point.

On the other hand, there exists a large literature on dynamic factor models including [1, 5, 9, 10], especially with focus on estimating price inflation and the effects of monetary policies, in particular, in the Euro area. Comparison with the factor models would be valuable in future work.

25.6 Summary

We demonstrate the following points in this paper.

1. The average frequency of price change of individual goods or service provides only a very limited information on the behavior of aggregate price because price change is not time homogeneous. The year 2008 is a good example. In the first half of the year, many prices were raised, but the bankruptcy of the Lehman Brothers turned the tide, and afterwords, many prices declined, some significantly. In other times, most prices simply remained unchanged for a long time period.
2. In order to fully understand the behavior of aggregate price, we must explicitly consider subsets or clusters of prices, not just a single macro-group of prices.

We have directly observed in Fig. 25.3 for the period 1995–2000, during which, except the raise of VAT, prices of some goods went up or down in clusters while others remained unchanged.

Micro information set on which firm sets its prices would not contain money supply, in particular. Rather, *prices affect prices with leads and lags*. The existing literature almost completely misses this important lead/lag relationships among micro prices. Our separate paper [11] shows the importance of such interactions of micro prices, and how the dynamics is not significantly related to money supply in an explicit way.

Acknowledgments We would like to thank anonymous referees for improving the paper. We are also grateful to Y. Ikeda, W. Souma, T. Watanabe and Y. Yajima for comments, and also K. Itoh and N. Shinozaki (both at NHK—Japan Broadcasting Corporation) for partial help in classification of the CPI data. We thank participants of seminars at the Bank of Japan, the University of Tokyo, and the Research Institute of Economy, Trade and Industry for comments.

This work is partially supported by Grant-in-Aid for Scientific Research (KAKENHI) Grant Numbers 24243027 and 25282094 by JSPS, and also in part by the Kyoto University Supporting Program for Interaction-based Initiative Team Studies: SPIRITS, as part of the Program for Promoting the Enhancement of Research Universities, MEXT, Japan.

References

1. Altissimo, F., Mojon, B., Zaffaroni, P.: Fast micro und slow macro: Can aggregation explain the persistence of ination? manuscript, fRB of Chicago Working Paper No. 2007-02 (2007)
2. Bank of Japan: Outline of statistics and statistical release schedule. https://www.boj.or.jp/en/statistics/outline/exp/pi/excgpi02.htm/ (2014)
3. Basawa, I.V., Prakasa Rao, B.L.S.: Statistical Inference for Stochastic Processes. Academic Press (1980)
4. Calvo, G.A.: Staggered prices in a utility-maximizing framework. J. Monetary Econ. **12**(3), 383–398 (1983)
5. Carvalho, C.: Heterogeneity in price stickiness and the real effects of monetary shocks. Front. Macroecon. **2**(1), article 1 (2006)
6. Gordon, R.J.: The history of the Phillips curve: consensus and bifurcation. Economica **78**(309), 10–50 (2011)
7. Klenow, P.J., Malin, B.A.: Microeconomic evidence on price-setting. In: Friedman, B.H., Woodford, M. (eds.) Handbook of Monetary Economics, vol. 3A, chap. 6. North-Holland (2011)
8. Statistics Bureau: Consumer price index. http://www.stat.go.jp/english/data/cpi/index.htm (2014)
9. Stock, J.H., Watson, M.W.: Forecasting inflation. J. Monetary Econ. **44**(2), 293–335 (1999)
10. Stock, J.H., Watson, M.W.: Handbook of Macroeconomics, chap. Business Cycle Fluctuations in U.S. Macroeconomic Time Series, pp. 3–64. Elsevier, Amsterdam, The Netherlands (1999)
11. Yoshikawa, H., Aoyama, H., Fujiwara, Y., Iyetomi, H.: Deflation/Inflation Dynamics: Analysis Based on Micro Prices. http://ssrn.com/abstract=2565599 (2015)

Chapter 26
Understanding the Diffusion of YouTube Videos

Mattia Zeni, Daniele Miorandi and Francesco De Pellegrini

Abstract In this paper we tackle several questions arising in the context of online content diffusion. In particular, we analyse the reason why some videos become viral, how popularity of a tagged video evolves over time and if there exist recurrent patterns in the dynamics of content popularity. Indeed, while the ultimate question is if it is even possible to predict the popularity dynamics of a newly published video, several interwoven factors impact the process of diffusion of online contents. In this paper we propose a framework able to put all the previous questions into a complex system science perspective. We first analyse the mechanisms that affect the popularity growth of a tagged video. We then illustrate why a multi-scale multi-level model appears the most appropriate to capture the effect of such phenomena. We finally present an open dataset of YouTube videos' popularity, which has been released with the aim to let researchers in the field validate their findings against real-world data.

26.1 Introduction

The study of how popularity of user-generated contents evolves over time has generated considerable interest from both the research and the business community in the past few years.

The connection with modern online marketing strategies comes from the fact that, having devised the ultimate early-stage prediction technique to identify contents undergoing popularity bursts [12], it would be possible to optimize for the resources

M. Zeni (✉)
Department of Information Engineering and Computer Science,
University of Trento, via Sommarive 14, 38123 Trento, Italy
e-mail: mattia.zeni@disi.unitn.it

D. Miorandi · F. De Pellegrini
CREATE-NET, via Alla Cascata 56/D, 38100 Trento, Italy
e-mail: miorandi.daniele@create-net.org

F. De Pellegrini
e-mail: depellegrini.francesco@create-net.org

S. Battiston et al. (eds.), *Proceedings of ECCS 2014*, Springer Proceedings
in Complexity, DOI 10.1007/978-3-319-29228-1_26

allocated, e.g., in a marketing campaign attaching a certain product/business advertisement to an uptaking popular video. Actually, several research papers have appeared in literature, each providing specific characterizations of the dynamics of popularity of online media contents.

However, from an abstract perspective, the process by which a certain video becomes popular has several features which qualify the resulting dynamics as the output of a complex multi-scale multi-level socio-technical system. In such system, individuals act as prosumers of content, platform owners set policies that impact how the content can be accessed and third parties provide services which generate a distortion of the natural ranking of popular contents by means of content "acceleration" tools (advertising, marketing, re-ranking, etc.).

Researchers have looked at various aspects related to how popularity (a multi-dimensional concept that accounts for how much a content spreads) changes over time. They also have tried to correlate popularity with various factors [1–13]. Businesses (in particular, marketing and web agencies) have tried to develop methods for enhancing the popularity of a given content, something that is typically sold to brands and professional content producers.

In this paper we develop a framework for studying and analysing how online contents can diffuse and become popular; in our case we focus on videos published on YouTube. We first analyse the mechanisms that affect the popularity growth of a video. We then illustrate a multi-scale multi-level model able to justify the effect of such mechanisms. We finally introduce an open dataset of YouTube videos' popularity and show how it can be used to validate findings in the field.

The remainder of this paper is organised as follows. In Sect. 26.2 we analyse the factors that drive the diffusion and popularity growth of contents in online video sharing platforms. In Sect. 26.3 we analyse the various scales and levels at play in this framework. In Sect. 26.4 we introduce the YOUStatAnalyzer open database (including the daily evolution of popularity of 1+ million YouTube videos) and show how the models developed can be used to purposefully analyse real-world data. Finally, Sect. 26.5 concludes the paper pointing out directions for future extensions and enhancements.

26.2 Driving Factors of YouTube Video Diffusion

YouTube videos are organised in channels: Fig. 26.1 shows a webpage providing access to a certain video channel. On that page, it is possible to visualize several videos which are featured on the specific channel. Any registered YouTube user can access any channel's videos and can create its own channels. Users can also manage multiple channels. YouTube videos are always published within a channel.

The platform maintains for each video a number of detailed metrics accessible only to the video's channel owner: they can be accessed through the YouTube portal

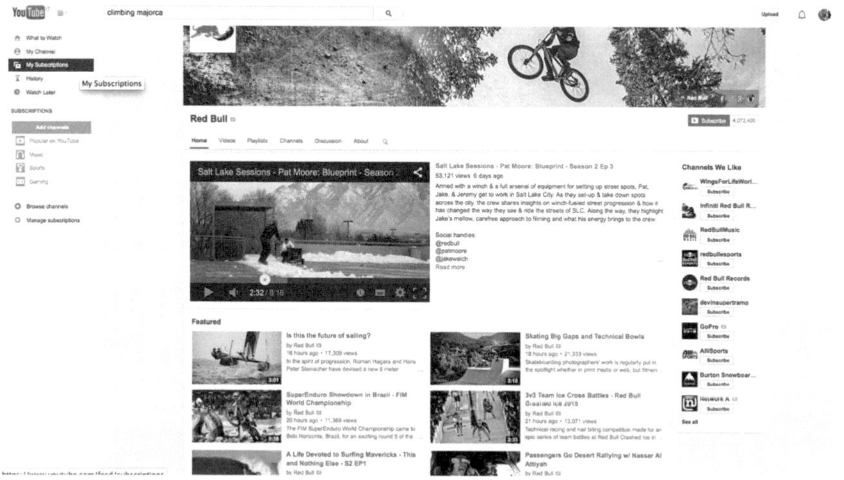

Fig. 26.1 A sample video channel: the icons on the *right* of the main window are recommenders for channels; the *central icons* should several videos on display

itself (once logged in) or through the YouTube Analytics API.[1] At the same time, daily statistics are typically visible to everyone below each video, unless the channel owner decides to block their display.

The popularity of a tagged YouTube video is a multidimensional concept: it actually accounts for a number of key performance indicators:

- *Views or viewcount*: number of times the URL corresponding to the video was opened;
- *Watchtime*: cumulative time spent by users watching the video;
- *Likes*: the number of preferences which users ascribed to the video;
- *Subscriptions*: the number of subscription to the channel generated by the tagged video;
- *Comments*: tracks the number of comments posted by YouTube users on the video.

Most studies on the popularity of videos focused on the number of views, namely the *viewcount*, as single metric of interest. However, there are at least two good reasons not to forget the other metrics:

- Since 2013, the YouTube recommender uses the watchtime as metric for understanding how a video is popular (and not the viewcount);
- The various metrics can be used to build a "conversion funnel", a commonly used tool in the marketing sector in order to characterize the impact of advertisement campaigns.

In Sect. 26.4, we are reporting on specific joint characterization of such performance indicators.

[1] See https://developers.google.com/youtube/analytics/.

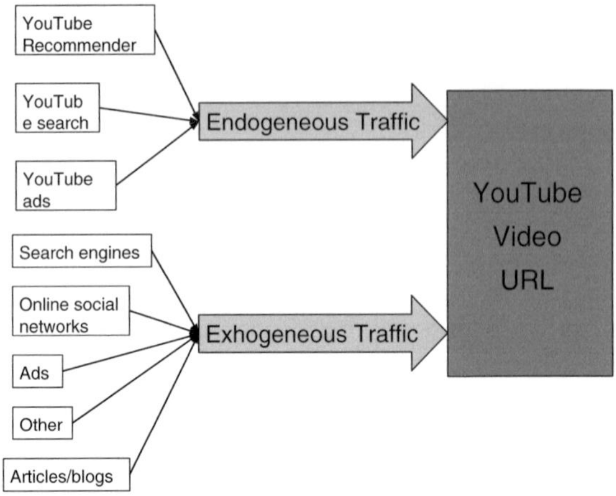

Fig. 26.2 Sources of traffic to a YouTube video URL: two main streams originate from exogenous sources, external to the platform, and from the inner recommendation and sharing mechanisms

Now, in general terms the study of the popularity of YouTube videos can be seen as a special case of web traffic analysis/web marketing. Each video is indeed characterised by a unique ID and a uniform resource locator (URL) of the form https://www.youtube.com/watch?v=XXXXX where XXXXX is the video ID.

The demand from YouTube platform users with respect to a certain tagged video can be interpreted as network traffic demand generated by different sources. Such traffic can be logically decomposed according to the tree represented in Fig. 26.2:

1. *Endogeneous traffic*: this is the traffic generated from within the YouTube platform itself. We can further classify this into:

   (a) YouTube recommender: the traffic generated by means of the YouTube recommender. Any user watching the YouTube frontend is presented with a number of 'Recommended' videos and with a number of 'Related Videos' once she is watching a video.
   (b) YouTube search: traffic generated by a user through the search engine embedded within the YouTube platform.
   (c) YouTube Ads: videos can be promote using Google AdWords. In particular, videos can be promoted through short video teasers, which are displayed before the content actually requested by the user.

2. *Exhogeneous traffic*: traffic generated from other web sites/platforms. We can further classify this into:

   (a) Search engines: traffic generated from commonly used search engines when users search for a given content.
   (b) Online social networks: one of the most significant sources of video traffic is the sharing by users of YouTube videos on social networking platforms, like Facebook, Twitter, Google+, Sina Weibo etc.

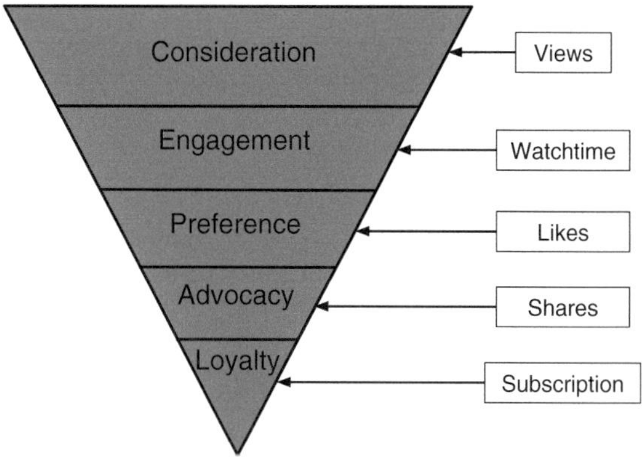

Fig. 26.3 Funnel representation of the various popularity metrics

(c) Advertisements: advertising bought by the publisher (e.g., through AdWords) to promote a given video content on various Web sites (excluding YouTube itself).
(d) Articles/blogs: online blogs/articles/magazines including an embedded version of the YouTube video.
(e) Other: any other source of exhogeneous traffic, e.g., TV or radio broadcasts referring to the tagged content.

All such possible traffic demands determine the interaction pattern of users with the tagged YouTube video, namely, how the key cumulative performance indicators (viewcount, shares, watchtime) increase over time. At a certain moment in time, it is possible to capture the notion of popularity of a tagged video by considering the five-stages funnel structure reported in Fig. 26.3:

1. *Consideration*: the first step in the funnel is the number of users who opened the video (impressions). This corresponds to the number of views.
2. *Engagement*: we can measure the interest in the content by considering the (normalized) watchtime.[2] If a content is uninteresting or of poor quality, it is likely that users who start watching it will stop the playback and navigate away within a very short time.
3. *Preference*: the number of likes is—at least in principle—a good measure of the feedback by users, in terms of whether they appreciated the video content.
4. *Advocacy*: the number of times a given video is shared by users is a good indicator of the fact that they recommend others (friends, aquaitances, followers) to watch it as well.
5. *Loyalty*: the number of subscriptions generated by a video is a good indicator of the fact that the user expresses a long-term interest in the type of contents generated by the author.

In Sect. 26.4, we illustrate the outfit of this multidimensional characterization using a set of videos belonging to sample video categories.

[2]The normalization over the video's length is a limitation but is the only way to compare the curves.

26.3 Scales and Levels in YouTube Video Diffusion

In general, the interwoven traffic paths described in the previous section determine the growth of key performance indicators of a tagged video at different time scales. Concerning time-scales, it is possible to identify two "natural" ones over which videos popularity dynamics takes place.

The first one (called hereafter the "fast" time-scale) account for fast dynamics. This represents, e.g., the cascading effect due one or more very popular Twitter users sharing a given video or to daily schedules of users who access the platform during specific peak daily hours. In order to study the dynamics over such a time-scale it is necessary to sample popularity indicators with sub-hour granularity. A sampling time between five and fifteen minutes appears appropriate for capturing most of the phenomena happening over such a time-scale.

The second type time scale operates over much longer time intervals. This type of evolution relates to the long-term popularity of a given content. In this case the dynamics can be observed over a period of months (when not years). A sampling time of the order of one day is appropriate to measure phenomena over such time-scale.

In terms of levels, we can consider two natural ones as well.

The first level is obviously that of a single video. Here all the factors identified in the previous section should be considered. The second one relates to the channel. As the channel aggregates different video contents, its popularity can be understood by measuring the aggregated popularity indexes of the videos it contains. Examples include, e.g., the total number of views, the average normalized watchtime or the average number of viewcounts/day for videos on the channel, or again the average ratio between shares and views (understood as a good proxy for the ability to generate advocacy). Another direct indicator of the popularity of a channel is the number of subscribers, whose evolution over time can also be studied. It is worth remarking that a specific recommendation list exists at the channel level, as can be seen in Fig. 26.1.

Yet, there is also a feed-forward effect to be considered: indeed channel subscriptions generate traffic, and videos generate channel subscriptions, so that a channel with a set of very popular videos is more likely to generate traffic to a newly published content.

In order to present a comprehensive analysis of the diffusion of YouTube contents, the aforementioned natural scales and levels should be jointly considered.

26.4 A Look into the Wild: The YOUStatAnalyzer Dataset

To characterize the reference framework we are proposing in this paper we have used a dataset of YouTube videos produced using the YOUStatAnalyzer software.[3] This dataset contains daily samples (from the publication time until early 2015) of the

[3]see https://github.com/mattiazeni/youstatanalyzer.

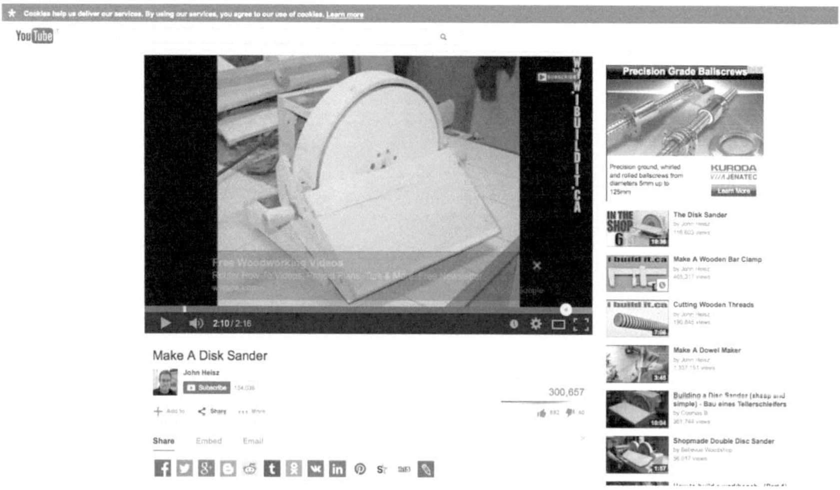

_ID	0ao1UwIYQCg
Category	Howto
Description	Older project, new uplo...
Title	Make A Disk Sander
Author	John Heisz
Published Date	2012-07-08T00:09:58.000Z
Access Control	['comment': ['allowed'], ...]
Comments #	50
Related Videos	[4kU2veNhyKI,A4PzyaoRYk0...]
Duration	207
Video Type	video/3gpp
Gplus Shares	activityID
Popularity Metrics	views, watchtime, subscribers, shares

Fig. 26.4 YOUStatAnalyzer dataset: information available for a sample video, namely the video with ID: `uPehlRGMtyw`

popularity metrics (views, watchtime, shares, subscribers) of more than 1 million YouTube videos, for a total of ~61 GB of data. Data is complemented by a set of additional meta-data information. Figure 26.4 describes the information collected for each video and stored in the YOUStatAnalyzer dataset.

To the best of our knowledge the YOUStatAnalyzer tool is the only one freely available to the research community that allows to create custom datasets of YouTube videos' statistics that allows to collect the daily and cumulative trends of Views, Watchtime, Shares, Subscribers. In fact, in June 2013 YouTube changed the backend system, blocking the endpoint that allowed to collect this kind of information as it was used in a number of papers, e.g., [7]. The authors found an alternative way to collect such data and decided to release it openly to the community in the form of an open source tool and an open, reference, dataset. Moreover, in the last version,the tool allows also collecting the shares (and reshares) of each video on Google Plus.

Using samples of such a dataset we represented a popularity vector in the sense described in Sect. 26.2. Figure 26.5 shows the same trend of the funnel chart proposed in Sect. 26.2 over the different popularity metrics for the Music and Technology

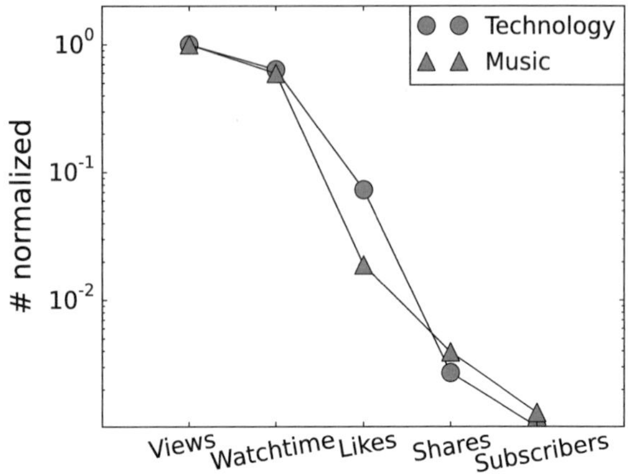

Fig. 26.5 The 5-dimensional popularity vector: comparing for the Music and Technology categories in the YOUTubeStatAnalyzer dataset. The normalization is done to be able to compare the metrics over all the videos

categories. The total number of videos interested in this analysis are 176627 and 66959, respectively. It is interesting to notice that even for two very different categories, namely Music and Technology, the trend is very similar. In particular, in the ratio between views, watchtime, shares and subscribers, with a substantial difference in the number of preferences expressed by the users. It appears that videos belonging to the Technology category are liked or disliked more than those belonging to the Music category.

Another interesting aspect to consider is the evolution curve of popularity for the videos in the dataset. We normalised the cumulative viewcount and run a clustering (k-means) algorithm to identify "typical" patterns (i.e., the resulting centroids). We considered the "Music" and "News" categories and used $k = 6$.[4] Results are depicted in Fig. 26.6a, b: each line represents a reference centroid waveform and its thickness is proportional to the size of the corresponding cluster.

It is possible to note from Fig. 26.6 that the News category has waveforms which concentrate around fast growing centroids. In particular, the uppermost centroid is almost vertical in the very first period after the video release; it becomes flat after only the 10% of the entire video lifetime. The distribution of weight and the shape of the centroids indicates that the videos belonging to the News category, compared to the others, are relevant to the YouTube community for just a very short period of time after the publication. Thus, we can infer that for such videos it is worth exploiting this very short period of time in terms of caching or advertisement.

A different behaviour is shown in the Music category: videos belonging to it are attractive to the users for a longer period of time. This can be understood from Fig. 26.6a, in which centroids do not present any critical vertical behaviour and in general they are more flat, thus demonstrating that the on average, views are distributed over a longer period of time.

[4]The choice of the parameter k was taken based on the analysis in [11].

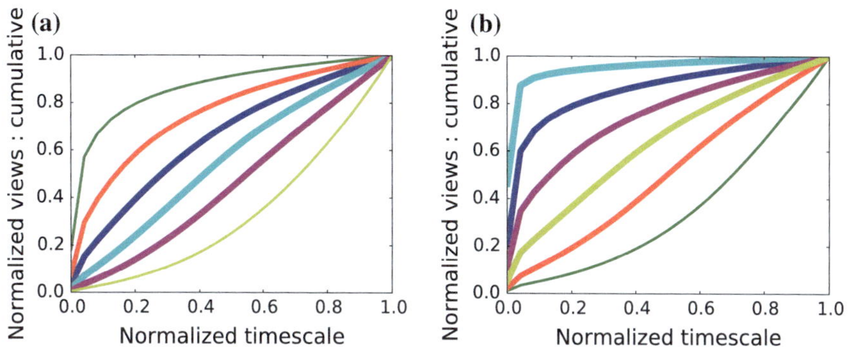

Fig. 26.6 The centroids over the views calculated using the k-means algorithm (with $k = 6$) over **a** 176K videos of the Music category, **b** 541K videos of the News category. They are plotted in different colours just for the sake of making them distinguishable

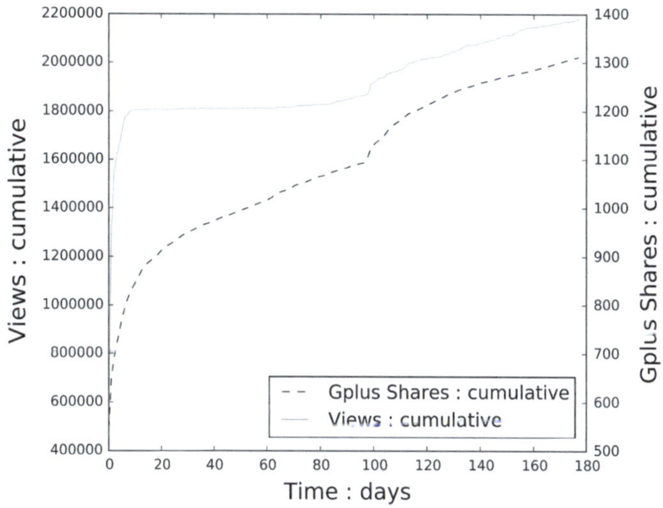

Fig. 26.7 A comparison of the trends of the views versus the Google Plus shares for the video ID: `uPehlRGMtyw`

With the latest version of YOUStatAnalyzer, we included also the possibility to collect the shares of a certain video generated on Google+. We used such an information to study to which extent this sharing process correlates with the diffusion of the video. Figure 26.7 shows the relationship between the popularity of the video, expressed by the cumulative viewcount and the shares on Google+. The strong coupling between the two curves can be detected by the fact that a fast change in the dynamics of the Google+ shares, occurring at fast time scale, is reflected by a corresponding jump in the viewcount. We remark that inspecting the shares operated directly on YouTube, we could not detect such fast time scale jump: we ascribe this behaviour to the fact that the re-shares operated on the social networks is what actually determines such discontinuity at the slow time scale.

26.5 Conclusion

In this paper we have provided a framework for answering a number of questions related to the diffusion and popularity of videos in online sharing platforms. Several fundamental questions exist in this field: why some videos do become viral and other videos do not, what is the definition of popularity able to best capture the way videos are perceived by the online audience, if there exist recurrent patterns, and whether videos from different categories behave differently.

Our framework is based on a funnel model that reproduces the behaviour of users which interact and engage with online video contents. We argue that scales and levels are a fundamental feature of the online diffusion process. We have shown cases in which the effect of the natural scales and levels can easily be inspected: the channel level and scales originating from the coupling of shares and viewcount. Tests for our general framework have been provided on a large, open database that can be used by scientists and practitioners in the field to develop new knowledge and test/validate their models and assumptions.

One limitation that requires additional work relates to the correlation and entangelement among actions carried out on different social networking platforms. We have provided evidence that the Google+ shares dynamics is indeed impacting the popularity dynamics in some specific cases. However, whether this sample is representative of the general sharing process on different platforms requires further assessment.

Acknowledgments The work of D. Miorandi and F. De Pellegrini has been partially supported by the European Commission within the framework of the CONGAS project FP7-ICT-2011-8-317672, see www.congas-project.eu

References

1. Ahmed, M., Spagna, S., Huici, F., Niccolini, S.: A peek into the future: predicting the evolution of popularity in user generated content. In: Proceedings of the sixth ACM international conference on Web search and data mining, pp. 607–616. ACM (2013)
2. Altman, E., De Pellegrini, F., El Azouzi, R., Miorandi, D., Jiménez, T.: Emergence of equilibria from individual strategies in online content diffusion. In: Proceedings of IEEE INFOCOM NetSciCOmm. Turin, Italy (2013)
3. Cha, M., Kwak, H., Rodriguez, P., Ahn, Y.Y., Moon, S.: I tube, you tube, everybody tubes: analyzing the world's largest user generated content video system. In: Proceedings of ACM SIGCOMM IMC, pp. 1–14. ACM, New York, NY, USA (2007)

4. Cha, M., Kwak, H., Rodriguez, P., Ahn, Y.Y., Moon, S.: Analyzing the video popularity characteristics of large-scale user generated content systems. IEEE/ACM Trans. Networking **17**(5), 1357–1370 (2009)
5. Chatzopoulou, G., Sheng, C., Faloutsos, M.: A first step towards understanding popularity in YouTube. In: Proceedings of IEEE INFOCOM, pp. 1–6. San Diego (2010)
6. Figueiredo, F., Almeida, J.M., Benevenuto, F., Gummadi, K.P.: Does content determine information popularity in social media?: a case study of YouTube videos' content and their popularity. In: Proceedings of the 32nd Annual ACM Conference on Human Factors in Computing Systems, pp. 979–982. ACM (2014)
7. Figueiredo, F., Benevenuto, F., Almeida, J.M.: The tube over time: characterizing popularity growth of YouTube videos. In: Proceedings of the ACM International Conference on Web Search and Data Mining, pp. 745–754. ACM (2011)
8. Li, H., Cheng, X., Liu, J.: Understanding video sharing propagation in social networks: measurement and analysis. ACM Trans. Multimedia Comput. Commun. Appl. (TOMM) **10**(4), 33 (2014)
9. Niu, G., Fan, X., Li, V., Long, Y., Xu, K.: Multi-source-driven asynchronous diffusion model for video-sharing in online social networks. IEEE Trans. Multimedia **16**(7), 2025–2037 (2014)
10. Pinto, H., Almeida, J.M., Gonçalves, M.A.: Using early view patterns to predict the popularity of YouTube videos. In: Proceedings of the Sixth ACM International Conference on Web Search and Data Mining, WSDM '13, pp. 365–374. ACM, New York, NY, USA (2013)
11. Richier, C., Altman, E., Elazouzi, R., Jimenez, T., Linares, G., Portilla, Y.: Bio-inspired models for characterizing youtube viewcout. In: 2014 IEEE/ACM International Conference on Advances in Social Networks Analysis and Mining (ASONAM), pp. 297–305 (2014)
12. Szabo, G., Huberman, B.A.: Predicting the popularity of online content. Commun. of the ACM **53**(8), 80–88 (2010)
13. Zhou, R., Khemmarat, S., Gao, L.: The impact of youtube recommendation system on video views. In: Proceedings of the 10th ACM SIGCOMM Conference on Internet Measurement, IMC '10, pp. 404–410. ACM, New York, NY, USA (2010)

Chapter 27
Free Energy Rate Density and Self-organization in Complex Systems

Georgi Yordanov Georgiev, Erin Gombos, Timothy Bates, Kaitlin Henry, Alexander Casey and Michael Daly

Abstract One of the most important tasks in science is to understand the self-organization's arrow of time. To attempt this we utilize the connection between self-organization and non-equilibrium thermodynamics. Eric Chaisson calculated an exponential increase of Free Energy Rate Density (FERD) in Cosmic Evolution, from the Big Bang until now, paralleling the increase of systems' structure. We term these studies "Devology". We connect the exponential growth of FERD to the principle of least action for complex systems leading to exponential increase of action efficiency. We study CPUs as a specific system in which the organization, the total amount of action and FERD are connected in a positive feedback loop, providing exponential growth of all three and power law relations between them. This is a deep connection, reaching to the first principles of physics: the least action principle and the second law of thermodynamics. We propose size-density and complexity-density rules in addition to the established size-complexity one.

G.Y. Georgiev (✉) · E. Gombos · T. Bates · K. Henry · A. Casey · M. Daly
Physics Department, Assumption College, 500 Salisbury St,
Worcester, MA 01609, USA
e-mail: ggeorgie@assumption.edu; georgi@alumni.tufts.edu

G.Y. Georgiev
Physics Department, Tufts University, 4 Colby St, Medford, MA 02155, USA

G.Y. Georgiev
Department of Physics, Worcester Polytechnic Institute, Worcester, MA 01609, USA

E. Gombos
Present Address National Cancer Institute, NIH, 10 Center Drive, Bethesda, MD 20814, USA

A. Casey
Present Address University of Notre Dame, Notre Dame, IN 46556, USA

M. Daly
Present Address Meditech, 550 Cochituate Rd, Framingham, MA 01701, USA

S. Battiston et al. (eds.), *Proceedings of ECCS 2014*, Springer Proceedings in Complexity, DOI 10.1007/978-3-319-29228-1_27

27.1 Introduction

After many years of study, the processes of self-organization of complex systems still do not have a satisfactory description. Non-equilibrium thermodynamics is essential to understand self-organization [1–3]. The driver towards higher levels of structure and organization in complex systems and in Cosmic Evolution has been recognized as the density (time and mass) of free energy flowing through a system [4–6]. Energy differences (gradients) create forces which move flows of matter, doing work to minimize constraints to motion, thus carving the flow channels along which the product of time and energy per event is minimized [7–10]. Other aspects of the principle of least action driving self-organization have been considered [11–14]. In our research program we set to investigate the entire chain of self-organizing events in systems spanning from the atoms to the society [7–10]. This is the goal of many scientists working in the fields of complexity, cosmic evolution, big history and others. This paper is one of the steps in this process answering the question whether there are other characteristics than level of organization and size of a system that are useful in describing the process of self-organization. Is free energy rate density, used by Chaisson, correlated to action efficiency as a measure of organization? Is the growth of free energy rate density exponential in time and does it fit in the positive feedback model between level of organization and size of a complex system? Will we be able to use those three measures as tools to characterize complex systems when partial information is available, and deduce them from each other?

Our approach is to study the minimization of average physical action per unit motion (action efficiency, quality) and the maximization of the total action for all motions in complex systems (quantity) [7–10]. We defined organization as action efficiency (α) and size of a system as the total amount of action in it (Q) and used the principle of least action as a driving force for increase of α [7–10]. We also posited a maximum action principle, where Q in a self-organizing complex system tends to a maximum, i.e. for the unit action to decrease, the total action has to increase in order to minimize the constraints to motion further. We have shown that the quality (α) depends on quantity (Q) and vice versa (the size-complexity rule) and when one is increased or decreased the other is affected in the same way, i.e. they are in a positive feedback loop [10]. This dependence is a major driving force and a mechanism of progressive development measured as the increase of action efficiency in complex systems.

Previously we described how in complex systems, elements cannot move along their least possible action paths that characterize their motion outside of systems, because of obstacles to the motion (constraints) [7–10]. We used variational approaches to optimization in complex systems, which are the least and most action principles mentioned above [7–10]. We extended the principle of least action as: complex systems are attracted toward a state with least average action per one motion given those constraints [7–10]. This is congruent with the Hertz's principle that objects move along paths with least curvature [15] and the Gauss principle that they move along the paths of least constraint [16]. We extended these principles for complex systems that the elements do work on the constraints to minimize them, reducing the curvature and the amount of action spent for unit motion. The tendency to move along geodesics drives the flows of elements in a system to remove obstructive constraints from their paths in order to achieve the state of smallest possible product of time and energy (action) for the processes. This is what we term self-organization. For this work the elements need energy, and the higher the free energy rate density, the more the work can be done, therefore the faster the self-organization. The new geodesics of the elements in the curved by the constraints to motion space are the paths with minimum action. The paths of least constraint are the flow paths in the system [7–10]. Therefore we defined organization (the action efficiency of the complex system) as the state of the constraints to motion determining the average action per one element of the system and one of its motions [10]. We posited a flow network representation of a complex system, where the flows are necessary to equilibrate any energy differences, attracted by their final state—that of thermodynamic equilibrium. Each element in a complex system is the smallest mobile unit in the system and usually moves in a flow channel along a network of paths (edges) between the starting and ending points (nodes) which are sources and sinks in the flow network. In CPUs, one unit of motion (event) is a single computation in which electrons flow from the start node to the end node [10].

The increase of FERD in complex systems is allowed by the increase of α driven by the principle of least action and of Q, driven by the principle of most action. In this paper, we understand the processes of progressive increase of level of organization, as a connected system of physics laws, which when put together yield complex systems that we observe around us. If any of those three principles is taken separately: the Least Action Principle (LAP), the Most Action Principle (MAP) and the principle of increase of Free Energy Rate Density (FERD), they do not lead to a self-organizing complex system. Only when they are connected in the same system in a positive feedback loop, acting together, they yield the amazing diversity of complex systems that we see in the world around us. We can use a new term, Devology (*dev* from "development", *-evo-* from "evolution" and *-logia* "study of") for a study of development of organization in complex systems in Cosmic Evolution, from the Big Bang to Humankind [4].

27.2 Model

Previously we connected in a positive feedback loop organization (α) and size (Q) of a complex system, leading to an exponential increase of both and to a power law relationship between the two, which matched well with data for CPUs [10]. We proposed that this feedback loop is the major mechanism of accelerated rate of self-organization and evolution of complex systems [10].

Here we show how non-equilibrium thermodynamics connects to this model, through measurement of what Eric Chaisson terms Free Energy Rate Density ($\boldsymbol{\Phi}$), as a distance from thermodynamic equilibrium, which he uses as a measure for Cosmic

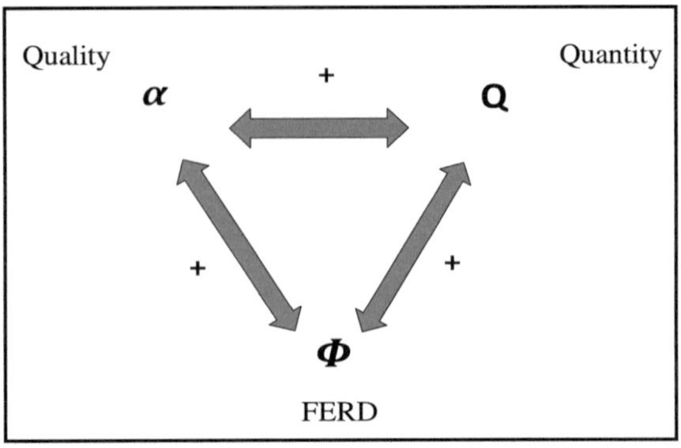

Fig. 27.1 A positive feedback model between α, Q and $\boldsymbol{\Phi}$. This loop can be described with a system of ordinary differential equations, as in [10], which solve to an exponential growth of each of the three and power law dependences between each two of them

Evolution [4]. On Fig. 27.1 we include $\boldsymbol{\Phi}$ in the model of positive feedback between α and Q developed earlier [10]. In this expanded model, all three are in a positive feedback, which as shown in [10] leads to exponential solutions of the differential equations for the involved quantities for each. When the exponential equations are combined, they yield a power law relation between them. Therefore for this paper it is enough to demonstrate with data whether the relationship of $\boldsymbol{\Phi}$ to either α or Q is a power law in order to connect all three in a positive feedback loop with the mentioned outcomes.

27.3 Data and Methods

CPUs are organized flow systems, where the events are well defined as computations and precise data for time and energy per event available over the entire period of their existence. CPUs are a good model system, because they are analogous to all other complex systems: they perform events, consume energy and increase their time and energy efficiency over time—evolve. Therefore they are an excellent system to test our positive feedback model. Free Energy Rate Density, $\boldsymbol{\Phi}$, is measured in MKS units. Those units are different than those used by Chaisson up to a constant, due to the proportionality of the area of the CPUs to their mass. We make the assumption that as a 2D system, the thickness of the silicon wafer is constant, or its change is negligible, and that there are equal amounts of mass per unit area across all generations of CPUs. Therefore, we calculate energy rate density in MKS units $J/s \cdot m^2$, which is analogous up to a constant to the units $J/s \cdot kg$ used previously for FERD and the trends in our data will not change if this constant is used. To calculate the mass per unit area constant, information about the thickness and density of the silicon wafers is necessary. Data only for processors for desktops or laptops were used for consistency, because some of the specialized processors, such as for phones or tablets, perform slower in order to use less energy and fall below this trend line.

Data were collected from Intel Corporation Datasheets [17]. The Instructions Per Second (IPS) for each processor was divided by the Thermal Design Power (TDP) as a measure of the total power consumption by the CPUs at maximum computational speed, for consistency. The result was multiplied by the table value of the Planck's constant, $h = 6.626\,10^{-34}$ Js, as the smallest quantum of action, to solve for, as the inverse of the average number of quanta of action per instruction per second [10]. To solve for Q, the TDP was divided by h to find the total number of quanta of action per second. To measure the FERD ($\boldsymbol{\Phi}$), we divided the TDP as the maximum rate of energy flowing through the CPUs, by the area (die size) of the CPUs.

27.4 Results

Figure 27.2 shows that $\boldsymbol{\Phi}$ is correlated with the size of the system Q by a power law, which we set to explore with our model. We do not observe large deviations from this power law relation, such as a system with small Q and a large density $\boldsymbol{\Phi}$ or a system with large Q and small $\boldsymbol{\Phi}$. Therefore the total amount of action in a system Q, or its size, is connected to the density of free energy $\boldsymbol{\Phi}$, but not just the amount of it. This is a Size-Density rule: size and density in a complex system are proportional. That means mathematically, that $\boldsymbol{\Phi}$ and α are also in a power law relationship, based on the positive feedback model on Fig. 27.1 and our previous results for the solutions of the differential equations describing this model [10]. Using this positive feedback cycle, the result is that $\boldsymbol{\Phi}$ is increasing exponentially in time, which matches with the observations by Chaisson [4]. CPU systems are not observed to have high action efficiency α at low $\boldsymbol{\Phi}$ and low α at high $\boldsymbol{\Phi}$. This proportionality of α with $\boldsymbol{\Phi}$, leads to

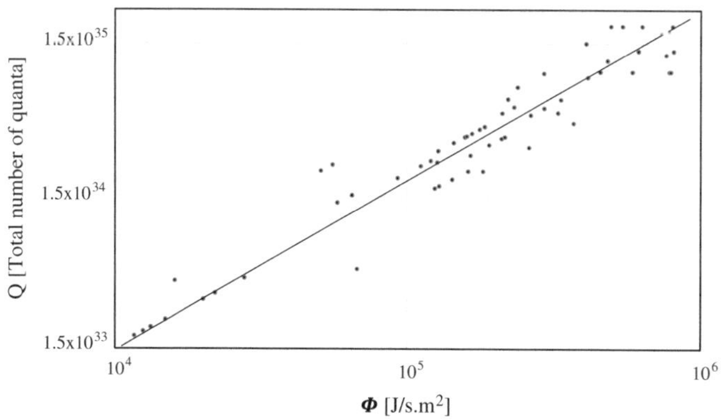

Fig. 27.2 A log-log plot of the total amount of action Q as a function of $\boldsymbol{\Phi}$. Data are *filled circles* and *solid line* is the fit. The data are from 1982 starting with Intel 286, to 2012, ending with Intel Core i7 3770k. There is a good agreement between the data and a power law fit. The two orders of magnitude change on each axis provide enough data to test the power law relationship between these variables

another, Complexity-Density rule: complexity (organization, level of development) α is proportional to the time and matter density of free energy, $\boldsymbol{\Phi}$, in a system. If one increases or decreases, the other increases or decreases as well.

27.5 Summary and Conclusions

The significance of these calculations is that FERD ($\boldsymbol{\Phi}$) is in a positive feedback loop with organization (α) and size (Q) of CPUs and all three reinforce each other, are related with power laws and increase exponentially in time. In order to become better organized, complex systems need larger energy flows to minimize larger constraints to motion of their elements. Connecting to non-equilibrium thermodynamics, we can say that the further a system is from equilibrium, the more work it can do to minimize constraints to flows, therefore increasing its organization in terms of action efficiency [4, 10]. Also, better organization means more action efficient flow channels, therefore higher α provides the necessary efficiency of the flow network to withstand and transmit larger energy flows. As Eric Chaisson points out, at a certain level of structure, the FERD level is an optimum [4]. Too low FERD will slow the system to a stop and too high level will destroy it. That is why we do not find data points much above or below the power law trend line. In order to move the optimum level of $\boldsymbol{\Phi}$ higher, the system needs to reorganize and grow in size. This correlations provide observational reason for connecting non-equilibrium thermodynamics with the principle of least action in order to explain progressive increase of organization in complex systems. It agrees with Chaisson's results for Cosmic Evolution, that $\boldsymbol{\Phi}$ grows exponentially in time paralleling the rise in organization [4]. The Least Action Principle (LAP), the Most Action Principle (MAP) and the principle of increase of Free Energy Rate Density (FERD) need to operate together in a positive feedback loop in order to produce an organized complex system.

It remains to be explored if the results are the same for $\boldsymbol{\Phi}$ outside this time interval, for other complex systems and in connection with other characteristics (parameters) of complex systems. If those dependencies hold in other complex systems, they can grow to universal Size-Density and Complexity-Density rules, in addition to the established Size-Complexity rule [10]. We term as "Devology" studies of self-organization in Cosmic Evolution and Development. Our future goal is to study other systems (stellar, physical, chemical, biological, social) for which we can obtain data for α, Q and $\boldsymbol{\Phi}$ and to compare with our observations for CPUs. This paper is one step in the further parametrization of the description of the processes of self-organization started earlier [10]. In following research, we plan to add other parameters such as the number of elements, density of elements, number of events and others in the description of the processes of self-organization, and find out if additional regularities exist. As shown in the model if one of the quantities α, Q or $\boldsymbol{\Phi}$ increases or decreases, the others increase or decreases predictably and lawfully as well, which is important to take into account in management of complex systems in ecology, engineering, economics, cities and elsewhere in society.

Acknowledgments The authors thank Professor Eric Chaisson, at the Harvard Observatory and Center for Astrophysics (CFA) at Harvard University, for fruitful discussions about Free Energy Rate Density and Cosmic Evolution and Professor Germano Iannacchione, Chair of the Physics department at Worcester Polytechnic Institute about discussions of non-equilibrium systems, as connected to self-organization and FERD. The authors also thank John Smart and Clement Vidal about discussions of the Evolutionary and Developmental processes in the Universe and Assumption College, for financial support and encouragement of this research.

References

1. Nicolis, G., Prigogine, I.: Self-Organization in Nonequilibrium Systems: From Dissipative Structures to Order through Fluctuations, Wiley (1977)
2. Onsager, L.: Reciprocal relations in irreversible processes. I. Phys. Rev. **37**, 405–426 (1931)
3. Onsager, L., Machlup, S.: Fluctuations and irreversible processes. Phys. Rev. **91**, 1505–1512 (1953)
4. Chaisson, E. J.: The cosmic Evolution. Harvard (2001)
5. Hübler Alfred W.: Predicting complex systems with a holistic approach: the "throughput" criterion. Complexity, **10**(3), 11 (2005)
6. Hübler, A., Crutchfield, J.P.: Order and disorder in open systems. Complexity **16**(1), 6 (2010)
7. Georgiev, G., Georgiev, I.: The least action and the metric of an organized system. Open Syst. Inf. Dyn. **9**(4), 371 (2002)
8. Georgiev, G.Y., Daly, M., Gombos, E., Vinod, A., Hoonjan, G.: Increase of organization in complex systems. In: World Academy of Science, Engineering and Technology 71. Preprint arXiv:1301.6288 (2012)
9. Georgiev, G.: Quantitative measure, mechanism and attractor for self-organization in networked complex systems. Self-Organizing Syst. LNCS **7166**, 90–95 (2012)
10. Georgiev, G., Henry, K., Bates, T., Gombos, E., Casey, A., Lee, H., Daly, M., Vinod, A.: Mechanism of organization increase in complex systems. Complexity (2014). doi:10.1002/cplx.21574,7,1
11. Annila, A., Salthe S.: Physical foundations of evolutionary theory. J. Non-Equilib. Thermodyn. 301–321 (2010)
12. Chatterjee, A.: Action, an extensive property of self-organizing systems. Int. J. Basic Appl. Sci. **1**(4), 584–593 (2012)
13. Chatterjee, A.: Principle of least action and convergence of systems towards state of closure. Int. J. Phys. Res. **1**(1), 21–27 (2013)
14. Gershenson, C., Heylighen, F.: When can we call a system self-organizing? Lect. Notes Compt. Sci. **2801**, 606–614 (2003)
15. Hertz, H.: Principles of mechanics, in miscellaneous papers, vol. III, Macmillan (1896)
16. Gauss, J.: Über ein neues allgemeines Grundgesetz der Mechanik (1831)
17. Intel Corporation. http://www.intel.com

MIX
Papier aus verantwortungsvollen Quellen
Paper from responsible sources
FSC® C105338

If you have any concerns about our products,
you can contact us on
ProductSafety@springernature.com

In case Publisher is established outside the EU,
the EU authorized representative is:
Springer Nature Customer Service Center GmbH
Europaplatz 3, 69115 Heidelberg, Germany

Printed by Libri Plureos GmbH
in Hamburg, Germany